Praise for *The Sugarmaker's Companion*

"*The Sugarmaker's Companion* is an amazing book. It uniquely fills the gaps in currently available maple reference material. It does not repeat what is in the *North American Maple Syrup Manual* but perfectly supplements it with up-to-date information, fresh ideas, and interesting examples. It also does not try to duplicate the details in the *New York State Maple Tubing and Vacuum Notebook* and the *New York State Maple Confections Notebook* but wonderfully adds to the ideas in these areas. It is the first book that integrates the making of other tree-based syrups, such as birch and various nut species, with the production of maple, improving the efficiency of equipment use and market opportunities. This is a great reference for all tree lovers."

—**Stephen L. Childs**, New York State maple specialist

"This most useful book for tree tappers is especially valuable because it includes the most up-to-date information on all the facets of the sap and syrup business. Particularly written for those in commercial sugaring, the book is also most interesting for the hobbyist or beginner who may not be aware of the amazing advances that have been made in the business lately. The book also gives detailed information on the tangential commercial and hobby possibilities in sugaring that are seldom examined in earlier books."

—**Gene Logsdon**, author of *A Sanctuary of Trees*

"*The Sugarmaker's Companion* is a delightful read. For the family with access to a few trees, or the larger sugarbush owner, the information provided in this book will be an affirmation of their efforts. The book delves into not only the maple harvest, but also many of our lesser-known tree resources, as well as fascinating facts like the nutritional benefits of maple sap. The message is clear. To preserve our forests we must enjoy them. . . .Yummy!"

—**B. Keith Harris**, BSc, owner/CEO, Troll Bridge Creek, Inc.

"Thomas Jefferson urged all farmers to plant maple trees, so that the colonies would not have to rely on imported sugar. In this spirit, Michael Farrell provides us with everything we need to know to produce America's own, natural, delicious sweetener. The section on maple sap—a healthy beverage that deserves a place in the American diet—is especially welcome."

—**Sally Fallon Morell**, president, The Weston A. Price Foundation

"Mike Farrell's *The Sugarmaker's Companion* should be on every maple producer's bookshelf. Along with the *North American Maple Producers Manual* and resource notebooks from Steve Childs, it is an essential reference resource. It contains a substantial amount of information not found elsewhere, especially marketing ideas, novel products, economic analyses, and creative ideas for expanding markets for pure maple products. Topics that are covered elsewhere receive updated treatment. Difficult concepts are explained well with attractive illustrations. The author's enthusiasm for this engaging business is displayed through an easy-to-read conversational style containing many personal anecdotes and opinions. There is a good amount of firsthand information from many hours spent in the woods, in the sugarhouse, and involving his community in enriching their lives with maple. Mike's positive point of view and creative ideas will encourage sugarmakers to engage a broader audience of potential customers."

—**Brian Chabot**, professor, Cornell University

"In *The Sugarmaker's Companion*, Michael Farrell presents both a philosophical and a practical look at today's tree sap and syrup industry. The book provides pertinent and useful information for both hobby and commercial tree-tapping operations. In a changing industry that is heavily shadowed by tradition, Michael combines the old tested methods with up-to-date research and science. This comprehensive book looks at both the big picture and many of the small details. Even for non-sugarmakers, it will be an informative and enjoyable read."

—**Gary Backlund**, author of *Bigleaf Sugaring: Tapping the Western Maple*

"Dr. Farrell's book is a must for any novice or beginning sugarmaker. It clearly explains the industry, products, and processes of maple sugaring. For experienced sugarmakers this is a book to join the Nearings' *The Maple Sugar Book* and the *North American Maple Syrup Producers Manual*—always on the shelf for ready reference. Thanks to Michael for a welcome addition to the maple library."

—**David Marvin**, president, Butternut Mountain Farm

The Sugarmaker's Companion

The Sugarmaker's Companion

AN INTEGRATED APPROACH TO PRODUCING SYRUP FROM MAPLE, BIRCH, AND WALNUT TREES

MICHAEL FARRELL

Chelsea Green Publishing
White River Junction, Vermont

Project Manager: Hillary Gregory
Editor: Makenna Goodman
Copy Editor: Laura Jorstad
Proofreader: Helen Walden
Indexer: Peggy Holloway
Designer: Melissa Jacobson

Printed in the United States of America.
First printing September, 2013.
10 9 8 7 6 5 4 3 19 20 21 22

Library of Congress Cataloging-in-Publication Data
Farrell, Michael, 1978–
The sugarmaker's companion : an integrated approach to producing syrup from maple, birch, and walnut trees / by Michael Farrell.
p. cm.
Includes bibliographical references and index.
ISBN 978-1-60358-397-8 (pbk.)—ISBN 978-1-60358-398-5 (ebook)
1. Sugar maple—Tapping. 2. Maple sugar. 3. Maple syrup. I. Title.

SB239.M3F37 2013
633.6'4—dc23

2013022231

Chelsea Green Publishing
85 North Main Street, Suite 120
White River Junction, VT 05001
(802) 295-6300
www.chelseagreen.com

This book is dedicated to my parents,
Marty and Kathy Farrell,
with deep gratitude for all they have done
for me and the entire Lynch/Farrell
family over many years.

Contents

Preface

I grew up in Albany, New York, and like most kids was only ever given Aunt Jemima or other fake syrup for my pancakes. It wasn't until I was 19 years old that I got my first taste of real maple syrup while eating brunch at a friend's house. Unfortunately my first introduction to pure maple was not a good one. I remember that the syrup was thin and runny with a bit of an off-flavor. I politely finished my pancakes but went back to using the fake stuff for the time being. Several years later I wound up in graduate school studying forestry at the SUNY College of Environmental Science and Forestry. During a class field trip to the Heiberg Forest, we attended a pancake breakfast with pure maple syrup produced from the university's sugarbush. This syrup tasted nothing like my previous experience—it was thick, flavorful, and unbelievably delicious. I couldn't believe what I had been missing out on and decided then that I wanted to try making my own syrup.

The following spring I tapped about a dozen trees at my parents' property in Lake George, New York. I had no idea what I was doing, and it certainly showed! My father helped me set up a collection system where most of the sap flowed into a black contractor bag nestled within a 30-gallon trash can (I do *not* recommend this!). I came home on weekends to boil the sap, and the entire family got involved in the process. After burning two of my mother's best pots, my brother and I were prohibited from entering the kitchen with sap, so we tried boiling it on an outdoor fireplace. Because we had no way to keep the ashes from getting into the boiling pot, to say the syrup took on a smoky flavor would certainly be an understatement! Unfortunately none of the syrup we produced was fit for human consumption. However, the one bright spot was that we discovered how delicious maple sap is, both fresh from the tree and boiled down partially. We gave up on trying to make syrup and instead just used all of the sap for drinking. We continued to tap a few trees each year to serve as a natural spring tonic, which we called Adirondack Sweetwater.

In 2004 I was working as a forester for Cornell and the New York State Department of Environmental Conservation when the position became available to manage The Uihlein Forest—Cornell University's Sugar Maple Research & Extension Field Station in Lake Placid, New York. Although my interest in maple had grown tremendously, I still had very little experience and did not expect to get the position. I applied anyway. During my interview, I remember telling the search committee that if they hired me, they would be making an investment for the future. I didn't know much about syrup production at the time, and my previous experiments were mostly a disaster. However, I was young, enthusiastic, had a solid background in forestry and economics, and promised to work extremely hard to learn about this new field. I must have made a convincing enough argument, because they hired me as director in January 2005.

I have now spent the past decade fully immersed in the maple industry. Being able to run a 5,000-tap sugaring operation while working on applied research and extension projects has given me a wide and unique perspective on syrup production. I've also spent the past five years pursuing my PhD at Cornell, allowing me to study the history, structure, and future growth potential of the maple industry. My thesis focused on the biologic, economic, cultural, and public policy factors that have impacted the North American maple industry to date and that will play a key role in future development. Much of this book stems from my experiences over the past decade working for Cornell and conducting the research for my doctorate.

I did my undergraduate work in economics at Hamilton College and had previously been destined for a career on Wall Street. After completing a couple of internships with Merrill Lynch, I soon realized that I had no interest in working for the financial sector. However, I do find basic economic theory interesting and relevant, since the main focus is on how to achieve the greatest benefit from scarce resources. Some of our resources are renewable and must be managed properly to ensure a sustainable supply for the future. Others are finite and must be conserved and allocated equitably over time. Our trees and forests are certainly a renewable resource and can provide us with many benefits for generations to come, but only if managed properly. One of the main reasons I wanted to write this book was to help others properly manage and utilize their forest resources, in particular the maple, birch, and walnut trees. I am continually amazed at the delicious and nutritious sap that these trees are able to produce every spring. My hope is that this book will allow you to start producing syrup yourself or help you expand upon your existing operation. Sugaring is truly an amazing process and provides us with some of the best food and beverages we could ask for. It's a shame that more people don't get the opportunity to produce and consume the sap and syrup of these incredible trees.

I have written this book more as a guide than a specific how-to manual on syrup production. Although this book is useful for hobbyists, you may want to pick up a more concise guide if your goal is simply to tap a few trees and boil the sap down on the kitchen stove or in your backyard. I have written this with the commercial producer in mind for two reasons. First, most people who start out on a small scale get hooked and wind up expanding into a larger operation. Furthermore, I think just about every tree should be tapped, so naturally I recommend developing as large an operation as you can support. My goal is to inspire you into action, providing information on many topics related to sustainable sap and syrup production from a variety of species. I couldn't cover everything there is to know about all topics, but I have tried to include the most pertinent and useful information that is difficult or impossible to find elsewhere. No person or book has all the answers, but I hope you find this reading useful and that it provides greater impetus and ideas to grow your sugaring operation. While the book is mostly focused on maple syrup production, nearly all the information presented can also be applied to tapping birch and walnut trees. The default setting is to discuss maple syrup production, but nearly all the same concepts can be easily transferred to birch or walnut sugaring. Where differences exist within and between these species, I have highlighted the main issues that you should consider.

My duties with Cornell are focused on research and extension for the maple industry. I help many people get started in or expand their sugaring operations, and I plan to stay in this position until retirement—which is a long way away! If you have any questions or need further advice as you develop your operations, feel free to contact me at mlf36@cornell.edu, or visit the Cornell Maple Program website at www.cornellmaple.com.

Acknowledgments

This book would not have been possible without the help of many people. First and foremost, I'd like to thank my colleagues at Cornell from whom I've learned so much over the years, in particular Brian Chabot, Peter Smallidge, Steve Childs, and Chuck Winship (Brian and Peter also supplied many of the photos for this book). I'm extremely grateful for the hundreds of maple producers I've met all over North America who have shared their insights and experiences with sugaring; my knowledge of the industry and this book have been greatly enhanced through these relationships. I am very appreciative of the efforts of other scientists who have conducted research and published hundreds of articles on tree sap and syrup production. Being able to talk with them and read their works has given me a deeper and richer perspective on sugaring, making this book much more informative and enjoyable. I am eternally grateful to Lew Staats for dedicating 35-plus years to building The Uihlein Forest from the ground up, and to Henry Uihlein II and The Uihlein Foundation for sponsoring Cornell's efforts in maple research and extension. Without their generous support and vision, I would never have had an opportunity to become engaged in such an incredible industry. Makenna Goodman, my initial editor at Chelsea Green Publishing and a small-scale sugarmaker herself, asked me to write this book and provided helpful guidance throughout the process. Nancie Battaglia spent countless hours going through her existing portfolio and taking many new pictures that added tremendously to the book. Finally, I would like to thank my wife, Andrea, for all her love and support over the years—I couldn't ask for a better partner in life.

CHAPTER 1

Why Maple Matters:
The Case for Increasing Production and Consumption of Pure Maple Products

It's amazing how the act of sugaring can change your outlook on pure maple syrup. Before I got into making syrup, I didn't really care what I put on my pancakes and never thought about the difference between pure and artificial syrups. Now I can't even fathom having that sort of attitude. If you don't already have a strong passion for pure maple, once you start tapping trees and making your own syrup I have no doubt that your feelings will change. There isn't a sugarmaker in the world who would think of using artificial syrups; many of us choose pure maple as our sole sweetener. The only reason anyone should buy the fake stuff is to use as teaching tool when explaining the differences between real and imitation syrups. I have to purchase a bottle of "pancake syrup" every year for our tasting trials with local students. It pains me to do so, but the benefits of showing kids how much better the real stuff is make it well worth it.

When something is pervasive in our everyday lives, it is human nature for us to start taking it for granted. Since maple trees are commonly found throughout eastern North America and sweet foods and drinks are also plentiful wherever we go, we don't always realize how lucky we are to have such an extraordinary maple resource. Picture for a minute that maple trees didn't already exist in the natural world and there was no such thing as tapping trees or boiling down tree sap to make syrup. Now imagine you read a news article about a new tree species that had been developed by scientists called "maples." They are relatively fast growing, have beautiful foliage and form, put on a spectacular fall foliage display, withstand extreme temperature fluctuations, and can reproduce naturally in the wild from seed. Chances are you would want to plant some of these new trees. Then imagine you read about the best part of all: In the late winter and early spring, the tree produces a delicious and nutritious sap that can be consumed as a spring tonic or boiled down into syrup and other confectionary products. Not only does it taste great, but it is also the healthiest sweetener in the world and commands a premium price in the marketplace. At this point you would probably be making plans to find additional land for planting. Of course we know that this species already exists, but we rarely take the time to appreciate how amazing maples really are. Whenever I stop to think about the incredible process of maple sugaring, it makes me wonder why so few people use their trees for sap and syrup production.

This first chapter explores the many virtues of maple syrup and the sugaring process; if you aren't already convinced that you should produce maple syrup, by the end of this chapter I hope you will already be making plans to buy your first evaporator. As I see it, the most important reasons why more people should get into sugaring are as follows:

- Sugaring is a time-honored tradition that connects us with our historical roots.
- Technological advancements have made sugaring *relatively* easy.
- Sugaring brings families and communities together around a common, worthwhile goal.
- Syrup produced from tree sap is one of the healthiest, all-natural sweeteners.
- Pure maple syrup tastes much better than artificial "pancake syrups."

- You can take pride and satisfaction in producing your own sweetener.
- There is a growing demand for pure maple products throughout the world.
- Syrup production allows for better conservation of trees and forests.
- Sugaring can be a profitable means of diversifying income streams from your land.

The rest of this chapter explores all these positive attributes in greater detail.

Historical Connection

There is a deep, rich history of maple syrup and sugar production in North America. By far the best account is found in Helen and Scott Nearing's *The Maple Sugar Book*. For a detailed history and firsthand accounts of sugaring in the early days, there is no better source than the Nearings' classic work. Native Americans had been collecting sap and making sugar for centuries before the Europeans arrived, and the early settlers learned the process from them. The Natives were using primitive technologies that relied on making large gashes in the trees to collect sap into birch-bark containers. They built fires to heat rocks and then dropped those rocks in hollowed-out logs full of sap. The heat from the rocks would transfer to the sap, and water would slowly evaporate away. With improved machinery and the use of metal drill bits, spouts, buckets, tanks, and evaporators, the European settlers gradually developed syrup production into a fairly large industry.

FIGURE 1.1. One of the many legends around how humans first discovered the sweetness of maple sap holds that they ate the delicious "sapsicles" that develop from broken branches.

Thanks to the efforts of the US Census Bureau, the USDA National Agricultural Statistics Service (NASS), and Statistics Canada, we have decent records of syrup production in the United States and Canada dating back to the mid-1800s. Figure 1.2 tracks total maple syrup production over the past 150 years. As you can clearly see, we used to produce a lot more maple syrup and sugar in the United States than we do today! Maple sugar was one of the primary sweeteners for farm families in the eastern United States and Canada during the 1800s. If a family was fortunate enough to have maple trees on their farm, they tapped them in the spring and produced maple sugar (very little of their production was kept as syrup). The height of maple production actually occurred during the Civil War era when northerners refused to import "slave sugar" from the southern states or Caribbean islands. The Nearings include several unforgettable quotes from the 1800s regarding the use of maple versus cane sugar. Two that stand out in particular come from the *Farmers' Almanac*:

> *Prepare for making maple sugar, which is more pleasant and patriotic than that ground by the hand of slavery, and boiled down by the heat of misery.*[1]

> *Make your own sugar, and send not to the Indies for it. Feast not on the toil, pain, and misery of the wretched.*[2]

Not only was there a great aversion toward "slave sugar," but cane sugar was also relatively expensive during that time period due to high shipping costs. We had not yet started making high-fructose corn syrup or beet sugar, so people mostly relied on maple sugar and

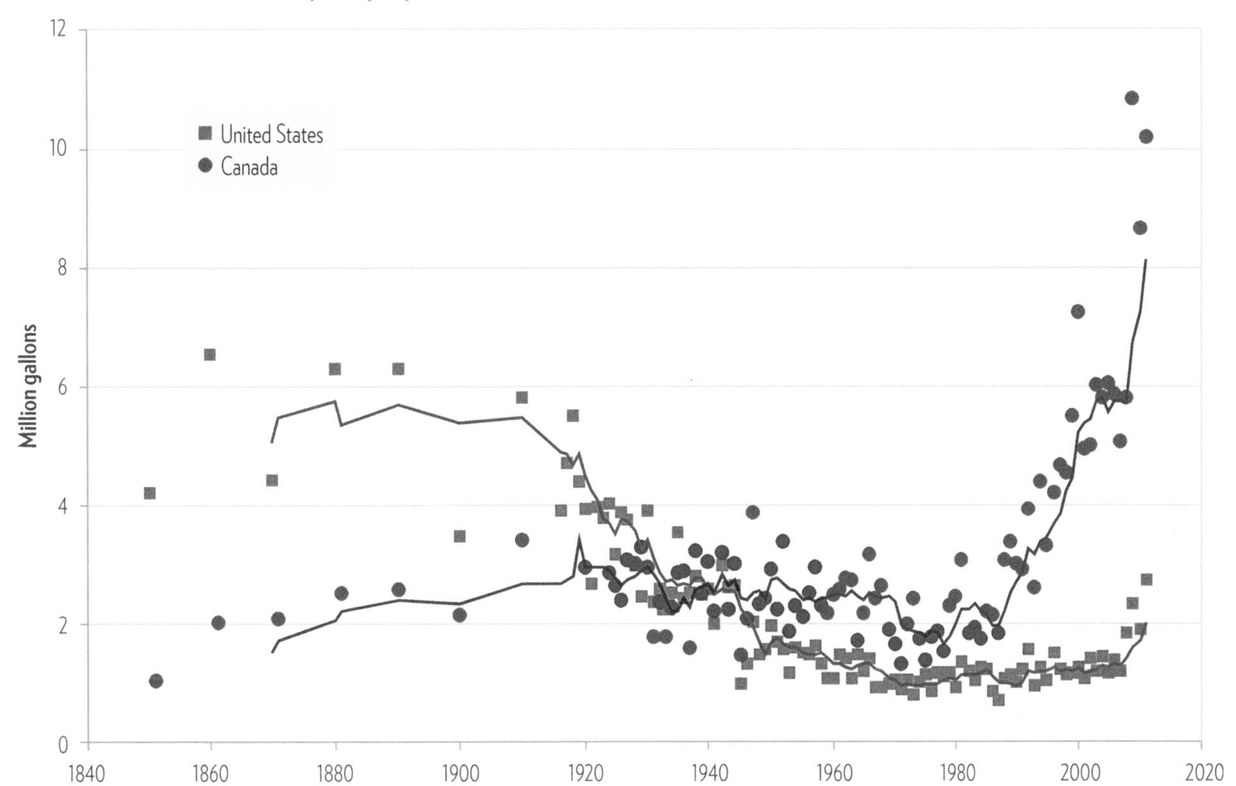

FIGURE 1.2. This graph is based on historical data from the US Census, the National Agricultural Statistics Service, and Statistics Canada.

FIGURE 1.3. Many of the Founding Fathers, including Thomas Jefferson, realized the virtues of maple syrup production in the newly formed republic. With a large grove of sugar maples planted near the Washington Monument on the National Mall, this iconic species still represents an important component of the DC landscape.

honey for their sweeteners. A great many farmers took pride in their ability to produce their own sugar and worked extremely hard to do so. When people started to move off farms during the Industrial Revolution, the remaining farms became larger and more specialized, and the production of maple sugar and syrup severely diminished. As production has declined, so has our consumption of pure maple products.

According to the Census Bureau, the United States produced a remarkable 6,613,000 gallons of maple syrup in 1860 from a population of 31.5 million people. If no maple products were exported or imported from Canada, the average consumption would have been 27 ounces per person—roughly 10 times the current consumption levels! Although our population has grown tremendously over the past 150 years, the production of maple products has fallen just as drastically, reaching an all-time low of 757,000 gallons in 1987. Production has picked back up again over the past decade, and we now produce roughly 2,500,000 gallons of syrup annually and production is growing at 5–10% per year. Our current consumption is just under 3 ounces per person, but if we didn't import the vast majority of our syrup from Canada, the average American would get about 1 ounce of maple syrup a year. This is certainly a sad statistic, especially for a country with the world's largest economy and over two billion tappable maple trees. If we could produce so much more maple syrup and sugar in the 1800s despite having far fewer people, trees, and processing equipment, surely we can increase the production and consumption of pure maple products today.

FIGURE 1.4. Former New York State Agriculture Commissioner Patrick Hooker drills a taphole with a brace and bit as part of a ceremonial tree tapping at Mapleland Farms in Salem, New York. These events are a common way of celebrating our maple heritage and remembering our historical roots in the maple industry.

There is promising news. The maple industry has been growing in recent years as more people get into it and technological advances allow sugarmakers to achieve higher yields per tap. In fact, maple syrup production is now one of the fastest-growing agricultural industries in the United States, and there was more syrup produced in 2013 than in any year since 1945. The extraordinary production of over 3 million gallons of maple syrup that year was driven by sugar rationing as part of the World War II efforts. In the good old days people produced maple syrup and sugar out of necessity since other options for sweeteners were limited. The current expansion of the maple industry has not been driven by necessity, but rather by choice. As people have begun to realize the benefits of eating healthier diets based on natural, whole foods, production and consumption of pure maple products has taken off. Producing your own maple syrup will allow you to rediscover this natural sweetener and its traditional roots.

In today's marketplace there is an overabundance of cheap, sugary foods available any time of year. It is more difficult to avoid these foods in your diet than it is to include them. However, people are increasingly rejecting the inferior, highly processed sweeteners and are willing to pay more for natural, healthier alternatives. While some may argue that we don't need any concentrated sugar in our diets,[3] people do enjoy sweet foods, and maple sugar is among the healthiest and most natural sweeteners that exist in the marketplace. Producing and consuming pure maple brings us back to simpler times when people lived off the land and primarily consumed natural, locally made goods. Moreover, to fully experience and appreciate maple in your life, you can't just consume maple syrup that you picked off a supermarket shelf—you need to make it yourself.

Technological Advancements

Making maple syrup certainly isn't easy—gathering sap and processing it into syrup usually requires long hours and significant effort. And hard as this work seems today, it was *much* more difficult to produce maple syrup and sugar back in the 1800s. At the peak of maple production in the United States, farmers had very primitive tools for collecting and processing the sap—it's really amazing how they were able to produce so much. I can't imagine how hard it would be to produce syrup without a cordless drill, metal buckets and/or vacuum tubing systems, reverse osmosis units, and efficient evaporators. The fact that people produced so much syrup with such primitive technology is a testament to the hard work and dedication of our ancestors. Consider how the process has changed over the years . . .

Rather than having to make V-shaped gashes in the trees or use a brace and bit to drill holes, we now have access to light, powerful cordless drills that can make a clean hole in a matter of seconds. Instead of having to build collection vessels out of wood and birch bark, we can buy metal or plastic containers relatively cheaply and use motorized vehicles to gather the sap. If gathering buckets is too much work, we can install a plastic tubing system and have all the sap flow into one large tank. By hooking up a vacuum pump, we can gather two to three times the amount of sap from each tree with far less effort and fewer tapholes.

When it comes to sap processing, gone are the days of heating rocks and putting them into a vat of sap to slowly evaporate the water. The settlers constructed evaporators with large metal pots and pans, which greatly improved efficiency at the time but would be considered impractical today. We now have extremely efficient evaporators that can boil off hundreds of gallons of sap every hour. In addition, reverse osmosis units remove at least 75 to 80 percent of the water from the sap before we start boiling—all we have to do is turn a few valve handles and press some buttons. Our ancestors couldn't even dream of the technology that we have at our disposal. If they had the same tools back then that we have today, maple sugar may have remained competitive with cane sugar for a very long time.

FIGURE 1.5. This combination bucket/pipeline was considered a huge advancement in the early 1900s as a means of getting sap to the sugarhouse. Earlier methods of gathering sap required even greater time and energy. PHOTO COURTESY OF THE ADIRONDACK MUSEUM

FIGURE 1.6. The largest sugaring operation in the world around the turn of the 19th century was located in the heart of the Adirondacks near Tupper Lake, New York. Although the series of large evaporators were extremely advanced at the time, modern reverse osmosis units and evaporators could process the same amount of sap with significantly less time and fuel. PHOTO COURTESY OF THE ADIRONDACK MUSEUM

Given how much easier it is for us to produce maple syrup today, you would think many more people would be doing it. However, nearly all survey research on landowners finds that aesthetics and privacy are the top two reasons why people own their land. Trees are mostly there for scenery and as a backdrop to their homes, not as practical sources of timber and food. Fortunately this attitude is changing, and many folks are now starting to grow their own food and make the most of their land. With technology that makes sugaring *relatively* easy, almost anyone with tappable trees can make their own syrup today.

Pride in Your Product

Producing maple syrup is something that you can take tremendous pride in. Every bottle of syrup that comes off your evaporator is something special and unique; the flavor and characteristics of the syrup are totally up to your processing capabilities and the trees that you collect the sap from. If you care about and take pride in the food you eat yourself and provide to your family and friends, this feature alone should make you want to produce your own syrup.

Nearly all the sugarmakers I know are extremely enthusiastic about their syrup and have a deep love of the production process. There are thousands of maple producers throughout the United States and Canada who go to great lengths to educate others about their products and beloved industry. Many farmers take tremendous pride in whatever crops they grow, but you would be hard-pressed to find farmers who are fanatical about growing the corn used to make artificial syrup. Corn growers may care deeply about their land and their farm operations, but it would be highly unusual to find any who take pride in opening a bottle of artificial pancake syrup that their corn goes into. They strive to get as many bushels per acre as possible, yet I'm sure none of them opens up a bottle of Aunt Jemima to revel in its unique qualities. On the other hand, it is hard to find a sugarmaker who isn't extremely passionate about producing pure maple syrup. This pride and passion is especially evident every spring when sugarmakers throughout the country host open houses to show others the incredible process of turning sap into syrup. To understand why this dichotomy exists, consider the following . . .

When a farmer plants a field of corn, harvests it, and sends it to the factory for processing, it gets blended with corn from thousands of other farmers. Once it has been processed and graded, one farmer's corn is indistinguishable from another farmer's corn. While this allows corn to be bought and sold in commodity markets throughout the world, it removes the sense of ownership and pride that many farmers have for their crops. Since only a small fraction of the corn grown in the United States is converted into corn syrup (most is fed to animals or turned into ethanol), and there are tens of millions of acres of corn planted each year, as a farmer you have a much better chance of finding a

FIGURE 1.7. Many sugarmakers are so proud of their syrup that they will enter it in contests at state fairs and related events. Pictured here is one of the dozens of awards earned by Clark's Sugar House in Acworth, in New Hampshire. PHOTO BY NANCIE BATTAGLIA

FIGURE 1.8. The best way to retain mature forests such as this one is to generate annual income through the production of maple syrup. When there is an economic incentive to retain forestland for syrup production, the likelihood of the land being harvested for timber or converted for development falls considerably.

needle in a haystack than of opening a bottle of pancake syrup and having even one molecule of it come from corn that you have grown yourself.

The situation with maple syrup couldn't be more different. As sugarmakers, we have a strong connection to our land, in particular the maple trees growing on our property. We have spent countless hours tapping trees, gathering sap, and boiling it down into syrup. This gives us a tremendous sense of ownership and pride in our finished products. When you sell a bottle of syrup or give it to friends and family, you can take pride knowing that it wouldn't be possible without your efforts.

Like many people, I would rather eat food that I produced myself or that was made by folks glad to show me around their farm. Whereas thousands of sugarmakers open their doors to the public each spring to show off their operations, you can't even get into a factory to watch the process of high-fructose corn syrup (HFCS) being made. In the documentary *King Corn*, the producers had to try making their own HFCS because no factories would allow them to see it being made. Judging by the clip from the film, it is quite obvious why the corn refining industry does not want the public to see the process. As someone who used to manage a slaughterhouse once told me, "If you want to keep on eating bacon, don't come visit me at work." The opposite sentiment is true with maple production—once people visit a sugarhouse and get to see and experience the sugaring process, they become pure maple fans for life.

Forest Conservation

While a small percentage of people object to tapping because they feel that it "hurts the trees," these feelings are often misguided. Tapping does remove some of the stored sugars and has a slight impact on the growth rate of trees, but it is very difficult to kill a tree by overtapping (people have certainly tried!). In fact, long-term research conducted in the 1990s as part of the North

American Sugar Maple Decline Project found that tapped trees live just as long as untapped ones.[4] I have seen a few cases of extreme overtapping, but most is done right and has little impact on the trees.

Rather than worrying about hurting the trees through tapping, I think the best way to save maple trees from real threats (being cut down for development or lumber) is to tap them for syrup production. When there is a strong economic, cultural, and personal connection to individual trees or forest stands, the chance that all the trees will be cut is very low. In fact, because syrup production is so culturally important in Quebec, they have laws prohibiting clearing of sugarbushes on private land. So if you are concerned about saving the maple trees—my advice is tap as many as possible.

There is a great deal of historical evidence to support this notion. We actually have many more maple trees growing in the United States today than we did during the peak of syrup production in the 1800s. At that time, most of the eastern forests had been cleared for agriculture and the countryside contained little woodland. However, sugar maple trees were often spared from the ax since the early settlers lived off the land and needed these trees to produce sugar. When almost all other trees were cut for firewood or building materials, many people wouldn't consider cutting the sugar maple groves. Our current forests are now largely an artifact of human preferences and management over time. When fields were abandoned and forests regenerated over the last 100-plus years, an abundant maple resource

FIGURE 1.9. A developing stand of sugar maples on abandoned land. Sugar maples have proliferated throughout the eastern United States over the past century as a result of our land management practices and preferences for these trees.

developed throughout the eastern half of the country. Sugar and red maples became much more prominent in the landscape and have continued to expand their range and distribution. If we want to ensure that future generations have the same maple resource that we are blessed with, maintaining a vibrant maple syrup industry will be essential.

If you become a large enough producer that you want to tap more trees than you have on your own land, you can help conserve maple stands in neighboring properties as well. By providing the owners with an annual payment to lease their trees for tapping, you may make it economically feasible for them to hold on to the land and pay the property taxes without having to subdivide or log the property. In fact, states such as New York and Wisconsin have laws that encourage maple sugaring and forest conservation by allowing landowners who produce syrup themselves or lease their land to a sugarmaker to qualify for a reduced property assessment for doing so. This policy has provided the impetus for some sugarmakers to get started and is the biggest incentive for landowners to let someone else tap their trees.[5] In fact, I've heard many stories of landowners asking maple producers to tap their trees so that they could qualify for reduced taxes under the agricultural assessment program in New York.

FIGURE 1.10. Sugaring is an excellent way to tie families together around a common goal. Earl Parker shows his grandson Joshua how to tap a tree that had been planted by Michael–Joshua's father and Earl's son. PHOTO COURTESY OF PAT PARKER

Sugaring Brings Families and Communities Together

For many people, especially those who consider themselves hobbyists, this is one of the best reasons to get into syrup production. Maple sugaring happens at a special time of the year when people are sick and tired of cold, snowy weather and can't wait for the warm, sunny days of late winter and early spring. After several months of gray skies and little outdoor activity—when cabin fever abounds—many people are looking for any excuse to get outside and be active. I can't think of a better way to celebrate this change in seasons than to be out in the woods tapping trees and gathering sap. Feeling the warm sun on your face, drinking fresh sap out of the buckets, and gathering around the evaporator at night making syrup is really an incredible way to spend the day!

There is a sense of community and togetherness that develops when people dedicate countless hours to achieving the same goal—collecting sap and producing delicious syrup! Kids love to tap trees and gather sap, yet most children never get to experience this and don't even know that there is a difference between pure maple syrup and its artificial competitors. If you are looking for an activity that the whole family can get involved with, then making maple syrup could be the right thing for you. In addition to bringing your own family together, sugaring also has the potential to strengthen community ties. This aspect of syrup production is so important that I dedicated an entire chapter to it. Chapter 9 describes some of the many communities that have used maple sugaring to bring people together around a worthy cause. You can read this chapter to get ideas about what others have done and then tailor a sugaring program to fit your particular situation. There are countless opportunities to develop these types of community operations—all you need are motivated people who take the initiative to make things happen.

Natural, Healthy Ingredients

When I teach kids about maple syrup, I often show them a bottle of artificial "pancake syrup" and ask them what it is. Invariably many yell out "maple syrup" very enthusiastically! I then explain that it's not real maple syrup but rather an imitation product that uses other sugars and chemicals to resemble the real thing as much as possible. One "lucky" kid gets to read the ingredients and always stumbles toward the end when trying to articulate terms such as *hexametaphosphate*. Several of the ingredients are difficult to pronounce (even for many adults) and are not what we would consider natural substances. On the other hand, the only ingredient in pure maple syrup is maple sap—one of the most pure and healthy liquids on the planet. As you will read about in detail in chapter 12, maple sap contains a wide variety of minerals and nutrients that our bodies need. Just as the sugar becomes concentrated when it is processed into syrup, all the minerals and nutrients also become concentrated in the final product.

There are a variety of imitation syrups on the market and most of them use corn syrup and high-fructose corn syrup for the base product. As I'm sure you know by now, the last thing most people need in their diet is more high-fructose corn syrup. If you make maple syrup, I can almost guarantee you that your intake of corn syrup and refined sugar will drop precipitously. Because you have made it yourself, you will use a lot more maple syrup than if you had to buy it. For instance, rather than buying salad dressing made with HFCS and/or refined sugar, you may discover how delicious making your own maple balsamic vinaigrette is. Instead of buying sweetened cereal for your kids, you will likely get the healthier options and just add some maple syrup to each bowl. You can also make your own soda by mixing maple syrup with seltzer or buying your own home carbonation machine. Given that most kids get a high percentage of their daily intake of calories from drinking soda, making this simple change could greatly improve the health of your children. These are just a few of the many ways that you can substitute pure maple into your diet and reduce the amount of refined sugar and HFCS you consume. Once you start producing syrup, there is no doubt your consumption of pure maple will increase dramatically as you find new ways to use it outside of the traditional pancakes and waffles.

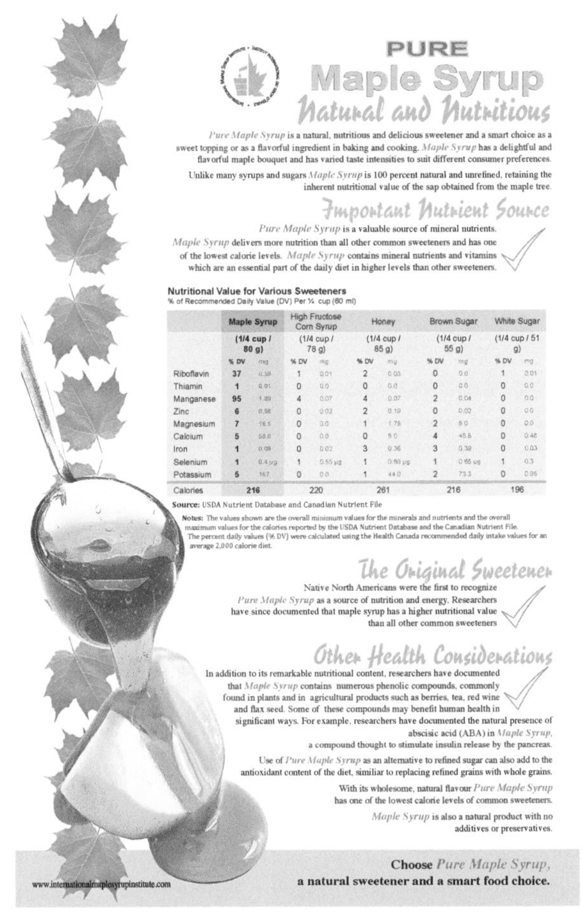

PURE
Maple Syrup
Natural and Nutritious

Pure Maple Syrup is a natural, nutritious and delicious sweetener and a smart choice as a sweet topping or as a flavorful ingredient in baking and cooking. *Maple Syrup* has a delightful and flavorful maple bouquet and has varied taste intensities to suit different consumer preferences.

Unlike many syrups and sugars *Maple Syrup* is 100 percent natural and unrefined, retaining the inherent nutritional value of the sap obtained from the maple tree.

Important Nutrient Source

Pure Maple Syrup is a valuable source of mineral nutrients. *Maple Syrup* delivers more nutrition than all other common sweeteners and has one of the lowest calorie levels. *Maple Syrup* contains mineral nutrients and vitamins which are an essential part of the daily diet in higher levels than other sweeteners.

Nutritional Value for Various Sweeteners
% of Recommended Daily Value (DV) Per ¼ cup (60 ml)

	Maple Syrup (1/4 cup / 80 g)		High Fructose Corn Syrup (1/4 cup / 78 g)		Honey (1/4 cup / 85 g)		Brown Sugar (1/4 cup / 55 g)		White Sugar (1/4 cup / 51 g)	
	% DV	mg	% DV	mg	% DV	mg	% DV	mg	% DV	mg
Riboflavin	37	0.59	1	0.01	2	0.03	0	0.0	1	0.01
Thiamin	1	0.01	0	0.0	0	0.0	0	0.0	0	0.0
Manganese	95	1.89	4	0.07	4	0.07	2	0.04	0	0.0
Zinc	6	0.58	0	0.02	2	0.19	0	0.02	0	0.0
Magnesium	7	16.5	0	0.0	1	1.75	2	5.0	0	0.0
Calcium	5	58.0	0	0.0	0	5.0	4	45.8	0	0.42
Iron	1	0.09	0	0.02	3	0.36	3	0.39	0	0.03
Selenium	1	0.4 µg	1	0.55 µg	1	[illegible] µg	1	0.65 µg	1	0.3
Potassium	5	167	0	0.0	1	44.0	2	73.3	0	0.06
Calories	216		220		261		216		196	

Source: USDA Nutrient Database and Canadian Nutrient File

Notes: The values shown are the overall minimum values for the minerals and nutrients and the overall maximum values for the calories reported by the USDA Nutrient Database and the Canadian Nutrient File. The percent daily values (% DV) were calculated using the Health Canada recommended daily intake values for an average 2,000 calorie diet.

The Original Sweetener

Native North Americans were the first to recognize *Pure Maple Syrup* as a source of nutrition and energy. Researchers have since documented that maple syrup has a higher nutritional value than all other common sweeteners

Other Health Considerations

In addition to its remarkable nutritional content, researchers have documented that *Maple Syrup* contains numerous phenolic compounds, commonly found in plants and in agricultural products such as berries, tea, red wine and flax seed. Some of these compounds may benefit human health in significant ways. For example, researchers have documented the natural presence of abscisic acid (ABA) in *Maple Syrup*, a compound thought to stimulate insulin release by the pancreas.

Use of *Pure Maple Syrup* as an alternative to refined sugar can also add to the antioxidant content of the diet, similiar to replacing refined grains with whole grains.

With its wholesome, natural flavour *Pure Maple Syrup* has one of the lowest calorie levels of common sweeteners.

Maple Syrup is also a natural product with no additives or preservatives.

Choose *Pure Maple Syrup*, a natural sweetener and a smart food choice.

www.internationalmaplesyrupinstitute.com

FIGURE 1.11. The International Maple Syrup Institute recently developed this brochure highlighting the health benefits of pure maple (available as a free download from the IMSI website). Handing these out to your customers is a great way to encourage more people to consume pure maple syrup. IMAGE COURTESY OF THE INTERNATIONAL MAPLE SYRUP INSTITUTE

FIGURE 1.12. Since ordinary sodas aren't healthy drinks for young children, Thomas Reynolds has never even tried a Pepsi or Coke, but he does get to enjoy a fresh glass of maple soda made with a home carbonation unit. PHOTO COURTESY OF BROOKS REYNOLDS

The Federation of Quebec Maple Producers has invested a lot of time and money recently to explore the health benefits of maple syrup. They provided funding for top researchers at the University of Rhode Island to chemically analyze maple syrup. This investment has certainly been fruitful. Dr. Navindra Seeram and his lab recently published several scientific articles discussing the health benefits of pure maple syrup.[6] Of the 54 compounds found in the syrup, 5 were brand-new discoveries not found in other foods. One of them was given the name Quebecol after the province that produces the majority of the world's syrup. According to Seeram, "Several of these compounds possess anti-oxidant and anti-inflammatory properties, which have been shown to fight cancer, diabetes and bacterial illnesses." They also discovered polyphenols in the syrup that inhibit the enzymes responsible for converting carbohydrates to sugars, presenting a new possible method for managing Type 2 diabetes.

This research has garnered a lot of media attention, and the Federation has been touting the newly discovered benefits of "Canadian" maple syrup. However, most consumers probably know that all maple syrup has these beneficial qualities, and the marketing efforts are helping to sell more syrup to health-conscious consumers. I found several articles on the web touting maple syrup as a new "superfood," but these claims may be a bit over the top. In fact, Dr.

Seeram has endured some criticism for advocating maple syrup (which is mostly sugar) as a health food. He now makes a point of stating that people should not consume prodigious amounts of syrup, but that if they choose to use sweeteners, pure maple syrup is a much better choice than artificial pancake syrups with HFCS. According to a press release on the University of Rhode Island website, Seeram states, "Pure maple syrup is not only delicious, it is so much better for you." I couldn't agree more. While it is important not to overindulge in sweetened, sugary foods and beverages, when you do want something sweet, pure maple makes an excellent choice.

Better Flavor

Although taste is subjective, most people who have grown up eating pure maple syrup will contend that the real things tastes much better than its artificial competitors. There are certainly many people who prefer the taste of imitation syrups to the real thing, but these are almost always people who grew up eating the fake stuff and are used to that flavor and texture. Those who are familiar with pure maple will vouch for its superior flavor. For example, in a consumer taste test carried out for the International Maple Syrup Institute, focus groups were given samples of a variety of pure maple syrups and artificial pancake syrup. The participants in Quebec and Ontario overwhelmingly rejected the artificial syrups, while a higher percentage in New Jersey actually preferred the fake stuff. Unfortunately the artificial syrups have covered America's pancakes and waffles for so long that many people have gotten used to them. However, it's not because they taste better, but rather because they are so much cheaper, that this trend continues. As an industry, we face an enormous opportunity and challenge in converting consumers of artificial syrup over to the real thing. Since about 85 percent of the shelf space in a typical grocery store contains artificial syrup, we have a long climb ahead, yet the potential is fantastic.

For people who say they prefer imitation syrups, one of the most common complaints I hear is that the real stuff is too thin and watery. Likewise, in the IMSI study, the number one positive comment on the artificial syrup tasted by the New Jersey focus group was that it was thick. Food scientists have been busy for many years creating pancake syrups with the texture and consistency that many people prefer. As sugarmakers, we don't need a degree in food science to make a delicious sweetener. Mother Nature provides the best-tasting natural product imaginable—all we have to do is boil the sap a little bit longer to get the syrup a tad thicker than usual. One of the reasons Vermont has been able to build its brand is the fact that its syrup needs to be at least 66.9 percent sugar concentration, whereas everywhere else the syrup only has to be 66 percent brix. Rather than settling for 66 percent minimum, I would encourage you to bring your syrup to 67 percent brix, no matter what state or province you live in. Just that 1 percent difference in sugar content gives the syrup a much thicker texture and feel. In chapter 6 you will learn much more on how to ensure optimum flavor in your syrup.

Growing Demand for Pure Maple Products

Maple syrup is only produced commercially in the eastern United States and Canada, yet there is a growing worldwide demand for pure maple products, especially once people are exposed to them through strategic marketing initiatives. Figure 1.13 tracks consumption of maple syrup in the US since the 1970s based on production and export/import figures from Canada and the US. As the production of syrup has spiked in Canada (and more recently in the US), Americans have developed a larger appetite for pure maple syrup. This is due in large part to marketing efforts and promotional campaigns aimed at selling all the additional syrup. There have been surpluses and shortages over the years due to weather patterns, yet the overall trend is that more maple syrup is being produced. Since maple syrup never goes to waste, much more maple syrup is also being consumed.

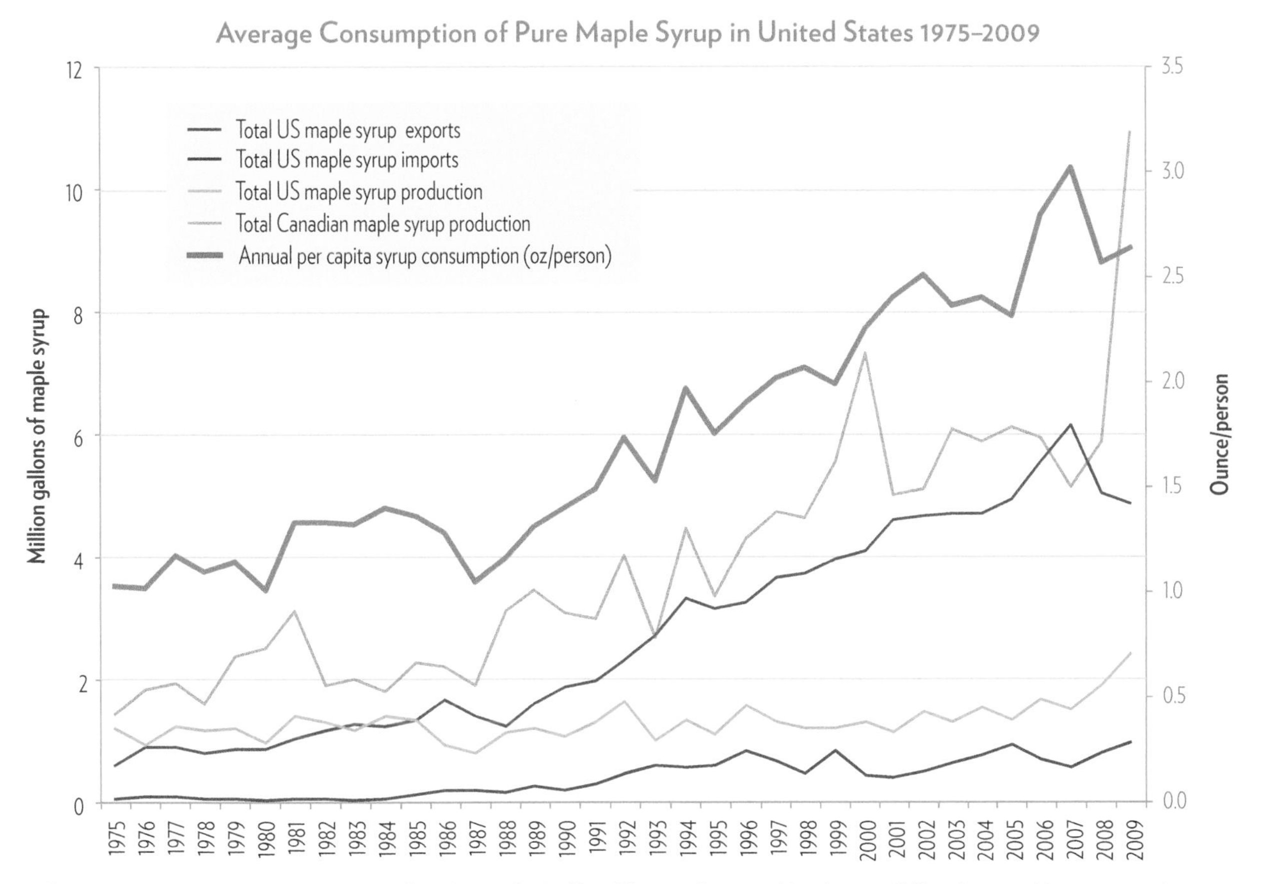

FIGURE 1.13. Average per capita consumption of maple syrup for the United States is determined by taking total US production, adding imports, subtracting exports, and then dividing by the population in a given year. Until recently, nearly all of the increase in United States consumption was driven by growth in Canadian syrup production. Consumption data is shown with the green line and references on the right-hand axis; import, export and production data references are on the left-hand axis.

These recent gains in production and consumption have been impressive, yet there is still enormous room for growth in both arenas. Although per capita consumption has nearly tripled in the United States over the past 35 years, it is still quite low at less than 3 ounces per person. To put this in perspective, I use about 3 ounces of pure maple syrup on a stack of pancakes. And even though the industry is rapidly expanding, we are still tapping less than 1 percent of the available maple trees in the nation. If more efforts were put into marketing pure maple syrup as local, healthy sugar in the regions where it is produced, the demand would rapidly rise and sugarmakers would have more incentive to tap more trees and produce more syrup. We need to do a better job of capitalizing on the growing demand for local, healthy, sustainably produced food. Once people realize all the benefits of pure maple, both production and consumption can easily grow together.[7]

In addition to the opportunities for pure maple to serve as the local sweetener for the Northeast, there is also a tremendous opportunity to supply international markets. Thanks in large part to the marketing efforts of the Federation, the largest market gains over the past decade have actually been overseas. Marketing campaigns have been extremely successful in other countries, as Canadian exports to Japan rose by 252 percent between 2000 and 2005 once the Federation initiated a marketing campaign there.[8] Markets have also been growing steadily in Western Europe, as

Canadian exports to Denmark and Switzerland each grew by more than 100 percent over the same time period. Whereas maple syrup is only produced commercially in the northeastern United States and Canada, people throughout the world want to consume this healthy, flavorful North American sweetener—especially when they are exposed to it! The more we promote maple syrup beyond our borders, the more pure maple we will sell to the rest of the world.

Some sugarmakers express concern about the possibility of overproduction driving down prices if supply exceeds demand. This has happened on several occasions over the past century, and many producers have suffered the consequences of falling prices for pure maple. However, I strongly disagree with the characterization of this problem as oversupply. With such low levels of consumption in the United States, in Canada, and throughout the world, we should never consider a short-term rise in supply as a case of overproduction. Rather, any surplus is simply the result of undermarketing. If we take the time and devote enough resources to teaching people about pure maple and the many uses for this all-natural, delicious, and healthy sweetener, we couldn't possibly make enough to meet market demand. Chapter 10 contains much more information on strategies for marketing your products to ensure a profitable and sustainable enterprise.

Income Diversification

There are very few sugarmakers who earn their entire livelihood from syrup production. Most sugarmakers who produce syrup to generate income do it as part of a larger farm operation or as a stand-alone business to supplement other income sources. Maple syrup is the first crop of the year for many farmers and generates income during a traditional slow period. If other crops fail or prices fall, income from maple production can help shoulder the blow. Producing and selling maple products also draws new customers to your farm stand or business, especially in regions where maple syrup is hard to come by. Any financial adviser will tell you it is a good idea to diversify your investments, and I believe adding syrup production to your farm portfolio could be one of the best investments you make.

CHAPTER 2

Is Sugaring for You?

After reading the first chapter, you may wonder why so few people actually produce maple syrup these days. At least that is what often goes through my head, especially when I travel around the Northeast and see so many untapped maples. Sugaring is an enjoyable, rewarding process that yields a delicious and nutritious product, so you would think more people would be doing it. In our survey research, we found that about 10 percent of landowners would like to produce syrup themselves and another 10 percent would like the trees on their property to be tapped by somebody else. Nearly half of the people we surveyed were not at all interested in syrup production while roughly a third were on the fence. Even though 10 percent of landowners would like to tap their trees and produce syrup, a much lower percentage actually follows through on this goal. There are a variety of reasons why people cannot fulfill their desires to become sugarmakers, so we set out to determine what the main barriers are to getting started.[1]

We were surprised to find that the number one obstacle identified was simply not having enough tappable trees. Since our survey took place in Maine, New Hampshire, New York, and Vermont—where maples are abundant—we did not expect a lack of trees to be the main barrier encountered by landowners. Certainly not all landowners are fortunate enough to have a large sugarbush on their property, but this doesn't have to be a limiting factor. Furthermore, I expect that many of the landowners who don't think they have enough tappable trees are probably just unaware of the resources they do have on their property. After all, sugar and red maples are two of the most commonly occurring species in the eastern United States. Even if you don't have many maples on your own property, chapter 8 provides much more information on strategies for leasing trees or buying sap from nearby landowners. If you live in the Northeast and really want to produce syrup, a lack of tappable maples should not be a significant barrier to getting started.

We were *not* surprised to find that the second greatest obstacle for landowners was a lack of time to devote to sugaring. This wasn't nearly as important for people who are retired, and in fact many people are taking up syrup production as an enjoyable and productive hobby in their retirement. But if you are working a full-time job, with kids to take care of and myriad other obligations, it's hard to find time to devote to maple sugaring. Still, we do make time for the important things in life, and there may be ways of incorporating a sugaring operation into an already busy life. Two of the

FIGURE 2.1. It's common to find hillsides of untapped maple trees when you're traveling through the Northeast. This chapter will help you decide whether tapping them is a good decision for you.

most important things in many people's lives are spending quality time with friends and family, and earning enough money to pay for living expenses. Sugaring can address both. After all, getting people involved in a worthwhile effort that brings everyone together around a common goal is an excellent way to spend quality time with friends and family. There are certainly long hours involved in gathering sap and processing it in the sugarhouse, and it is much more enjoyable when the work is shared among many people. Furthermore, depending on the size and scope of the operation, syrup production can also generate a significant revenue source for a family. I know many sugarmakers who take time away from their normal jobs to produce syrup in the spring.

FIGURE 2.2. Collecting sap can be a tiring affair, especially if you are only four years old! I know many parents who have their kids gather sap after school—it gets them fresh air and exercise, it keeps them out of trouble, and they sleep a lot better at night. PHOTO COURTESY OF JEREMY FARRELL

If they kept their normal work schedules during that time of the year, they wouldn't have enough time to fully devote to the sugaring operation. However, since sugaring can be a good source of seasonal income and an enjoyable activity, they make time for it. Everyone has 24 hours in a day—how you decide to spend that time is largely a matter of your priorities and the demands that life brings. If you can incorporate friends and family into your sugaring operation and generate enough money to make it pay for itself, not having time becomes less of an issue.

The other major impediment for landowners who would like to get involved in syrup production is the fact that the initial cost of equipment is so high (47 percent selected this option in our survey). There are several ways to get around this issue. If you have good credit and can convince a bank to loan you money, you may be able to finance the upfront costs and make the payments with the revenue generated by selling your products. There are banks such as Farm Credit that specialize in lending for agricultural ventures, so you may want to check with them first. Another option is to start out small and simply sell sap to a nearby producer. The cost in materials for buying buckets and spouts or setting up a tubing system usually ranges from roughly $6 to $15 per tap, and if you go with plastic bags, it can be as low as $1 per tap. Your sap collection costs can be much lower than building a sugarhouse and buying an evaporator, reverse osmosis, filter press, and all the other equipment items necessary for producing high-quality syrup in a cost-effective manner. Simply selling the sap and not processing it yourself also greatly reduces the time you need to invest in the process. Once tubing is installed and the trees are tapped, it only takes a limited amount of time to maintain the tubing lines and gather/transport sap, though gathering buckets or bags does require a lot of time and effort. Either way, just focusing on gathering the sap saves you considerable time and money in processing the sap into syrup.

Another no-cost option is to "purchase" the tubing materials with the syrup that will be produced from the sap you collect. I know several large producers who are also equipment dealers and regularly buy in sap from neighboring landowners. In some instances, these dealers have provided the tubing materials in advance to the landowner in exchange for sap or syrup the following sugaring season. This can be an attractive option if you would like to tap your trees but don't want to make any financial investments yourself. If you happen to live close to an equipment dealer who buys sap, you may want to look into this option.

Finally, if your sugaring operation will be small enough to be considered a hobby, you may be able to justify spending money on the equipment just for that reason. There are probably more people producing maple syrup for pleasure on a small scale than there are sugarmakers doing it as a profitable business enterprise on a large scale. Although hobbyists rarely make any money producing syrup, that is not the ultimate goal for many sugarmakers. Most people actually spend a lot of money on their hobbies without any chance of financial return. For instance, how many people do you know who actually make money golfing, skiing, or traveling? Yes, there are some professional golfers, skiers, and travel writers, but these are the exceptions, certainly not the rule. Sugaring is one of the few hobbies where you may eventually break even or earn a bit of money while having a great time producing an incredible product.

There are certainly many ways to get around the perceived barriers to getting started in maple production, yet it's important to realize that sugaring is not for everybody. It often requires long hours under difficult conditions, and the financial rewards don't always justify the efforts. Most sugarmakers consider their efforts a labor of love and greatly look forward to the coming of sugaring season every year. It's not about making money, but rather the experience and tradition that brings thousands of people into the woods year after year as winter fades into spring. However, sugaring is not always enjoyable, and sometimes it can be downright frustrating and tiresome, especially when your equipment fails and the weather isn't cooperating. Most sugarmakers love what they do, but are realistic about the fact that it is hard work. A poem by Albert Southwick addresses the difference in perceptions between those who produce syrup and the "city folk" who would like to be sugarmakers . . .

Sap Time ✦ by Albert B. Southwick

The city man drives past and sees
the sap pails on the trees.
He stops his car and steps outside
and sniffs the fragrant breeze.
He sees the happy farmers with their
maple trees on tap.
He breathes a sigh of envy but
he's never gathered sap.

He sees the sled and team come in—
it looks like so much fun.
The farmers look so healthy
and he wishes he was one.
But in his logic there is apt
to be one major gap.
For all his vim and eagerness,
he's never gathered sap.

He wanders to the sap house
with its clouds of fragrant steam.
He watches how the rising foam
is quelled by drops of cream.
He sees the golden syrup pour
and fill the thick felt nap.
He thinks it's simply super but
he's never gathered sap.

He never slogged for hours at
a stretch through mud and slush.
He's never emptied buckets
'til his mittens turned to mush.
He's never slipped and fallen down
and spilled it on his lap.
He thinks it's wonderful because
he's never gathered sap.

He doesn't go to bed to dream
of maples row on row.
With miles and miles of buckets
just about to overflow.
He thinks it's quite romantic.
He's a very pleasant chap.
But the brutal fact, my friends is that
he's never gathered sap.

The thing I like about this poem is that it presents some of the challenges sugarmakers can encounter. Sugaring is often romanticized, yet it can be extremely frustrating and tiring at times. However, I don't like the insinuation is that if the "city man" actually did the hard work of gathering buckets full of sap every day, he would not enjoy it and would think less of the sugaring process. While most people wouldn't choose to collect sap from hundreds of buckets in deep snow, those who do gather sap tend to enjoy it. I think it helps us appreciate the sugaring process and the final product even more. It's easy to lose sight of the blessings we are given; too often we take things for granted. While it may seem tedious and tiring to gather sap some days, it is useful to step back and realize how fortunate we are to be healthy enough to gather the delicious sap and produce such an incredible product.

Of course, tapping trees, gathering sap, and making syrup just isn't for everyone. Thus, before you venture into sugaring or invest a lot of time and money into expanding your sugaring operation, there are many questions you should ask yourself. The purpose of this chapter is not to scare you away from syrup production, but rather to encourage you to think of all the options and scenarios that could affect if and how you develop your operation. There are many questions that you should ask yourself as you plan your next moves.

FIGURE 2.3. If trudging through snow to collect buckets isn't something you are interested in doing, then I recommend setting up a tubing system that will let the sap flow to your collection tank—even when everything looks like a winter wonderland. PHOTO COURTESY OF PAM MASTERSON

How Much Time Do You Have?

As discussed above, not having enough time is one of the major obstacles for folks who would like to get into syrup production. Figure out how much time you *realistically* have to devote to sugaring during different parts of the year and plan accordingly. For instance, if you are very busy during sugaring season but have greater availability at other times, you may want to use a tubing system instead of buckets to gather sap. Tubing can be installed and maintained at any time, whereas buckets require daily attention and significant time when the sap is flowing. Another option, if you would really like to make syrup but don't have much time, is to invest in time- and energy-saving technology. For example, instead of spending sleepless nights putting wood into an evaporator, buy a reverse osmosis unit and let that remove about 80 percent of the water from the sap while you are at work or engaged in other activities. There are ways around a perceived lack of time—you just have to do your homework and make realistic plans.

Do You Like Collecting Sap and Making Syrup?

Although I would strongly recommend sugaring to just about everybody, I also suggest starting out slow and making sure you truly enjoy syrup production before getting in too big. Most people who start out small get hooked and continue to grow their business year after year. However, there are certainly others who find out how much work it is and scale back or stop making syrup altogether. One possibility is working with another sugarmaker for a season before branching out on your own. This will provide you with great insight into the process—you'll see what aspects you really like to do and what chores you would rather leave to someone else. If you really like gathering sap but don't care too much for the processing end of things, you may decide to just collect and sell sap. On the other hand, you may really enjoy running the evaporator and making syrup, so perhaps you could structure your operation to include a lot of sap buying (see chapter 8 for more information on these options).

How Many Trees Do You Have, and What About Your Neighbors?

If you have enough forestland, I would strongly suggest having a forester conduct an inventory of your woods and develop a management plan based on your goals and objectives. In addition to other benefits, this will give you a relatively accurate count of the number of tappable trees, providing valuable information on how to structure your sugaring operation. If you don't have many maple, birch, or walnut trees, but would like to develop a larger operation, then it would also be very useful to explore options with leasing trees or buying in sap from your neighbors (again, see chapter 8).

How Will Sugaring Fit into Your Overall Farm Enterprise?

During the 1800s, at the height of maple production in the United States, nearly all farmers with a decent number of maple trees tapped them in the spring to make maple syrup. Maple was the first crop of the season, and the timing fit well with other agricultural pursuits. Since sugaring season often coincides with mud season, there is little field work that can be done at this time. It is also the worst time of the year to be in the woods cutting trees and dragging out logs, so sugaring can occupy that brief window of time when it's too warm and muddy to do any logging yet too cold to grow any crops.

What Kind of Equipment Do You Have or Need?

Syrup production requires a lot of tools and equipment, but you may already have many items that could be used for your sugaring operation. Do you own a cordless drill? Hammers? A good source of clean, food-grade buckets? Large tanks? A garage, barn, or shed that could be renovated as a sugarhouse? Before making any purchases, see what you already have yourself, or what could be borrowed from friends and family.

How Many People Are Willing and Available to Help?

Talk to your friends and family to get a sense of who is interested in helping with various aspects of the

FIGURE 2.4. Maple sap dripping out of a spout is a familiar sight throughout the Northeast. PHOTO COURTESY OF STEVE CACCAMO/NEXT GENERATION MAPLE PRODUCTS

FIGURE 2.5. Collecting buckets is one of the best ways to get young people involved in a sugaring operation. Here students from Northwood School in Lake Placid head out to the woods to gather sap at Heaven Hill Farm. PHOTO BY NANCIE BATTAGLIA

enterprise. If you are planning a large-scale operation, you will need an abundant and reliable source of labor. Everyone has their own skills and talents, and a successful sugarmaker will be able to identify the unique qualities of each person and assign him or her to tasks that utilize these natural talents.

How Much Money Are You Willing to Invest?

The amount of money you are willing to spend will greatly determine how large an operation you can start and whether you will be purchasing new or used equipment. While it is usually true that you need to spend money to make money, when starting a maple enterprise it is important to control costs if you want to receive a decent return on your investment in a timely manner. Decide on how much money you are willing to invest and stick to it. It is easy to spend tens of thousands of dollars even on just a small hobby operation. Before shopping around for equipment, it is a good idea to have a reasonable budget in mind and to explore all options before you make any major purchases.

Are There Nearby Producers Who Could Boil Sap?

If you are considering just collecting sap and bringing it to someone else to be processed into syrup, then you should see who else is making syrup nearby and if they would be interested in a collaborative arrangement. In some areas there are a lot of sugarmakers and it would be relatively easy to find someone to purchase sap or boil it on shares, whereas in other regions you may be the only one tapping trees within a 20-mile radius. While collecting sap and having someone else boil it for you may sound appealing, if there is nobody nearby willing or able to do that for you, it's simply not a feasible option.

What Are the Local Markets for Pure Maple Products?

There are some areas where sugarmakers are a dime a dozen and you can find maple products for sale everywhere. However, there are a lot more areas where nobody is making syrup, and if you started your own operation, there would be little to no competition. From a production perspective, I'd rather be located in a place with lots of other sugarmakers. You will likely form lasting friendships with other sugarmakers, and you will quickly gain a foothold in the sugaring community. There will be plenty of other folks to bounce ideas off and troubleshoot when you are having problems. However, strictly from a marketing perspective, I would much rather make syrup where I'm the only game in town. When you are unique and produce something that no one else has locally, the chances of selling your products at a profitable level are much greater.

To illustrate the opportunities and challenges in making maple syrup outside of the typical production areas, consider the case of the Benson family in southern Illinois. I recently came across a story in which someone reported them to the police for operating a crystal meth lab. The person had no idea why else there would be large amounts of steam coming off a property and buckets hanging on trees. The county sheriff went out to investigate and of course was pleased to find nothing but pure maple syrup being produced at the Benson household. The story got a lot of media attention and quickly became the talk of the town. An editorial in the *St. Louis Dispatch* gave the Bensons some great free publicity, noting that it is comforting that there are still places that are more like "the little house on the prairie" rather than the "little meth lab in a trailer." What started out as a small hobby may develop into a new business venture, as the Bensons now have far more demand for syrup than they can produce themselves.

CHAPTER 3

Sugarhouse Design and Construction

Unless you are planning on selling all your sap to another sugarmaker, one of the first things you need to think about is where you are going to process all your sap into syrup. Building your sugarhouse is one of the most exciting and rewarding things you will ever do as a sugarmaker. I have never seen two sugarhouses the same—each one is a unique reflection of the sugarmakers' creativity and preferences. They are often referred to as "sugarshacks," but nowadays many sugarhouses don't look anything like a "shack." Before you build your sugarhouse, I highly recommend visiting as many other sugarhouses as you can. You will undoubtedly see many things that you want to do (or don't want to do) in your own operation. Talk to the sugarmakers and ask them what they like about their sugarhouse, what they don't like, what they are glad they did, and what they wish they hadn't done. Hindsight is 20/20, and most people only get one chance to build a sugarhouse, so spending a lot of time exploring various options will be time well spent. This chapter certainly won't teach you everything you need to know about building a sugarhouse, but it will provide some of the general principles you should consider in designing your unique operation.

FIGURE 3.1. If you have the resources available, I recommend constructing a timber-frame sugarhouse. Although it may cost more than traditional framing, the added beauty and structural stability make it an ideal location for making syrup. You may not need a fancy new building to produce maple syrup, but having one sure helps in marketing maple as the gourmet food that it is. PHOTO BY NANCIE BATTAGLIA

Location

Everyone has heard the expression that the three most important features of real estate are "Location, location, location." Indeed, where to locate your sugarhouse is of the utmost importance in determining what type of sugaring operation you will have. If you don't already have a sugarhouse built or your location in mind, there are several aspects that you should consider.

Accessibility to Trees

Traditionally sugarhouses were located in a central area within the sugarbush in order to facilitate sap collection with buckets. This was especially true in relatively flat areas where the sap was collected with horses or carried in by hand. With the advent of tubing, sugarhouses were naturally placed at the bottom of a hillside to allow gravity to deliver all the sap to one central location for processing. It often makes sense to locate the sugarhouse at the lowest point on your property, but this may not work for everyone. Depending on your particular situation, it may be more advantageous to put sap collection

FIGURE 3.2. An ideal sugarhouse location is at the base of a hillside where all the sap flows directly to the feed tanks for your RO or evaporator. Not all sugarmakers are that lucky! PHOTO BY NANCIE BATTAGLIA

facilities at the lowest point and then transfer the sap to another location for processing into syrup.

Proximity to a Road

Sugarhouses built today are often located along a well-traveled, easily maintained road for a number of good reasons. First of all, if you would like to use your sugarhouse for marketing maple products, it makes sense to be as visible and accessible to the general public as possible. While some people may like the adventure of going deep into the woods to visit a sugarhouse and watch maple syrup being made, many more people would rather visit a place that is on a paved road with nearby

FIGURE 3.3. It would be unusual to see a sugarhouse located right along a railroad track in today's world. This picture was taken in the early 1900s at a large sugaring operation in the Adirondacks. PHOTO COURTESY OF THE ADIRONDACK MUSEUM

parking. Being on a well-traveled road will also draw visitors who only discovered your operation because they could see it as they are driving by. Nothing draws people into a sugarhouse like steam coming out of a cupola and an open sign clearly visible by the entrance.

You might also want to place your sugarhouse by the road if you are collecting sap from many different locations. Having easy access to receive sap is essential to growing your operation through leasing agreements or buying sap. It would be very difficult to collect sap from nearby sugarbushes or purchase sap from your neighbors if getting the sap to the sugarhouse is challenging. Making the delivery of sap as easy as possible will increase your ability to grow your operation in the future.

FIGURE 3.4. If you aren't located on a main road, it is important to build a nice gravel road to your sugarhouse. This is especially important if you will be hauling sap, whether with horses, a tractor, or a pickup truck.

Proximity to Utilities

Although people produced plenty of maple syrup and sugar for centuries before electricity was harnessed, having reliable electric service will make your operation much more efficient and enjoyable. You'll be able to use various electronics (including an automatic drawoff), lighting will be simplified, and you'll be able to easily operate an RO and electric vacuum pump. Even the most efficient wood-fired evaporators now require electricity to run the fans for gasification purposes. Although generators can provide electricity, they require additional maintenance and the noise tends to take away from the peaceful nature and tranquility of sugaring. Solar panels and windmills are certainly possibilities for generating your own electricity, although being attached to the grid will make them much more useful. Sugaring requires a lot of electricity for a brief period—probably more than you can generate yourself. Being tied to the grid will allow you to sell the excess capacity during the rest of the year and draw additional power during sugaring season.

Indoor plumbing with running water is also a huge plus. This will allow you to have real bathrooms in lieu of an outhouse or port-a-potty and will make it much easier to clean and maintain all your floors, surfaces, and expensive stainless-steel equipment. For small, hobby-scale operations, you can get away without running water, but if you believe that a clean home is a happy home, you'll be much happier with it. If government regulations or the large syrup buyers eventually dictate that syrup be produced in a facility that meets certain minimum guidelines, it is highly likely that running water will be required. Even without mandatory regulations, if you plan on selling products and hosting visitors to your sugarhouse, running water and indoor plumbing are almost necessities. Having a good answer for "Do you have a bathroom?" will lead to much happier visitors and will make your life as a sugarmaker much more convenient.

Possible Conversion of Existing Building

I know many sugarmakers who have chosen their sugarhouse location by simply repurposing a building they already had. Converting an existing barn, garage, or other structure into a sugarhouse may make sense, depending on the features of the building and what you desire for your sugarhouse. This can be a much more cost-effective strategy than building a new structure, so it's a good idea to consider what you already have before building something new.

Design Considerations

Once you have chosen your location, the next step to consider is how you are going to design the sugarhouse.

Going Big While Going Green

Despite what some people might claim, becoming a large-scale maple producer does not necessarily mean compromising your values or being wasteful with energy. On the contrary, consider the example of Tom Gadhue of Solar Sweet Maple Farm in Lincoln, Vermont. On the way back from picking up his son Ryan from graduation at architecture school in Georgia, they had a long ride to contemplate their next moves in life. Tom had made syrup in the past and wanted to get back into it. Ryan was fresh out of school with a lot of new ideas on how to construct environmentally friendly and energy-efficient buildings. Shortly thereafter the Gadhue family embarked on an ambitious project to start a 15,000-tap sugaring operation with a net-zero carbon footprint. They installed solar panels on the roof and are able to supply more power over the course of the year than they use in sugaring season. They use much more power than they can generate from February through April, when they are making syrup, yet in the summer months they use very little power and are able to supply the grid with additional capacity on those hot and sunny days when lots of homes and businesses have their air conditioners running. The Gadhues have put in variable-frequency drive (VFD) and slow-start motors on all their electric pumps and machinery. Although the initial cost was higher, it will save them money in the long run.

Not only is their 4,000-square-foot sugarhouse energy-efficient, it is also a handsome structure. They used reclaimed timbers from an old post-and-beam barn and the siding came from some old Vermont barns, adding to the beauty and charm of the building. The Gadhues' operation has drawn a lot of interest among other sugarmakers and the general public. The *Maple News* ran a front-page story on them in 2012, and their operation will soon be featured in a video about Vermont maple syrup production at a rest area on the interstate. The Gadhues have proven that building green not only saves energy and money, but is also a very useful marketing tool!

This will depend a lot on the size and scope of your sugaring operation, your budget, and your desire to accommodate visitors.

Size

The main consideration that will determine the size of your sugarhouse is the size of your evaporator. It is typically placed in the center of the building (though it doesn't have to be) and should allow for plenty of operating space around it. I recommend a minimum of 5 feet around all the sides—more if you plan on having visitors in your sugarhouse. If you will have a wood-fired evaporator, it is important to have even more space in front for loading wood into the firebox.

I have never met a sugarmaker who complained that his or her sugarhouse was too big! When you are deciding on what size to build it, always err on the side of making it too large, especially if you have the time and resources to build it that way. If you have to build a smaller sugarhouse than you would prefer, be sure to design it such that it is easy to add on to. When your operation grows and you want to do more with your sugarhouse, you'll be glad that you planned ahead.

Cupola

The rule of thumb is to make your cupola as large as your evaporator. This ensures that steam can easily get out of the sugarhouse without condensing and dripping water back onto the evaporator, yourself, or your visitors. Cupolas come in all shapes and sizes, so I strongly recommend checking out as many sugarhouses as you can to see what design you like the best. I prefer cupolas that can be easily opened and closed to keep the critters and weather out when you aren't making syrup. Including screens and wire cages will also help to keep out curious wildlife species attracted to the sweet smells and warm cover.

FIGURE 3.5. Having ample room around your evaporator makes your sugaring operation much safer and easier to manage. Although few of us will ever build a sugarhouse of this magnitude, it helps to make it as big as you can afford to. PHOTO BY NANCIE BATTAGLIA

FIGURE 3.6. This cupola opens up completely on the sides when the evaporator is in operation and can be easily closed off when not in use. PHOTO BY NANCIE BATTAGLIA

FIGURE 3.7. Rather than installing a cupola, this sugarhouse discharges the steam directly through the roof with large piping. A pulley system allows the lids to be closed when the evaporator is not in use. PHOTO COURTESY OF PETER SMALLIDGE

A lot of modern sugarhouses are moving to a cupola-free system in which the steam is collected with a steam hood and piped through the roof. These systems work well when installed properly, but they sacrifice the traditional look of the cupola on a sugarhouse. If you plan on bringing visitors to your sugarhouse and you want to project the traditional image of a sugarhouse, I highly recommend installing an attractive cupola.

Sap Storage

It is important to have plenty of sap storage on an elevated structure above the evaporator. Since sap weighs about 8 pounds per gallon, it is equally important to make sure the platform that holds the tank is sturdy. Ideally sap storage would be located in a well-ventilated area that stays cool and is easily accessible for cleaning. Sap can spoil rapidly, especially once it has been concentrated with reverse osmosis, so keeping your sap cool is essential for producing high-quality, light-colored syrups.

Access Doors

If you make syrup for enough years, there is a good chance that you will want to move different evaporators

FIGURE 3.8. In this sugarhouse, the RO is placed in a well-insulated, heated room that can easily be seen from the drawoff valve on the evaporator.

and other equipment in and out of your sugarhouse at various times. Technology is always improving, sugarmakers are usually expanding, and it's only a matter of time before you wind up getting yourself a new evaporator. To accommodate this, make sure that you put in doors wide enough that you can easily move equipment in and out of your sugarhouse.

Electricity

Since sugarhouses are extremely moist environments, you want to make sure that all your outlets are of the GFCI (ground fault circuit interrupter) variety. ROs and vacuum pumps can require significant electricity draws, so oversizing your service is a good idea. It is cheaper and easier to put in additional capacity the first time than to go back and redo it in the future.

Lighting

Proper lighting, both natural and artificial, will make working in your sugarhouse much more enjoyable, efficient, and safe. All your lights should either be shatterproof or come with protective covers to prevent them from breaking and possibly getting broken glass in your sap and syrup. When somebody calls to let you know that they found broken glass on the bottom of a jug of syrup, you want to be able to tell them with 100 percent confidence that it is actually sugar crystals that precipitated out because the syrup you sold them was extra thick.

Heat

Very few sugarhouses are heated, but if you do want to put heat in your sugarhouse, I recommend radiant floor heat in your concrete slab. Most sugarhouses have poured concrete floors, and it doesn't cost much money to lay down some PEX tubing before you pour concrete. Furthermore, when standing near your evaporator for extended periods on cold nights, you'll greatly appreciate having warm feet.

Drains

Working drains are very important to keep your sugarhouse clean and safe for yourself and visitors. Sugaring season usually coincides with mud season, and spraying down floors is usually much easier and more effective than sweeping. We also generate a lot of wastewater when cleaning evaporator pans, barrels, filters, and other equipment, so it is worth the effort to put in high-quality drains when you are pouring the concrete floor. Although this should be obvious, be sure that the floor slopes to the drains! I'm sure we have all seen poorly installed concrete pads that don't actually do so. When water gets on the floor and doesn't drain, it can create a dangerous slipping hazard if it turns to ice. This is a valid concern since most sugarhouses are unheated and naturally dip below freezing at night during sugaring season.

Parking

If you plan on having visitors to your sugarhouse, be sure to carefully plan for parking. The parking area should be well drained, free of ice and other hazards, and well lit in the evening.

Reverse Osmosis Room

Most of the ROs being sold on the market today require a heated room to prevent them from freezing. Some models can be drained when not in use and therefore do not need to be in a heated space, but the vast majority still require warm rooms to keep ice from destroying your pumps. Ideally you should place the RO in a location that is easily seen from the drawoff area on your evaporator. Being able to watch the evaporator and keep an eye on your RO at the same time will help you run a smoother operation. It pays to make sure the room is well insulated, as losing power and heat for an extended period in the winter could do serious damage to your RO. With good insulation, however, the RO may be fine for several days without heat while you wait for power to be restored, even when temps outside are below freezing.

Final Thoughts

Although very few of us have all the resources we would need to build the sugarhouse of our dreams, we should all try to make our sugarhouses as nice as possible. If you invest sufficient time and resources to build a comfortable, attractive, and functional sugarhouse, you will want to spend more time there and will probably get more enjoyment out of it. Remember that maple syrup is a luxury, gourmet food and should be made in a reputable facility. If you want to exemplify high-quality maple syrup, you'll need a high-quality, attractive sugarhouse. The last thing you want is for someone to walk in to your sugarhouse and say, "They make food here?"

An Untapped Resource:
The Sap-Producing Trees of North America

Did you ever wonder where the phrase *untapped resource* came from? Or why people often say that we should *tap into* something? According to Merriam-Webster's dictionary, the first known use of *untapped* in the English language was 1779, many years after Europeans had come to America and learned how to tap maple trees to produce sugar. While maple trees that aren't used for syrup production are certainly an untapped resource, this expression has also made its way into other segments of our society. Webster defines *untapped* as "not drawn upon or used," and the example they give is "untapped reserves of coal." While coal miners certainly aren't installing taps in mountainsides to gather coal, sugarmakers do tap into maple trees to collect sap. It is possible that the popular use of *tapping* and *untapped* dates back to the days when many American farmers produced sugar from their maple trees. As our language and culture have evolved over the centuries, we have begun referring to many underused resources as untapped.

Given the vast resource of maple, birch, and walnut trees throughout the temperate forests of the world, there are two widespread myths that I would like to dispel in this chapter:

1. Maple syrup can only be produced in the northeastern United States and Canada.
2. Maples are the only trees that can be tapped to produce syrup.

These are common misconceptions based on the fact that the vast majority of maple syrup does come from eastern North America and there is very little syrup made from other tree species. However, there are trees growing in every state and province that can be tapped for sap and syrup production, including a wide variety of maple, birch, and walnut trees. The only state that doesn't have any of these species is Hawaii, though people do collect sugary sap from palm trees in tropical regions. Eastern North America may offer the best climate and forests for making maple syrup, but as long as you live somewhere that has distinct seasons and freezing temperatures in the winter, you can probably find trees to tap between January and April.

This chapter describes the main species of maple, birch, and walnut found in the United States and Canada. I have included some pictures that will help you in identifying the tappable species, but this chapter should not be used as a field guide for identifying individual trees. If you think you may have some tappable trees, but are unsure how to identify them, I suggest purchasing a field guide for trees in your region. There are excellent resources on the web, in particular the Virginia Tech dendrology course website, available at http://dendro.cnre.vt.edu/dendrology/main.htm. The SUNY College of Environmental Science and Forestry also has great videos of Professor Donald Leopold talking about 100 tree species of the Northeast, including all the sap-producing species discussed in this book. Simply type "youtube dendrology ESF" into a search engine and it is bound to come up at the top of the list.

The Maples (*Acer* spp.)

Although tapping maples for syrup production is a North American tradition, this process could

Sugaring in Denmark

My first summer working for Cornell I got a visit from Oluf Steen Carstensen and his Danish family when they were traveling on vacation in Lake Placid. I was particularly surprised when Oluf told me he that he had just started making maple syrup in his home country of Denmark. Oluf's father had planted about 25 sugar maple trees back in the 1960s when he was only about 8 or 10 years old. Unfortunately the trees were not well taken care and most of them died, but the five that did survive grew quite rapidly. Oluf has been tapping these trees and producing syrup for the last decade with fairly good results. He is a scientist by trade and therefore keeps great records of weather conditions and sap flow—it has turned into one of his favorite hobbies. Denmark has a relatively mild climate thanks to the Gulf Stream, so he usually starts tapping in late January or early February and his season lasts until sometime in March (occasionally early April). Having benefited from his father's efforts in planting trees, Oluf wanted to continue the tradition and brought back seed from our "sweet trees." He has since planted dozens of seedlings and will eventually have a much larger sugaring operation for his own kids to carry on this new tradition. Oluf's experience is a testament to the fact that maple syrup can be produced in unusual places, especially if you put some effort and long-term planning into it.

FIGURE 4.1 A few of Oluf's sugar maples being tapped for syrup production with the Danish flag waving proudly in the wind. PHOTO COURTESY OF OLUF STEEN CARSTENSEN

theoretically be carried out in many other parts of the world. There are over 100 species in the genus *Acer* spanning five continents, with the greatest diversity occurring in China.[1] A wide variety of maples are also found in the very northern border of Africa, the Middle East, the Himalayas, Europe, and Southeast Asia. There are actually a few tropical and subtropical maples that keep their leaves year-round, and *A. laurinum* is the only species of maple found in the Southern Hemisphere. I have heard of people planting sugar maples and successfully producing syrup in New Zealand and Chile—of course their optimum sap flow period would be August through October. One of the most unusual places that I have heard of finding maple trees is Abbottabad, Pakistan. The city made famous for the capture of Osama bin Laden was once known as the City of Maple Trees since all the roads had been lined with maples. There are also remnant populations of a sugar maple subspecies (*Acer saccharum* subsp. *skutchii*) in the high-elevation cloud forests of Mexico and Guatemala.[2]

Part of my dissertation research at Cornell involved looking at the growth potential of the maple industry in the United States, in particular the number of sugar and red maple trees that could be tapped in each state. I performed analyses using the latest USDA Forest

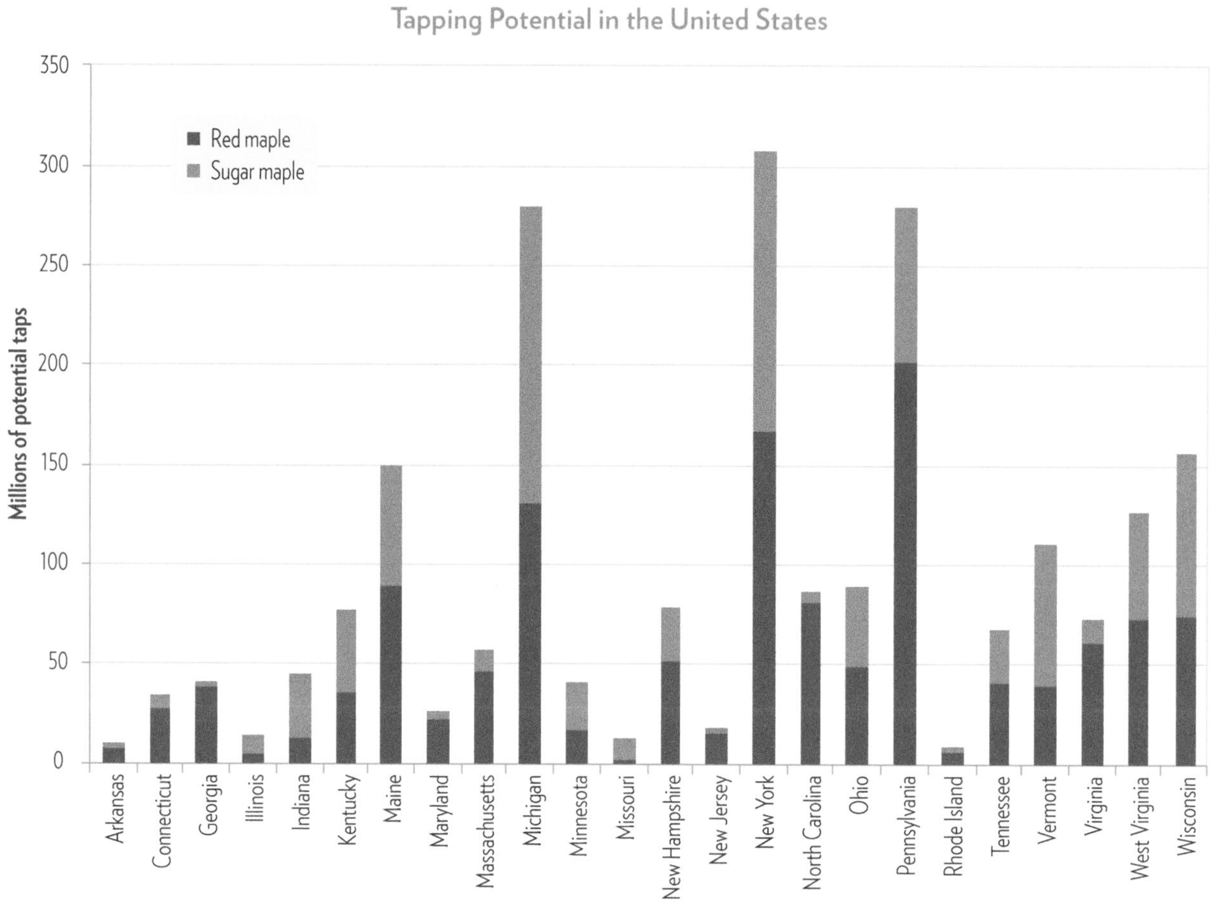

Tapping Potential in the United States

FIGURE 4.2. Total number of potential taps from sugar maples and red maples in United States.

Service Forest Inventory & Analysis (FIA) data from 25 states that contain a significant number of sugar and/or red maples. The FIA program measures permanent forest inventory plots on a periodic basis throughout the country, providing the most comprehensive assessment of what is happening to our forests and trees over time.[3] This is an invaluable database that allows researchers to answer a variety of questions about our diverse forests.

As seen in figure 4.2, I estimated the number of potential taps by summing all the live sugar and red maple trees greater than 10" diameter at breast height (dbh) and applying conservative tapping guidelines: one tap for a 10" to 17" tree and two taps for trees 18" and greater. I further classified the data according to ownership status, public or private. For the public land, the trees were divided between the tappable (non-reserved) and non-tappable (reserved) trees, as the reserved forestlands that are legally prohibited from timber production are also likely to be restricted from syrup production. An example of this is the Adirondack Park in New York, where state guidelines prohibit any trees from being tapped.

With the vast hardwood forests of the Upper Peninsula, Michigan has the largest resource of sugar maples in the United States. Pennsylvania is also a heavily forested state and contains the most red maples. When considering sugar and red maples combined, New York has the greatest number of total potential taps. The more southerly and western states tend to have more red maple than sugar maple trees, though there are exceptions to this "rule." For instance, Connecticut, Maine, Massachusetts, and Pennsylvania all have significantly

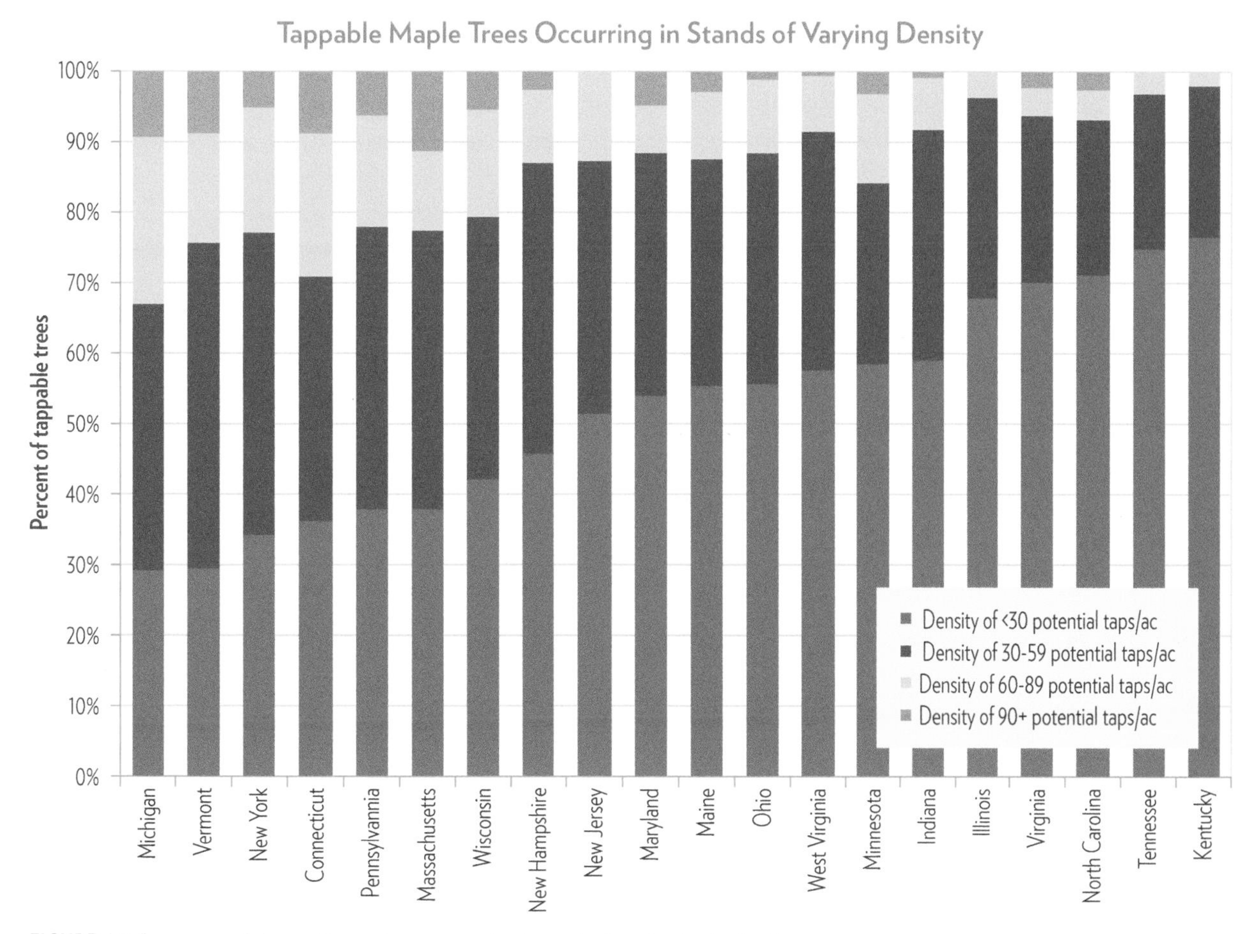

FIGURE 4.3. Percentage of all tappable maple trees occurring in stands of varying density for 20 states.

more red maples than sugar maples, whereas Illinois, Indiana, and Kentucky have more sugar maples than red.

It is important to realize that the figures presented here overestimate the realistic tapping potential for several reasons. To economically tap maples, the trees must be accessible, located close enough to an access road, and the density of trees must be high enough to justify installing a tubing or road system to collect the sap. Whereas the FIA data include all sugar and red maple trees growing on non-reserved forestland, many of these trees are growing in locations that are not suitable for tapping. Some of them are in stands that have a low density of maples, are too far from an access road, or are otherwise inaccessible. Thus, I performed further analyses of the FIA data that takes into account the density of maples in a stand and the distance of the trees to a road.[4]

The main limiting factor as to whether or not trees can be realistically tapped is the density of tappable trees within a stand, so I developed four categories of sugarbushes based on the number of potential taps per acre. Maples occurring in low-density stands that have less than 30 taps per acre are not really suitable for installing a tubing system or putting in roads to gather the sap. Stands with 30 to 60 taps are suitable for sugaring, though not ideal, and stands with 60-plus taps per acre are definitely worth tapping. Those that have more than 90 taps per acre are ideal and would likely even benefit from thinning out some of the trees to develop the "perfect sugarbush." Figure 4.3 shows the percentage of all tappable maple trees occurring in stands of varying density for the 20 states in the eastern United States that contain a significant number of maple trees.

The vast majority of maple trees in Appalachin states such as Kentucky and Tennessee occur in stands where maple is a minor component of forests that are primarily dominated by oaks and hickories. On the other hand, northern states such as Michigan, New York, and Vermont have the majority of their maple trees occurring in stands where sugar and red maples are the dominant species. In most woodlots containing maples in these states, the density is sufficient to justify tapping. If you happen to have maples on your property where the density is not ideal, you could still theoretically tap them; just remember that your costs of running tubing or installing roads or trails to gather sap will be much higher than normal.

Finally, it is important to note that the FIA data I used to develop these figures only deal with forestland and therefore do not account for a significant percentage of the trees that are actually tapped. Maples growing in yards, parks, hedgerows, and along roads are favored by producers who collect with buckets due to the easy access and large volumes of sweet sap they generate. To quantify the potential for tapping these trees, much more detailed inventory data must be collected and analyzed through urban and community forestry research initiatives. The Forest Service FIA program is hoping to expand research plots to include more urban areas, so we may have more detailed information on the urban forest resource in coming years.

Even if you don't have much woodland or any maples growing in your woods, given the popularity of planting maples for shade and ornamental purposes there is a good chance that you have an abundant supply in your area that you could tap. This is especially true in the eastern half of the country, though maples are often planted in the Midwest and western states where suitable climates exist. As soon as you start to look, you will probably start seeing them everywhere. The challenge and opportunity exists in persuading the people who own the trees to let you tap them—that subject is covered extensively in chapter 8.

Utilization of the Maple Resource for Tapping

The National Agricultural Statistics Service has been keeping track of maple syrup production for 10 states in the eastern United States since the early 1900s. Based on the Forest Service FIA data and the latest tap counts

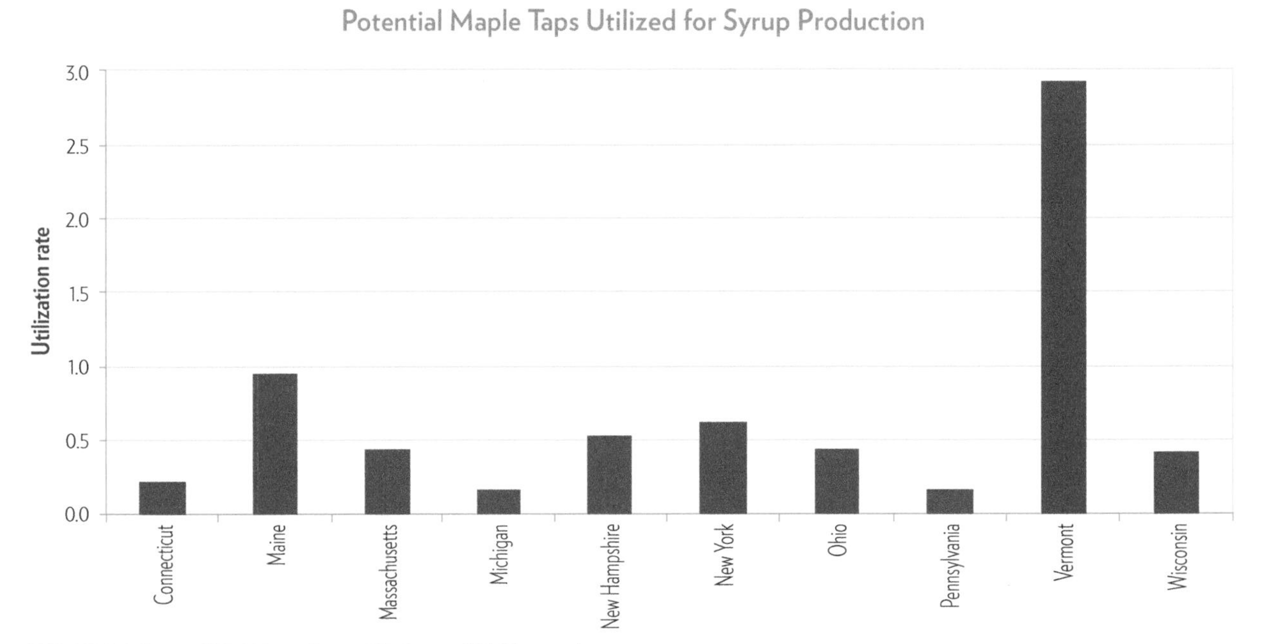

FIGURE 4.4. Source: USDA Forest Service FIA data and NASS report from 2011.

TABLE 4.1: Percentage of Trees Tapped Based on Four Categories of Potential Sugarbushes

		Category 1 Sugar maples on private land Stands with 60+ taps/acre Within 0.5 mile of access road		Category 2 Sugar and red maples on private land Stands with 60+ taps/acre Within 0.5 mile of access road		Category 3 Sugar maples on private land Stands with 30+ taps/acre Within 1 mile of access road		Category 4 Sugar and red maples on private land Stands with 30+ taps/acre Within 1 mile of access road	
State	Thousands of taps in 2011	Thousands of potential taps	Utilization rate	Thousands of potential taps	Utilization rate	Thousands of potential taps	Utilization rate	Thousands of potential taps	Utilization rate
Connecticut	71	–	–	7,550	0.9%	1,029	6.9%	17,613	0.4%
Maine	1,470	5,572	26.4%	10,548	13.9%	26,519	5.5%	55,334	2.7%
Massachusetts	245	–	–	5,718	4.3%	1,314	18.7%	19,989	1.2%
Michigan	495	24,786	2.0%	44,103	1.1%	66,258	0.7%	128,251	0.4%
New Hampshire	420	3,370	12.5%	6,566	6.4%	9,316	4.5%	30,634	1.4%
New York	2,011	15,126	13.3%	48,429	4.2%	63,486	3.2%	171,361	1.2%
Ohio	405	1,290	31.4%	8,589	4.7%	11,830	3.4%	35,717	1.1%
Pennsylvania	503	7,195	7.0%	32,033	1.6%	26,902	1.9%	116,446	0.4%
Vermont	3,300	6,360	51.9%	12,100	27.3%	32,375	10.2%	56,493	5.8%
Wisconsin	660	6,487	10.2%	13,537	4.9%	28,546	2.3%	55,969	1.2%
GRAND TOTAL	**9,580**	70,187	13.6%	189,174	5.1%	267,574	3.6%	687,808	1.4%

identified by NASS, I determined the percentage of all tappable maple trees being utilized for syrup production for the 10 largest producing states. As you can see in figure 4.4, the utilization rate in Vermont is well above every other state, yet Vermonters are still tapping only 3 percent of the sizable maple trees. All other states are using less than 1 percent of the available trees for syrup production. Michigan and Pennsylvania are ranked at the bottom of the list, even though these two states contain the second and third largest maple resource. Clearly there is an enormous potential to increase the amount of syrup being produced in the United States.

Although the percentage of maple trees being utilized for syrup production seems extremely low, there are good reasons why some maple trees shouldn't be used for syrup production. Therefore, I calculated utilization rates for all these states based on different categories of potential sugarbushes that take into account the density of maples in a stand as well as the distance of that stand to an access road. The first category of an optimal sugarbush contains over 60 taps per acre of sugar maples that fall within 0.5 mile of a road. The second category is nearly identical, but includes sugar and red maples to get the 60 taps per acre. Categories 3 and 4 allow for fewer taps per acre (30) and can be as far as 1 mile from a road. These are feasible to tap, yet not optimal for sap collection purposes.

Table 4.1 provides the utilization rates for 10 states based on four categories of potential sugarbushes. There are broad differences among states in the percentage of available trees being used for syrup production. If we consider all of the feasible taps (Category 4), Vermont sugarmakers would be tapping 5.8 percent of their available trees. However, by considering only the most optimal sugarbushes in Category 1, the utilization rate would be over 50 percent. Since a large percentage of the optimal sugarbushes are already being utilized, for Vermont to increase its output of maple syrup, sugarmakers will have to tap trees that may not be the most economical to gather sap from. Maine and New York regularly trade places as the number two ranked syrup production state, although New York is utilizing a much lower percentage of its tapping potential than Maine. If all the taps were located in optimal sugarbushes, Maine would be using 26 percent of its capacity whereas New York would be using far less (13 percent). Considering all the feasible taps, Maine's utilization rate is 2.7 percent whereas New York's is 1.2 percent. Michigan and Pennsylvania have the lowest utilization rates and greatest potential for growth. Only 2 and 7 percent, respectively, of their most optimal taps are used for syrup production, and only 0.4 percent of all the feasible taps are utilized in each state. In particular, with nearly 25 million potential taps in the optimal density category, Michigan has the greatest growth potential of any state.

Overall, with 9.6 million taps spread out over the 10 major producing states, US sugarmakers are utilizing 2.5 percent of all the potential taps in Category 2 and 0.8 percent of all the potential taps in Category 4 sugarbushes. To put this in perspective, Quebec puts out 40 million taps each year out of a total potential of 110 million potential taps in their Category 2 sugarbushes, resulting in a utilization rate of 36 percent. Quebec dominates the maple industry not because it has the largest resource of tappable trees, but rather because sugarmakers there use such a large percentage of the tappable trees for syrup production. We can certainly do a better job of utilizing the maple resource in the United States.

Suitable Maples for Tapping

There is clearly an enormous resource of sugar and red maples in the eastern United States and Canada, yet there are also many other species of maple that are suitable for tapping in our part of the world. In fact, every state within the continental United States contains tappable maple trees, even states such as Florida, Texas, and Arizona that most people would never associate with maple trees. This section highlights the eight different species of maples found in North America that you would be most likely to tap—sugar maple, black maple, red maple, silver maple, Norway maple, canyon maple, bigleaf maple, and boxelder. I've also mentioned species from the American West and one species of maple found in Asia, *Acer mono*, to demonstrate the international importance of maples.

Sugar Maple (*Acer saccharum*)

Sugar maple was given its name since it generally yields the highest volumes and concentrations of sweet sap. It is without a doubt the best tree to produce syrup from in the world. It grows naturally throughout the eastern United States and thrives in moist, well-drained, fertile soils. If the soil is too wet, too dry, or not very fertile, sugar maples may not grow at all, and if they do occur, chances are they will be stunted or exhibit symptoms of decline. Although many people consider it to be a northern, cold-climate species, sugar maples actually do quite well in the warmer, more southerly states, provided there is adequate rainfall and soil moisture. Sugar maples are extremely shade-tolerant and will grow slowly in the understory of an established forest for many years. However, when given more sunlight, they can grow rapidly, sometimes achieving growth rates of 0.5" or more in diameter per year.

Sugar maples are also widely known for their fall foliage and lumber. Sugar maple leaves turn a brilliant mix of yellow, red, and orange throughout the autumn, making them the premier species for leaf peeping throughout the Northeast. Their lumber is the hardest of all the maples, so it is also commonly referred to as hard maple or rock maple. The fact that the lumber is so valuable causes many landowners to hesitate about drilling holes in them for sap production; however, as you will see in chapter 15, tapping for syrup production is usually more lucrative than traditional timber management.

Black Maple (*Acer nigrum*)

Originally thought of as a subspecies of sugar maple, black maples are virtually indistinguishable from sugar maples in their appearance and uses. They have a smaller range than sugar maples and are usually found in states surrounding the Great Lakes. The most distinguishing characteristic is the fact that black maples tend to

FIGURE 4.5. It is hard to find a nicer scene than a row of sugar maples with buckets hanging on them in the spring. PHOTO BY NANCIE BATTAGLIA

FIGURE 4.6. With five rounded lobes and the perfect shape, sugar maple leaves are an iconic symbol of the Northeast. PHOTO COURTESY OF PETER SMALLIDGE

FIGURE 4.7. The buds of sugar maple are brown and pointed. PHOTO COURTESY OF ROB ROUTLEDGE, SAULT COLLEGE, BUGWOOD.ORG

FIGURE 4.8. The leaves of black maple look very similar to sugar maple, but usually only have three distinct lobes. PHOTO COURTESY OF PAUL WRAY, IOWA STATE UNIVERSITY, BUGWOOD.ORG

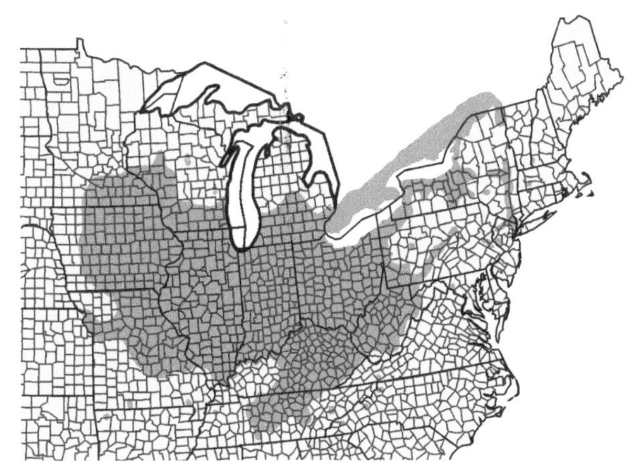

FIGURE 4.9. The range of black maple is centered along the Great Lakes and Upper Midwest.

have three major lobes in their leaves, as opposed to five in sugar maple. Although you may never be able to tell black maples and sugar maples apart, don't let that bother you—the black maples will produce just as much sweet sap as sugar maples.

Red Maple (*Acer rubrum*)

Red maples are quickly becoming the dominant tree species throughout the eastern United States. They grow fast in a wide array of soils and climate, tolerating both extremely wet and dry conditions. I have seen red maples growing well from the southern tip of Florida all the way into Canada, providing one of the broadest geographic ranges of any tree species. Another name for them is soft maple, as their lumber is not nearly as dense as that of sugar maple. The prices for soft (red) maple lumber are usually significantly less than those for hard (sugar) maple, so there is not nearly as much pressure to log these trees.

The sap sugar content of red maples is generally lower than that of sugar maples; they also bud out earlier in the spring. Although this has the potential to reduce syrup quality toward the end of sugaring season, the sap is often done flowing by the time this happens. Red maples have always been considered inferior to sugar maples, and many sugarmakers don't bother tapping the red maples in their sugarbush. Although red maples might not be as productive as sugar maples, they do provide an excellent source of sweet sap and should not be disregarded, especially for larger sugaring operations with reverse osmosis units. Since ROs remove 80 to 90 percent of the water content before boiling, the initial sugar content of the sap does not matter nearly as much. Red maples produce syrup of comparable flavor and quality

FIGURE 4.10. Red maple leaves are easily distinguished by their serrated margins. PHOTO BY NANCIE BATTAGLIA

FIGURE 4.11. The large, round red buds are very distinctive in the winter, making it easy to tell them apart from sugar maples. PHOTO COURTESY OF PETER SMALLIDGE

to sugar maples—it is nearly impossible to tell the difference between them. I know many producers who tap primarily red maples; their syrup is delicious and they get good yields per taphole. Although you could do even better with sugar maples, if you happen to have red maples, you might as well use them for sugaring!

Silver Maple (*Acer saccharinum*)

Silver maples are extremely fast growing and have been widely planted as street trees throughout the United States. They naturally occur along streambanks and in wet areas but are rarely found in upland forests. Just like red maples, silver maples tend to have lower sugar content in their sap and will bud out fairly early in the season. They are also referred to as soft maple and are commonly confused with red maples, though the leaves are much different. It is certainly possible to make high-quality syrup from silver maples, but they are not the ideal species for syrup production.

A study was conducted in Illinois to evaluate the potential of utilizing silver maples for syrup production.[5] Although the mean sugar content of silver maples was only slightly less than that of sugar maple (1.7 versus 2 percent), the sugar maples produced nearly twice as much sap per taphole as the silver maples (44.6 versus 23.5 liters per tap). However, there is tremendous variation in the sugar content and volume of sap produced between individual silver maples, so it is certainly possible to only tap the most productive trees, if desired. Finally, if you are interested in planting trees and have wet areas where sugar maples would not do well, there are genetically improved seedlings with high sugar content available. Based on the research and findings of Cedric Larsson in Canada, Bill MacKentley from St.

FIGURE 4.12. Older red maples tend to have shaggy bark that peels in strips running vertically along the trunk.

FIGURE 4.13. The bark of red maples is highly variable and can often be smooth grey, especially on smaller trees. PHOTO BY NANCIE BATTAGLIA

FIGURE 4.14. Silver maples are also called swamp maples since they occur naturally along riverbanks and in floodplains. Collecting sap from these locations could be quite challenging! PHOTO COURTESY OF STEVEN KATOVICH, USDA FOREST SERVICE, BUGWOOD.ORG

FIGURE 4.15. Silver maples were planted extensively for landscaping purposes due to their fast growth rate. However, they often have poor form and the weak wood is prone to breakage, so large silver maples are now a hazard in many urban and suburban environments. PHOTO COURTESY OF JOSEPH O'BRIEN, USDA FOREST SERVICE, BUGWOOD.ORG

Lawrence Nurseries in Potsdam, New York, grows "sweet sap" silver maple seedlings that have been bred for higher sap sugar concentration. Since the seedlings are produced via tissue culture, they are genetically identical to the parent tree that consistently produced sap between 3 and 5 percent in the wild. Silver maples are an excellent choice for planting on low-lying, poorly drained, wet areas where sugar maples would fail. If you put them in the right spot and take care of them, you may have tappable-size trees in 10 to 12 years.

Norway maple (*Acer platanoides*)

Although this species is native to Europe, it has been widely planted throughout the United States and Canada for ornamental purposes and has naturalized into urban and suburban forests. It grows fast, is a prolific seeder, and thrives in poor soils that will not support sugar maples. Due to its aggressive growth habit, Norway maple is now considered an invasive species throughout much of the US. Some municipalities have even banned it from being planted. One of the most popular cultivars is 'Crimson King', which many people wrongly confuse with red maple since the leaves are a deep burgundy color throughout the growing season. The leaves of most Norway maples look very similar to sugar maple, though the bark is much different. Norway maples often have black dots on their leaves, a symptom of tar spot fungus that is found almost exclusively on this species.

I used to have a strong aversion to Norway maples since they are replacing sugar maples in many areas of the Northeast. However, since they are already here and not going away, we might as well make the best use of them. The sap of Norway maples may not be quite as sweet as sugar maples and can become milky fairly early in the season, so it's not as desirable for drinking. However, I have tasted maple syrup produced entirely from Norway maples on several occasions, and it was quite delicious. If you live in an urban or suburban area with an abundant supply of Norway maples, I would definitely recommend using them to make syrup.

FIGURE 4.16. Black spots from tar spot fungus are usually found on Norway maple leaves. PHOTO COURTESY OF LESLIE J. MEHRHOFF, UNIVERSITY OF CONNECTICUT, BUGWOOD.ORG

FIGURE 4.17. 'Crimson King' is one of the most popular varieties of Norway maple planted. It is mistakenly called red maple because the leaves stay red all summer long, turning a dull brown in the autumn. PHOTO COURTESY OF LESLIE J. MEHRHOFF, UNIVERSITY OF CONNECTICUT, BUGWOOD.ORG

FIGURE 4.18. Boxelders are often found in urban area and waste places, taking on a contorted and shrub-like appearance. PHOTO COURTESY OF KEN FOSTY

FIGURE 4.19. Ken Fosty makes great uses of his own name and the unique nature of the trees he is tapping with his original Frosty's Manitoba Maple Syrup. PHOTO BY NANCIE BATTAGLIA

Boxelder (*Acer negundo*)

This species is also known as Manitoba maple since it is the only maple tree commonly found on the Canadian prairies. Some people also call it ashleaf maple since its leaves look more like an ash tree than a maple. Boxelders are typically short, multistemmed, unattractive trees and are almost never planted for ornamental purposes in the East. However, since they grow rapidly, produce a tremendous amount of wind-borne seeds, and will survive under harsh conditions, they are often found growing in urban areas and along roadsides. If you try to cut them down, they will likely sprout back with even more vigorous stems. While it is certainly possible to make syrup from these trees, I would only use them as a species of last resort in the eastern United States and Canada. However, since they do not have any sugar or red maples on the Canadian prairies, the maple producers in this region rely almost entirely on boxelder for their source of sap.

Recent research found that the average boxelder may only yield up to half the amount of syrup of a typical sugar maple.[6] Even though boxelders aren't ideal for tapping, there are enough producers tapping these trees to warrant the attention of the Canadian government. In fact, four pages of a recent government document on agriculture in Canada were devoted to the Manitoba maple syrup industry.[7] The interest is definitely growing, as the number of taps rose from 2,100 in 1991 to nearly 20,000 in 2001 spread out over 119 producers in Manitoba and Saskatchewan. Although more recent figures are not available, I got a chance to talk with Ken Fosty about the current status of the industry. Ken served as an educator with the Manitoba Forestry Association for 20 years and now teaches various workshops and seminars on Manitoba maple (and birch) syrup production. He believes that there are over 1,000 people now producing syrup from Manitoba maples (or planted silver maples in the Manitoba region), yet only a few are doing it on a commercial scale. The largest producer has approximately 3,000 taps and is the only person using vacuum tubing and reverse osmosis technologies. Although the yields from Manitoba maples are not as great as sugar maples, the higher prices resulting from the unique origin of this syrup make it a viable enterprise.

Bigleaf Maple (*Acer macrophyllum*)

Bigleaf maple is the main species of maple native to the Pacific Coast, ranging from Central California up to British Columbia. It sometimes forms pure stands in riparian areas, but is usually found as a smaller component of coniferous forests. Based on USFS FIA inventory data (figure 4.21), there are nearly 75 million potential taps from bigleaf maple trees in the forests of Washington, Oregon, and California (and many more in British Columbia). Bigleaf maple has often been thought of as a weed species that can outcompete valuable western conifers for growing space, so foresters have generally tried to get rid of bigleaf maples in their management activities. Until recently, hardly anyone had considered tapping this species for syrup production. In fact, the *Bigleaf Maple Manager's Handbook* for British Columbia does not mention the sap production potential of this species anywhere in the 54-page, otherwise thorough manual. Although Native people had been collecting the sweet sap from bigleaf maples for centuries, tapping bigleaf maples is a "new" idea today that is becoming increasingly popular, especially on Vancouver Island.

Much of the development of the bigleaf maple syrup industry can be attributed to the efforts of Harold Macy and Gary Backlund. Harold worked for many years as a research forester on Vancouver Island and regularly taught classes on woodland management. In the early 1990s, he started dabbling in bigleaf syrup production, and whenever he mentioned it to landowners their eyes and ears always perked up. One of Harold's star students was Gary Backlund, who was initially a bit skeptical of the idea, but once he tried some of the syrup he was immediately hooked. Upon completion of the course Gary immediately picked up some tapping supplies, and within the first 24 hours of tapping, he got roughly 10 gallons of sap from only three spouts. This stroke of beginner's luck gave him the impetus to develop a very successful enterprise and help thousands of other people get started as well. Together with his daughter Katherine, he wrote a guidebook on

FIGURE 4.20. In a few short years, Gary and Katherine Backlund have already sold 3,000 copies of their guidebook on making syrup from bigleaf maples. PHOTO BY NANCIE BATTAGLIA

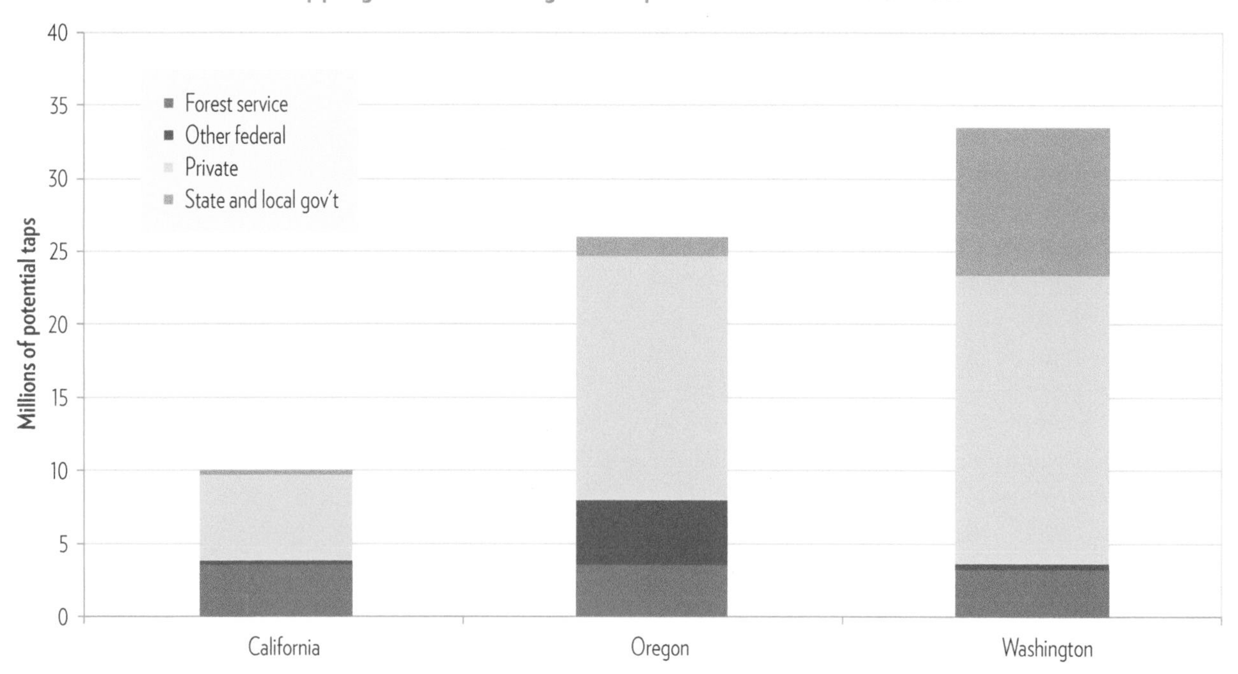

FIGURE 4.21. Number of potential bigleaf maple taps in the Pacific Northwest (data are classified according to ownership category; only trees growing on land where tapping is legally allowed are included in these calculations).

FIGURE 4.22. Most bigleaf maple sap is collected with small sections of tubing connected to plastic buckets. Trees located next to streams are preferred for tapping. PHOTO COURTESY OF GARY BACKLUND

FIGURE 4.23. Since maple syrup is currently in short supply on Vancouver Island, Gary Backlund makes his limited harvest go farther by blending some of it with apple cider and a hint of cinnamon to be processed into his maple apple cinnamon syrup. PHOTO COURTESY OF GARY BACKLUND

how to make syrup specifically from bigleaf maples.[8] They have since sold over 3,000 copies, a testament to the growing enthusiasm surrounding bigleaf maple syrup production. Gary helped organize a group of bigleaf maple tapping enthusiasts called "the sapsuckers," and they started a maple festival at the British Columbia Forest Discovery Centre. The festival is highly successful and draws in thousands of people every year. Gary estimates that there are now at least 10,000 people producing bigleaf maple syrup in British Columbia, nearly all of whom would be considered hobbyists or backyard sugarmakers by eastern standards.

The first scientific research on bigleaf maple syrup production was undertaken by the USDA Forest Service in the 1970s.[9] They tapped 13 trees in Oregon, kept track of sap production characteristics, and processed the sap into syrup with a steam kettle. Although they found that the flavor was different from sugar maple syrup produced in the East, this could have been due to the method of boiling. More recently, a graduate student at the University of Victoria named Deirdre Bruce conducted her master's thesis on bigleaf maple syrup production.[10] She was originally going to do a project on community forests, but during her initial exploratory phase she got to visit Gary Backlund's sugaring operation. After tasting some partially boiled-down sap and getting to tap a bigleaf maple tree, Deirdre was hooked. She decided to change her master's thesis to focus entirely on the characteristics of bigleaf maple sap and syrup production. Both studies found that the sugar content and sap flow volumes of bigleaf maple are less than those exhibited by sugar maple in the East, though bigleaf maples can still provide a commercially viable source of locally produced syrup for the West Coast. There are many people in this region willing to pay a premium for locally produced syrup, and *bigleaf maple syrup* has a great ring to it. The demand far exceeds supply and there are millions of untapped trees, so I expect this industry will continue to grow at a rapid pace.

The only major downfall to producing bigleaf maple syrup in the Pacific Northwest is the weather. This region has a temperate climate without massive

FIGURE 4.24. The tallest maple tree in the world is a bigleaf maple recently discovered by Mario Vaden in Humboldt Redwoods State Park in California, topping out at an amazing 157.8 feet (although it is *only* 3 feet in diameter). The "Humboldt Honey" grew so tall in order to fight for sunlight amid the towering redwood trees. PHOTO COURTESY OF MARIO VADEN

temperature fluctuations, so winter sap flows are highly unpredictable. The limiting factor is the lack of freezing nights, especially for lower elevations that are closer to the shore. Just like eastern maple syrup producers, the bigleaf maple "sapsuckers" hope for a good freeze followed by a nice sunny day. The peak flows can happen anytime between December and February, so it's difficult to know when to tap. One way to overcome the weather issues is through the use of vacuum, as this allows producers to gather sap even without sufficient freezes. I have not found any research on the use of vacuum for collecting sap from bigleaf maples, nor have I heard of any producers who have tried it, but I suspect it would boost yields tremendously. If I had access to a large stand of bigleaf maples in the western United States or Canada, there is no doubt that I would be installing high vacuum tubing.

Canyon Maple, Bigtooth Maple (*Acer grandidentatum*)

As the name suggests, canyon maple is a small, shrubby tree found primarily in moist soils and canyons throughout the Rocky Mountain states. Maples require plenty of water, and in much of the West the only places that have sufficient summer moisture to support maples are canyons and ravines. Research conducted in the 1970s by the USDA Forest Service examined the feasibility of producing maple syrup from canyon maples in Utah.[11] Although the sugar content of the sap was comparable to that of sugar maple, the volume produced was significantly less. The researchers were able to produce light-colored syrup whose flavor was characterized as "a delicate fruitiness." Part of the study included a taste test between their canyon maple syrup and regular light amber syrup; 57 percent of participants preferred the sugar maple syrup whereas 43 percent preferred the canyon maple. They concluded that while production of canyon maple syrup may not be commercially viable, it did possess strong potential as a family-based recreational experience. Indeed, there are enough native maples growing in the wild and other maples planted for ornamental purposes throughout the Rocky Mountain states to make syrup production a viable activity for many people.

Not only do canyon maples thrive out west, but this species can also be found in Texas, where it goes by a different common name: Uvalde bigtooth maple. There is even a state park called Lost Maples State Natural Area that features a large, isolated stand of these trees. This species came down to Texas during the last ice age about 6,000 years ago, but when the glaciers retreated, certain stands of trees were able to survive in this relatively cool portion of Texas. I learned about these trees from Dr. Mark Vorderbruggen, a research chemist who spends his free time foraging in the wild lands surrounding Houston. He also goes by the name Merriwether and posts a great deal of information on his website, www.foragingtexas.com. He has tapped maples planted in the Houston area and collects a small amount of sap in the winter months. In fact, the maples will often yield sap even without going through a typical freeze–thaw cycle. Mark reports that it does not produce enough to make syrup, but the sap does quench his thirst as a refreshing drink. Unfortunately many of the maples that Mark had been tapping in the Houston area died in the summer of 2011 due to the extreme drought in that part of the country. This reinforces the point that if you are going to try to plant and tend maples south or west of their native range, it is imperative to have a good water supply. Maples can usually tolerate high heat, but not if it is coupled with persistent drought.

FIGURE 4.25. A group of dedicated volunteers helps grow, plant, and distribute bigtooth maples throughout the town of Boerne, Texas. PHOTO COURTESY OF JACK MORGAN

Rocky Mountain Maple, Douglas Maple (*Acer glabrum*)

If you live in western North America and can't find any other maples to tap, you may be able to find some Rocky Mountain maples large enough. The species is usually found in shrub form but will occasionally grow to the size of a small tree. In addition to the main species of *Acer glabrum* found in the Rocky Mountains from Montana to New Mexico, there are five other varieties with their own specific and limited ranges.

- *Acer glabrum* var. *diffusum*, Rocky Mountain Maple: eastern California, Nevada, Utah
- *A. glabrum* var. *douglasii*, Douglas Maple: Alaska south to Washington and Idaho
- *A. glabrum* var. *greenei*, Greene's Maple: endemic to Central California
- *A. glabrum* var. *neomexicanum*, New Mexico Maple
- *A. glabrum* var. *torreyi*, Torrey Maple: endemic to Northern California

Gorosoe (*Acer mono*)

Although maple syrup is primarily a North American tradition, there are dozens of other species of maple found throughout the world. *Acer mono* (gorosoe) is the most commonly tapped maple in the mountains of Korea and recently gained a lot of attention in the United States based on a 2009 article highlighting the gorosoe sap industry in South Korea. *Gorosoe* means "the tree that is good for the bones," and hardly any of the sap is boiled down into syrup. Rather, nearly all of it is consumed fresh as a natural beverage and spring tonic (more about this in chapter 12). Gorosoe does not usually grow to very large sizes or occur in vast stands, but it does appear in sufficient quantities to merit tapping for the beverage industry in Korea.

FIGURE 4.26. A relatively large *Acer mono* being tapped for sap production in Korea; roughly half of the maple trees in Korea occur on government-owned land, and residents must apply for permits to extract the sap. PHOTO COURTESY OF KIYOUNG LEE

The Walnuts (*Juglans* spp.)

There are 21 species within the *Juglans* genus that occur throughout the world. The two most common ones found in the eastern United States are black walnut (*Juglans nigra*) and butternut (*J. cinerea*). Very few people know that walnuts produce sap and hardly anyone is currently tapping these trees, though I expect that will change in the future. There has been some interest in "sugarbushing the black walnut" among nut growers to yield yet another great product from these amazing trees, yet very little walnut syrup has actually been produced.[12] I highly encourage you to try making syrup from these trees, with one word of caution. Walnut syrup comes from nut trees, and while this may seem obvious to you, some folks may not make that connection. Since there are many people allergic to tree nuts, it is possible that someone could also suffer allergic reactions from consuming walnut syrup. Therefore, be sure to take extra precautions if you plan on providing your walnut syrup to others, either as a gift or for sale. I was not able to find any regulations pertaining to walnut syrup, but food safety specialists in Ontario believe that the boiling process would denature and destroy any allergenic nut proteins if they are present in the sap. Without any peer-reviewed literature documenting this, I would still just use common sense and make sure that anyone who consumes walnut syrup is fully aware that it comes from a nut tree and may cause problems for those with severe nut allergies.

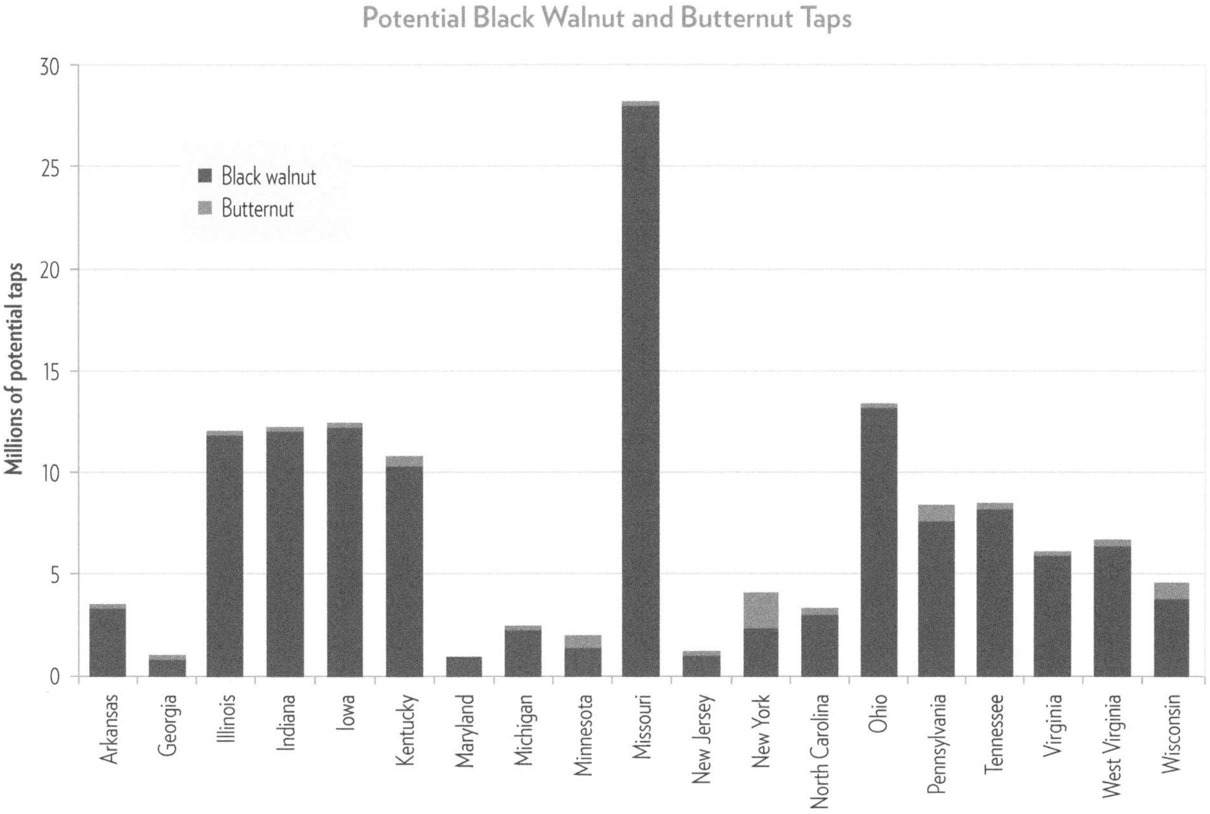

Potential Black Walnut and Butternut Taps

Millions of potential taps

Black walnut
Butternut

Arkansas, Georgia, Illinois, Indiana, Iowa, Kentucky, Maryland, Michigan, Minnesota, Missouri, New Jersey, New York, North Carolina, Ohio, Pennsylvania, Tennessee, Virginia, West Virginia, Wisconsin

FIGURE 4.27. Number of black walnut and butternut trees of tappable size in United States.

Figure 4.27 shows the number of potential black walnut and butternut taps in the eastern United States. Although there are not nearly as many walnut trees as there are maples, there are enough black walnuts growing in many states to develop cottage industries in walnut syrup production. In fact, many of the mid-Atlantic and midwestern states that don't have a lot of maples do have a significant resource of black walnuts. Butternuts are much less common than black walnuts, so the opportunities to tap this species will be quite limited. It is worth noting that these figures only include walnut trees growing on forest stands that are at least 1 acre in size and do not include those found in yards, parks, and hedgerows (where I often see them growing).

Butternut (*Juglans cinerea*)

Also known as white walnut, this species is normally recognized for its edible nuts and beautiful lumber. One of its lesser-known qualities is that it produces a delicious sap that can be drunk fresh or boiled down into syrup. Butternut sap usually contains about 2 percent sugar and also flows at the same time of the year as maples. The syrup is quite delicious—many people have described it to me as having nutty and almost fruity overtones.

I could only find one research article on butternut syrup—it came from the Michigan Agricultural Experiment Station way back in 1925.[13] In their limited trials, researchers found that the timing of sap flow, sugar content, and total volume of sap produced were very similar to those of sugar maple. In fact, the overall yield of syrup from their butternut trees was only 6 percent less than their sugar maples. I have not been able to find any other research on tapping butternuts; nor have I been able to find anyone who is currently making this syrup. Thus, for his fourth-grade science fair project a few years ago, I suggested to my nephew Owen that he tap the butternut tree in their backyard. He had great success, but I never got a chance to try any of the syrup as it was all used for the science fair. Once I decided to write this book, I figured I had to try making butternut syrup myself. After all, you really shouldn't write about something unless you have done it yourself! The last couple of years I have made a few gallons of butternut syrup from a grove of 16 large, healthy trees in my hometown of Albany, New York. There is even a video of me tapping one of the butternut trees—just do a search for "youtube tapping butternut." I have been pleasantly surprised by the amount of sap flow and the taste of the finished syrup over the past couple of years and plan to continue tapping these trees in future years.

There is one thing that you must be keenly aware of when tapping butternut trees. Both the Michigan research and my own experience found that a jelly-like substance regularly appears in butternut sap. I found it

FIGURE 4.28. The compound leaves of butternut have 11 to 17 leaflets and are yellowish green with fuzzy undersides. PHOTO COURTESY OF PETER SMALLIDGE

FIGURE 4.29. The black splotches with whitish margins are indicative of cankers on butternut trees. PHOTO COURTESY OF MANFRED MIELKE, USDA FOREST SERVICE, BUGWOOD.ORG

floating on the top of my sap buckets and discovered that it becomes even more concentrated in the syrup if it is not filtered out. I asked Dr. Randy Worboro, a microbiologist at Cornell University, to analyze a sample of butternut sap, and he determined that it contains a great deal of naturally occurring pectin. This pectin is not necessarily something to be concerned about; you just need to adequately deal with it. We add pectin when making jelly and jams to act as a gelling agent, and it even has some beneficial properties since it is considered a dietary fiber. However, unless you want to make butternut jelly, I can't stress enough the importance of filtering the sap and syrup as much as possible to remove the pectin. I attempted boiling some sap without first filtering it, and the resulting product simply concentrated the pectin and took on the consistency of jelly. It was an edible spread on toast, but the texture was far from desirable, at least from my point of view. Apparently it can also be found in the sap of other *Juglans* species, and I have found it in limited quantities in black walnuts, but not nearly to the same extent as the butternuts.

Butternuts are an increasingly rare species and have been declining largely due to the spread of a fatal disease called butternut canker that is brought upon by the fungus *Sirococcus clavigignenti-juglandacearum*. It was first discovered in Wisconsin in 1967 and has since decimated butternut populations throughout eastern North America.[14] If you are lucky enough to still have healthy trees growing on your property, great care should be taken to ensure their continued growth and survival. The fungus can infect trees through buds, leaf scars, and wounds on the trunk or exposed roots, so purposefully drilling holes for sap collection could provide an entryway for the fungus. If you have very healthy butternut trees, you may consider not tapping them so you don't create any openings for the fungus to penetrate.

If you do decide to tap your butternuts, I recommend only one 5⁄16" spout per tree and minimizing any vehicular traffic around the root system. In fact, you may only wish to tap a small fraction of your butternut trees to see how they respond before tapping the rest of them. If you have trees that are already infected with the canker, I wouldn't hesitate in tapping them. The trees will soon be dead, thus any additional wounding and stress brought upon by tapping is unlikely to have much impact. However, if the canker infestations are severe enough you may wind up wasting your time, as trees that have had cankers for many years may not yield much sap.

If you do find healthy butternut trees on your land, I recommend gathering as many nuts as you can in the early fall—before the squirrels take them all! Even though butternuts may eventually get the canker, don't let that discourage you from planting these trees. Butternuts are fast growing and can be easily started by planting a nut in the ground and protecting the emerging seedling from rodents, deer, lawn mowers, and the like. If planted in the right place, they can grow quite fast. You can start tapping them when they are about 12 to 15 years old and then collect sap for many years before they eventually get the canker and go into decline. At that point you can saw the logs into valuable, unique lumber and use the rest of the tree for firewood. If you are lucky, perhaps some of the trees you plant will be resistant to the canker and you will help perpetuate this important yet declining species.

FIGURE 4.30. At 225" in circumference (nearly 6 feet in diameter), this butternut tree likely served as the mother tree for a large grove of smaller trees on neighboring properties in Albany, New York.

Black Walnut (*Juglans nigra*)

Although they are somewhat rare in the Northeast, black walnuts are much more common in Appalachia and the Midwest. They have a reputation as one of the most valuable timber species due to the beautiful color, strength, and durability of their lumber. They also produce large crops of edible nuts that are prized by both people and wildlife. What is not well known about black walnuts is that they yield a sweet sap at the same time of the year as the maples. Even if you don't like the taste of black walnuts, chances are you'll love black walnut syrup. In 2003–04, researchers at Kansas State University experimented with producing black walnut syrup from 30 trees that were scheduled to be thinned out of a plantation. They boiled all of the sap into syrup using a steam kettle and then performed research on consumer preferences for black walnut versus maple syrup. They found no significant differences on the likability scale between these two syrups and concluded that black walnut syrup could develop as a niche market in the Midwest.[15] Despite this promising research, hardly anyone is currently tapping black walnuts for syrup production.

Because walnut lumber is so expensive, many people balk at the idea of drilling holes in otherwise valuable logs. If you are concerned about this, you may want to hire a professional forester to come examine your trees before you start tapping them. It is possible that you have extremely valuable trees that could generate much more money as sawlogs than they ever could as sap sources. However, it is also possible that your trees are not prime sawlog or veneer trees, in which case you should not hesitate in tapping them. If your trees do have potential as sawlogs but you still want to tap them, I recommend placing the taps as low as possible on the trunk to reduce the impact on future lumber quality. However, since tapped maple lumber can command premium prices in niche markets (more on this later), it is quite possible that walnut boards with evidence of tapping could also appeal to the right market. Tapping maple trees causes long columns of stained, compartmentalized wood that is surrounded by mostly white sapwood. On the other hand, the dark heartwood is what is valuable in black walnut trees, so when you drill holes in the outer white sapwood, you

FIGURE 4.31. Maple syrup is not only the only species that has a fake syrup substitute. This "black walnut syrup" is actually just corn syrup, cane sugar, and walnut extract with pieces of black walnuts added. PHOTO BY NANCIE BATTAGLIA

may wind up with even more heartwood. This could potentially increase the value of the log, especially for niche markets. However, if you have any intention of selling your black walnut logs on the open market, I wouldn't recommend drilling any holes in them, as most log buyers are unlikely to agree with you that the tapholes add value.

With black walnuts, you may be better off tapping smaller-diameter trees with a lot of sapwood rather than large old trees that contain mostly heartwood. Only the white sapwood near the bark will produce sap—hence the name *sapwood*. Small-diameter trees between 6 and 10" dbh cannot be harvested for sawlogs, yet you can gather sap for many years until they eventually grow into sawlog size. Furthermore, the sapwood will eventually turn to heartwood anyway (even

Michael Jaeb: The Largest Black Walnut Sugarmaker in the World

There are currently very few people making syrup from black walnut trees. Michael Jaeb of Simple Gourmet Syrups in Ohio may be the largest producer, and he only taps about 150 trees. Michael worked as a graphic designer for many years and always produced maple syrup on a small scale. When the recession hit and his regular business started to dry up, he decided to focus on turning his hobby into a business. In addition to his maple syrup, he started producing black walnut syrup and a variety of other naturally flavored syrups (including shagbark hickory). Michael says that the black walnut syrup is a customer favorite; he sells out of it very quickly every year. In 2012 one person bought every last bottle he had when she realized there was only a limited amount available. Given the success that Michael has had with tapping his black walnuts, I expect there will be a lot more people making this gourmet syrup in the future.

FIGURE 4.32. A bottle of black walnut syrup produced by Michael Jaeb with Simple Gourmet Syrups. PHOTO BY NANCIE BATTAGLIA

without tapping), so it makes sense to gather sap out of that portion of the tree before it becomes useless for sugaring purposes. I believe that much more research is necessary to determine the optimum tapping size and sap yields from black walnuts of different sizes. However, until we have sufficient data to make solid recommendations, I wouldn't hesitate to tap the smaller trees. Since they grow so fast, black walnuts can be tapped at a younger age and put on enough new wood to allow for sustainable tapping, even when starting with small trees. No matter what size of black walnut trees you tap, you should expect the sap yield to be much less than maples (or butternuts). This may be due to the fact that black walnuts contain a lot of heartwood and very little sapwood on the outside of the tree. The sugar content is comparable to that of maples, yet the volume seems to be significantly lower. We will be conducting research to determine the yields from different-sized trees, so stay tuned for further information.

The timing of tapping is another unknown with walnut syrup production. Research in France on *Juglans regia* (English walnut) found significant sap flow in autumn, winter, and spring.[16] The authors concluded that the winter sap flow was based on stem pressure

FIGURE 4.33. Black walnut trees planted next to a sugarhouse. Most maple producers are not aware that walnut trees also produce a sweet sap that can be boiled down into delicious syrup.

FIGURE 4.34. Although black walnuts are widespread in Appalachia and the Midwest, they are not nearly as common in the Northeast. Some of the trees that I found to tap were located behind the neighborhood library and had been planted back when this part of Albany was still farmland.

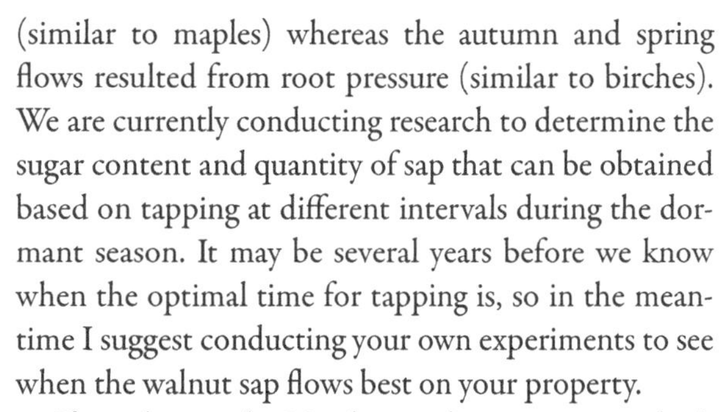

(similar to maples) whereas the autumn and spring flows resulted from root pressure (similar to birches). We are currently conducting research to determine the sugar content and quantity of sap that can be obtained based on tapping at different intervals during the dormant season. It may be several years before we know when the optimal time for tapping is, so in the meantime I suggest conducting your own experiments to see when the walnut sap flows best on your property.

If you live in the Northeast, chances are you don't have enough black walnut trees to make syrup from your own property. Still, in my travels along the East Coast, I often see black walnuts growing along roadsides, yards, and hedgerows. If you can convince the landowners to let you tap their trees, you will likely be able to have a small walnut sugaring operation. If you happen to live in the Midwest, the situation is much different. There are many black walnut trees in Missouri and the surrounding states, and even more black walnut orchards that could be tapped for syrup production. The trees are already laid out in nice rows, so it would be relatively easy to install a tubing system to gather the sap. Parts of the tubing may need to be

FIGURE 4.35. A bottle of maple-walnut syrup with 90 percent light amber maple syrup mixed with 10 percent black walnut syrup. PHOTO BY NANCIE BATTAGLIA

taken down and put back up each year due to orchard maintenance and nut harvest, but that would not be an insurmountable problem. Also, since the trees are being grown primarily for nuts, they are short and have wide-spreading crowns to maximize nut yield. These trees are not likely to be harvested for timber, and if they are eventually cut down, the trunks will not yield any commercial sawlogs; thus the impact on timber value from tapping is negligible. If you happen to live near a walnut orchard, it is worth the effort to ask the landowner if you could lease some of the trees for a fee (see chapter 8 for more information on leasing taps).

I think one of the best opportunities for producing walnut syrup could be through collecting sap from small branches when pruning young trees in the winter months. Many people will prune small branches on the main trunk to increase the future value of the trees for sawtimber. If you can find a way to economically gather the sap, you could make great use of it rather than letting it just drip away on the ground. More research is necessary to determine the potential yields and optimum collection strategies from pruning cuts; it may be many years before we figure this out. In the meantime, I strongly encourage anyone growing walnut trees to try collecting the sap from winter pruning cuts and see what kind of results you get.

The flavor of black walnut syrup is surprisingly similar to a light or medium amber maple syrup, but with more butterscotch and nutty overtones. The first run of black walnut syrup I made tasted just as good as any maple syrup I've ever had. In fact, many people that I have given samples to, including many sugarmakers, feel the same way. I gave a workshop on black walnut syrup production in Vermont recently, and after everyone got to learn about the process and taste some of the syrup, nearly everyone was making plans to go plant some walnut trees!

If you are only able to make a small amount of walnut syrup and would like to stretch it farther, you may want to try mixing it with maple syrup. I would use at least 75 percent maple syrup and choose the lightest you have, so as to not overpower the walnut syrup. You could also simply mix the walnut sap in with maple sap and boil them down together to make maple-walnut syrup. Maple-walnut is a distinctive, well-known flavor that many people enjoy, and although your syrup won't taste exactly like the ice-cream flavor, I'm sure it will be delicious and much better! Producing maple-walnut syrup provides a unique marketing opportunity that will appeal to many high-end clients.

Heartnut (*Juglans ailantifolia*)

Heartnuts are a cultivar of Japanese walnuts that have gained popularity throughout eastern North America over the past century. Similar to butternuts, yet resistant to the canker disease, they have been planted as a replacement species in some areas. As you might expect, the nut is heart-shaped in cross section, easy to crack, and has a sweet flavor without the bitter aftertaste that can be found in black walnuts. Heartnuts are supposedly hardy to Zone 5, so if you live in a warm enough climate, you may want to try planting these. Todd Leuty, an agroforestry specialist for the Ontario Ministry of Agriculture, experimented with tapping a

FIGURE 4.36. A row of Japanese walnuts tapped at the University of Guelph in Ontario. PHOTO COURTESY OF TODD LEUTY

stand of heartnuts growing at the University of Guelph over two years. The weather in 2012 was too warm to obtain reliable data, yet in 2013 the weather was more favorable and they got the taps out early enough to collect sap to make some walnut syrup. The sap sugar content and flow rates were highly variable. Most trees exhibited sugar contents comparable to sugar maple, though the amount of sap was *much* less on average. Some trees did produce large quantities of sweet sap, but these were the exception, rather than the norm.

A buartnut is a cross between heartnuts and butternuts. I first discovered these at St. Lawrence Nurseries in Potsdam, New York, a nursery that specializes in Zone 3–hardy fruit and nut trees. Heartnuts would not survive the frigid winters of northern New York or many other places, but by crossing them with butternuts, you wind up with a nut almost identical to butternut that is resistant to the canker and hardy to Zone 3. If your climate is particularly cold, but you would like to experiment with walnut syrup production, I highly recommend planting these.

English Walnut (*Juglans regia*)

If you have ever eaten walnuts or anything with walnuts in it, chances are you were eating English walnuts. Also known as Persian walnut (based on its native origin), this species is grown throughout the world for its delicious nuts. They have a much milder flavor than our native black walnut, the percentage of nutmeat is much higher, and the shell is easier to crack open. Although nearly all the English walnuts in America are grown in California, there has been a lot of interest in growing this species in the eastern United States. English walnut is rated as hardy to Zones 5 to 7, so depending on where you live and your own microclimate, you may be able to give it a try. Those that are grafted onto black walnut rootstock tend to be hardier and more robust. One of the main reasons they are not grown commercially in the eastern US is that late-spring frosts kill flowers and severely diminish nut production. The trees can survive, but they usually don't produce any nuts. However, if you are interested in tapping them, our freezing temperatures throughout the winter and spring would create the ideal conditions for collecting sap.

The Birches (*Betula* spp.)

Tapping birch trees is another potentially lucrative activity that you can include in your overall sugarmaking operations. Whereas sap flow from maples and walnuts relies on freeze–thaw cycles to produce adequate stem pressure, birch sap flow is based on root pressure that develops once the soil warms to about 50°F. Toward the end of the maple sugaring season, when sap flow is starting to slow down and there are few freezing nights, birches start to yield sap. In fact, when you start to hear the spring peepers, this familiar sound heralds the end of maple season and the beginning of birch sap flow. Once it starts flowing, the sap in birch trees will flow almost continuously for several weeks until the leaves pop open. Even freezing nights don't stop the sap from running. The sugar content in birch sap is much lower than that of maples, and the flavor of the syrup is much more distinct and intense. While many people do not like the taste of birch syrup, don't let that discourage you. Birch syrup is a novelty product that commands much higher

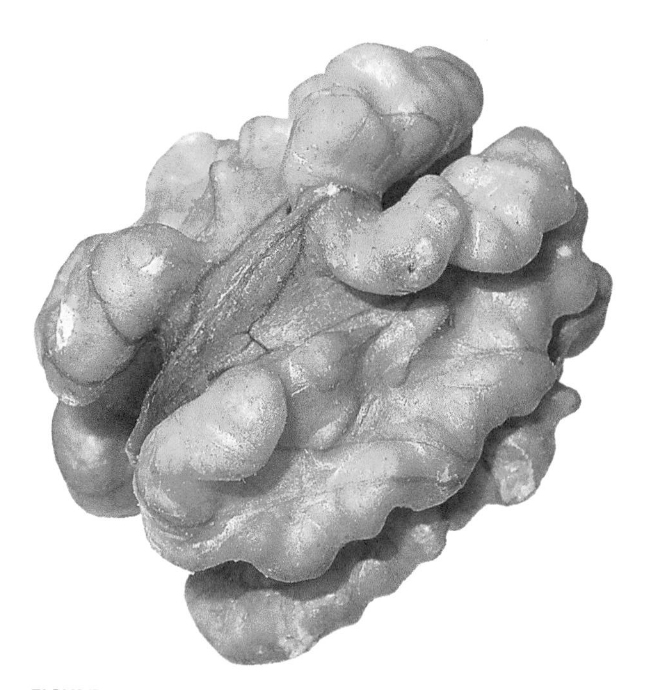

FIGURE 4.37. Although English walnuts are prized for their delicious nutmeat, they may also be tapped to produce a gourmet syrup. PHOTO COURTESY OF H. ZELL, WWW.GNU.ORG/COPYLEFT/FDL.HTML

prices than maple—there are many people that appreciate its unique flavors and are willing to pay a premium.

The difference in timing of sap flow provides for a great complementary relationship between maple and birch sugaring operations. If you aren't completely worn out by the time the maple season ends, then you can use the same equipment from your maple sugaring operation to also produce birch syrup. The processing equipment is identical, and your sap collection materials may also be used for both species. The only caution comes in using galvanized buckets, spouts, and collection tanks, as the betulinic acid contained in the birch sap can possibly corrode the metal and result in the sap and syrup developing a metallic flavor (or possibly even leaching out lead into the sap). Stainless steel is the preferred material for all sugaring equipment and plastic bags and buckets are now available to replace galvanized ones, so these issues can be avoided. For tubing systems, you could utilize the same mainlines and simply run different lateral lines to the birch trees. If you do so, it's important to thoroughly clean the tubing before collecting any birch sap. Even if you attempt to clean your tubing, it may still have some microbial contamination left over from the maple season that could ruin your birch sap. You may want to use an entirely different tubing system altogether. Although you can theoretically use the same evaporator, most of us have many more maples than birch trees. Thus, the evaporator that you use to process your maple sap will likely be oversized for the amount of birch sap you collect. If this is the case, I recommend getting a smaller evaporator to make birch syrup.

Even though birch syrup production is a growing enterprise in the eastern United States, very few people are currently doing it. At The Uihlein Forest in Lake Placid, we tap about 700 trees and are probably one of the largest birch syrup producers in the eastern United States. I'm surprised that more people have not gotten into birch syrup production, especially given all the advantages we have producing here in the East. The problems that have hindered birch syrup production in Alaska—lack of infrastructure and knowledge of syrup production, high cost of raw materials, and extreme weather conditions—do not exist for eastern sugarmakers to the same

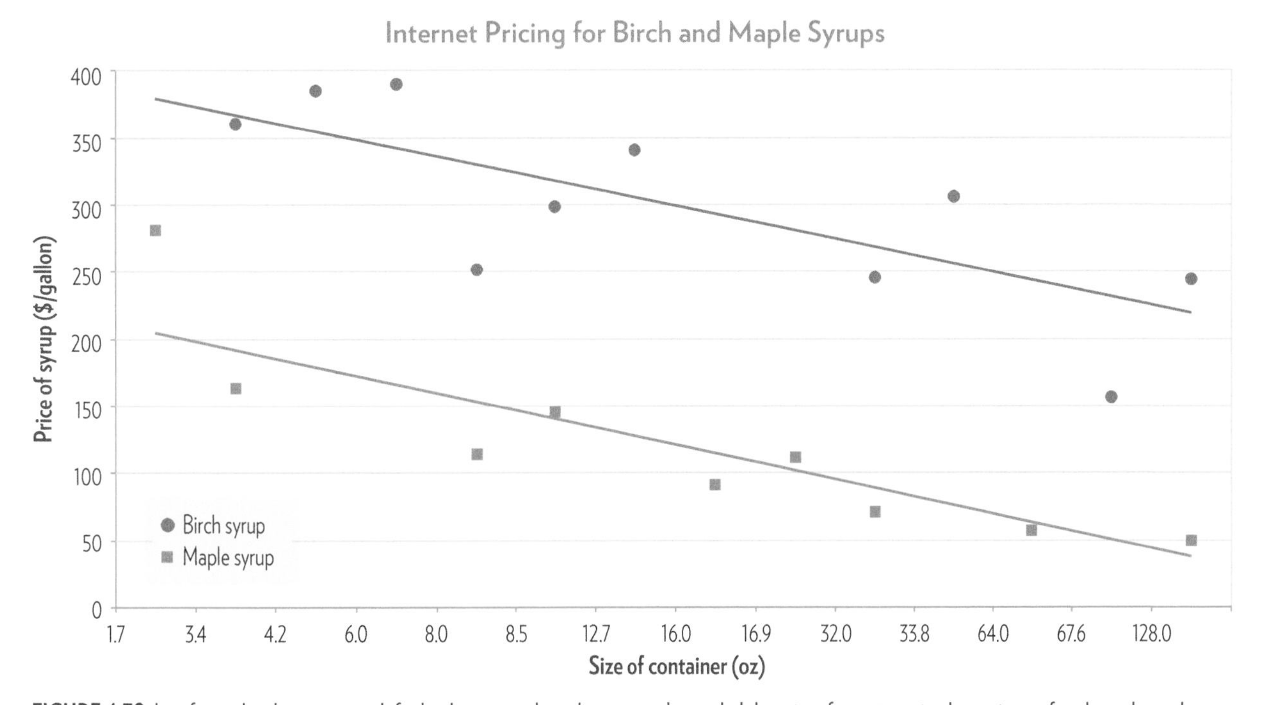

FIGURE 4.38. I performed an Internet search for birch syrup and maple syrup and recorded the prices for various-sized containers of each product—they are significantly higher for birch syrup. Most of the birch syrup is sold in 8-ounce containers for roughly $20 each, which equates to over $300 a gallon.

FIGURE 4.39. We put all our birch syrup into 40ml glass bottles and affix a nice tag with a "birch-bark" background, which works well for the Adirondack market. At $5 a bottle, our retail prices equate to $473 per gallon. PHOTO BY NANCIE BATTAGLIA

FIGURE 4.40. Maple and birch trees share the same mainlines in this tubing system. Some of the lateral lines are solely devoted to yellow and paper birch trees; other lines have birch mixed in with maple. If you combine your tubing systems, it is important to keep the birch capped off during maple season and to properly wash the tubing before tapping in the birch trees after maple sap flow has ceased.

extent. By taking advantage of the existing infrastructure for maple syrup production, many sugarmakers could expand their operations by producing birch syrup as another valuable crop once the maple season has ended.

The three most common birches in North America that can be tapped are paper birch, yellow birch, and black birch (also known as sweet birch). Other birches include river birch, gray birch, and European white birch. The following pages discuss each of the birch species in greater detail. Pictures of the bark are provided for each species, as this is the easiest way to distinguish among them.

Paper Birch (*Betula papyrifera*)

Paper birch, also known as white birch, gets its name from the pure white bark that peels off in a paper-like fashion. Since they do not have many maples in Alaska and western Canada, the people in these regions tap paper birch, or its close relative Alaska birch (*Betula neoalaskana*). Paper birch is a fairly short-lived pioneer species; it requires full sunlight and bare mineral soil to get established. Following land clearing and fires in the mid-20th century, birches seeded in many areas of western North America have restored the forest cover and ecological integrity. Whereas finding a maple is a rare sight in this part of the world, it's not uncommon to find pure stands of paper birch as far as the eye can see.

The birch syrup industry has been growing rapidly in Alaska over the past 20 years. The Alaskan syrup producers got organized in the early 1990s, started the Alaska Birch Syrupmakers' Association, and originally put a mobile sugarhouse at the Alaskan State Fairgrounds.

Given its success, they now have a permanent building there to promote their products. Several of the producers have formed sap collection cooperatives, buying sap from their neighbors and processing it in a central sugarhouse. While Alaskan birch syrup is popular among the locals and tourists, it has also gained worldwide recognition, especially in Italy. The limited number of producers haven't been able to supply the growing demand for birch syrup, so there is little doubt that the production and marketing of birch syrup will continue to grow.

Although paper birch has much lower sap sugar content than maple trees, it may be the sweetest of the birch trees. David Moore, a birch syrup producer in New Hampshire who runs The Crooked Chimney, tested the sugar concentrations of his paper birch, yellow birch, and black birch trees for two years as part of a university research project. The sugar content of paper birch was the highest whereas yellow birch and black birch were significantly lower. In several years of tapping birch trees in the Adirondacks, I've found that our paper birch trees rarely get to 1 percent sugar content, and the yellow and black birch are roughly half as sweet as the paper birch. Because the sugar concentrations can vary greatly from site to site and tree to tree, I recommend getting a reliable refractometer and measuring sugar concentrations on your own trees. If you don't have a reverse osmosis unit, you may decide to just tap the sweetest trees, thereby reducing the amount of time and fuel spent processing the sap. However, I wouldn't even consider trying to make birch syrup without concentrating the sap first. If you don't have an RO, I highly recommend just drinking and cooking with the birch sap (see chapter 12 for more information on the health benefits of birch sap).

FIGURE 4.41. Paper birches often get established after fires, agricultural abandonment, or clear-cutting, so it's not uncommon to find pure stands like this one that are ideal for tapping. PHOTO COURTESY OF JOSEPH O'BRIEN, USDA FOREST SERVICE, BUGWOOD.ORG

FIGURE 4.42. Heading out to the birch woods with a sled full of tubing ready to be put up. Although most of the birch trees are tapped with buckets in Alaska, the Kahiltna Birchworks started installing vacuum tubing in 2012 with favorable results. PHOTO COURTESY OF KEVIN SARGENT

Yellow Birch (*Betula alleghaniensis*)

Yellow birch is a common associate with sugar maples in northern hardwood forests; could be utilized to a much greater extent for sap and syrup production. The inner bark contains a distinct wintergreen flavor that is easily discovered when you scratch the bark off a twig. If the sap and syrup also contained this wintergreen flavor, I have no doubt that there would be millions of taps in yellow birches throughout the Northeast.

FIGURE 4.43. The yellow-bronze exfoliating bark is very pronounced on young trees, though the bark on older stems becomes rough and patchy. PHOTO COURTESY OF JOSEPH O'BRIEN, USDA FOREST SERVICE, BUGWOOD.ORG

FIGURE 4.44. Yellow birch seeds readily germinate on moist, decaying stumps. Once the stump has decomposed, the yellow birch remains in its elevated state.

Unfortunately the oil of wintergreen is an aromatic compound that escapes during the boiling process, leaving the syrup without the slightest hint of a wintergreen flavor. Nevertheless, the sap and syrup have their own unique flavor that many people enjoy as a unique ingredient in their kitchens.

In the northern hardwood forests, yellow birch grows well on rich, well-drained soil and can also occupy sites that are too wet for sugar maple. You can often find them growing on mossy boulders and decaying stumps, as the tiny wind-borne seeds require a moist surface to germinate. When the stump eventually rots away, the roots remain, encompassing the skeleton of the former stump and making the tree appear to be up on stilts. Yellow birch produces a valuable lumber that can command stumpage prices rivaling those of sugar maple. For this reason, you may be better off tapping other birch species besides yellow birch if you have valuable sawtimber or veneer-quality trees.

There is a company in Canada that has recently exploded into the birch syrup market by tapping yellow birch trees. As a way of diversifying its large maple sugaring operation, in 2009 Érablière Escuminac conducted limited trials with birch syrup. After experiencing favorable results, they have quickly expanded to 6,000 taps and could reach up to 25,000 taps in the near future. According to their website, their syrup has a distinct caramel flavor that is featured in a number of beers, wines, and spirits; it's prized by chefs for desserts and pastries. Given these positive developments, the l'Institut des Nutraceutiques et des Aliments Fonctionnels (INAF) in Quebec has been conducting research to determine the chemical constituents and health benefits of yellow birch syrup. The INAF found that birch syrup has many of the same properties as maple syrup with enhancements in some areas. For instance, birch syrup has higher mineral concentrations and lower sugar content while also possessing a higher oxygen radical absorbance capacity (ORAC).

Black Birch (*Betula lenta*)

Also known as sweet birch and cherry birch, this species is best known for its use in making birch beer. Just like yellow birch, it contains oil of wintergreen in the twigs but not the sap or syrup. Black birch tends to grow in more southerly and warmer areas whereas yellow birch is mostly found in northern states with colder climates. However, black birch is expanding its range and becoming much more prevalent throughout the eastern United States. It is fairly shade-tolerant and germinates on exposed mineral soil, so it often comes in after logging events.

FIGURE 4.45. Older black birch trees of tappable size have black, plate-like bark that is sometimes confused with black cherry. PHOTO COURTESY OF PETER SMALLIDGE

River Birch (*Betula nigra*)

River birch can be found throughout the southeastern United States and is also planted as an ornamental throughout the Northeast. Certain cultivars, which are often planted as clumps of three, have pinkish white bark that peels off in sheets similar to paper birch. It is a fast-growing, beautiful, strong tree that is also resistant

FIGURE 4.46. Often found in clumps, river birch has wide variations in bark appearance. Ornamental cultivars often have pinkish white bark that exfoliates in curly papery sheets.

FIGURE 4.47. Gray birch bark is chalky white with black triangular patches called chevrons where the branches meet the trunk. Unlike paper birch, it does not readily exfoliate.

FIGURE 4.48. The showy white bark makes European white birch a popular ornamental.

to the bronze birch borer. In a similar fashion to the white birch, it does not contain any oil of wintergreen in the stems but is reported to produce a fine sap.

Gray Birch (*Betula populifolia*)

I hesitated about including gray birch in this book because it is more of a shrub than a tree. It looks very much like white birch, but has a more limited range and does not grow as tall or as sturdy as the white birch. Gray birch is a pioneer species that will colonize an open area and provide an initial tree cover while slower-growing, longer-lived species get established. It certainly wouldn't hurt to tap these trees once they're large enough and while they are still alive, but I wouldn't base a business plan on tapping gray birch for an extended period of time.

European white birch (*Betula pendula*)

As its name suggests, this species is native to Europe but is commonly planted as an ornamental tree species in urban and suburban areas. It will often be grown in a clump of 3 (similar to the river birch) and the stems rarely get much larger than 12" in diameter. Some people prefer this species over our native white birch due to its glistening white bark. It is a relatively short lived tree and is susceptible to the bronze birch borer. If you happen to have some of these trees available, it would be worth tapping them just to drink the sap.

Sycamore (*Platanus occidentalis*)

Sycamores are easily recognized by their distinctive bark, which peels off in an assortment of different colors. They are often planted in urban areas and occur naturally along streambanks and floodplains. Whereas some people realize that birch and walnut trees produce sap, hardly anyone knows that sycamore trees also produce a sweet sap in the spring. The only person I've met who has produced sycamore syrup is David Moore from The Crooked Chimney in New Hampshire. The sap has a lower sugar content than maple, and the syrup

Taste Testing Different Syrups

FIGURE 4.49. Visitors to The Uihlein Forest in Lake Placid sampling a variety of tree-sap-derived and imitation syrups.

During our Maple Weekend open house event at The Uihlein Forest in 2013, we put out samples of 10 different syrups for people to try. In the blind taste test, people took a small cup of each syrup and then marked on their scorecard the degree to which they liked it. They were told that most of the syrup came from tree sap, though not necessarily from maple trees. After they handed in their scorecard, we explained a little bit about each of the syrups. I was glad to discover that our two pure maple syrups scored the highest, with comparable ratings between the light amber and extra dark. People seemed to like the butternut syrup more than the black walnut, though this may have been due to the fact that the black walnut syrup sample we had wasn't the best quality (we've made much better-tasting black walnut syrup in the past). Our maple-walnut syrup, a blend of 85 percent light amber maple syrup with 15 percent black walnut and butternut syrup, scored equally with both pure maple syrups. I was surprised that the bigleaf maple syrup from Oregon wasn't as well received—it had a very distinctive caramel flavor that I really enjoyed, but was probably much too different from the flavors people were used to and expecting. Nearly everyone expressed their strong disapproval of the birch syrup, but we did have several people who really liked it and bought some bottles as a result. Almost everyone who purchased the birch syrup said they planned on using it as a unique flavor in their cooking. Mrs. Butterworth's did surprisingly well, as many people were surprised (and disappointed in themselves) that they actually liked the fake stuff. The agave syrup with "organic maple flavor" performed poorly—I found this one particularly objectionable! Whereas some people really liked the sycamore syrup (including myself), overall it received low scores. In general, this experiment gave us some interesting data on people's preferences and provided an excellent means of teaching people about pure maple and other tree syrups. Everyone was really interested in doing the tasting and thoroughly enjoyed this part of the Maple Weekend experience. I highly recommend doing a variation of this experiment at your next open house event.

FIGURE 4.50. Occasionally found along streams and riverbanks in the eastern United States, sycamores have exfoliating bark that makes these a popular ornamental in many urban and suburban neighborhoods.

FIGURE 4.51. Ironwood, or hophornbeam, develops shaggy bark as it gets older and is easily distinguished by the hop-like seed clusters that ripen in the fall. PHOTO COURTESY OF ROB ROUTLEDGE, SAULT COLLEGE, BUGWOOD.ORG

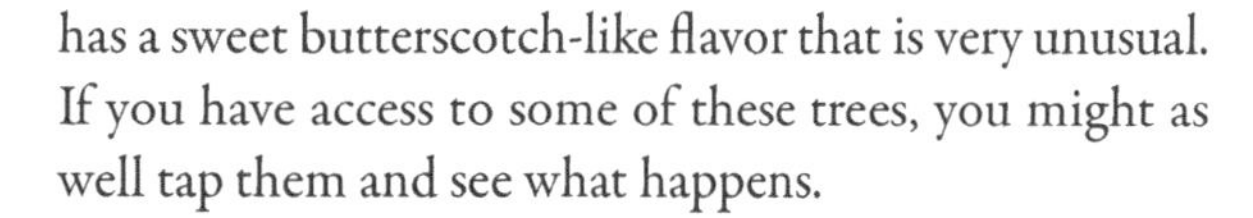

has a sweet butterscotch-like flavor that is very unusual. If you have access to some of these trees, you might as well tap them and see what happens.

Ironwood/Hophornbeam (*Ostrya virginiana*)

Depending on where you come from, you may call this tree ironwood or hophornbeam. The name *ironwood* comes from the fact that it has the densest and hardest wood you've ever come across, whereas some call it hophornbeam due to the hop-like seeds hanging from its branches. These trees are usually relegated to the understory and rarely grow larger than 5" or 6" in diameter, but occasionally you will find large specimens that extend into the main canopy. They also produce sap later in the spring when the birches start to run—the sugar content is fairly low and they don't seem to yield nearly as much sap as birches. While it may not be worth the trouble to try producing syrup from this species, I would think a hophornbeam ale would make for a great beer if you can gather enough sap to brew with.

Tree Bark Syrups

Although we normally consider tree-based syrups as originating from the sap, shagbark hickory (*Carya ovata*) syrup is actually made from the bark. The long shags that peel off are collected and then used to add flavoring to simple syrup made with cane sugar. The process starts by washing the pieces of bark in cold water, toasting them in an oven, and making an amber-colored tea with the bark. You then add sugar to the simmering tea until you have achieved the proper consistency. I have come across about a dozen people

FIGURE 4.52. The shagbark hickory is a striking tree with beautiful exfoliating bark. PHOTO COURTESY OF PETER SMALLIDGE

who are making and selling this syrup with great success. They take great pride in their syrup and are strong proponents of its smoky, earthy flavors. A similar process can also be carried out with the bark from tulip poplar (*Liriodendron tulipifera*) trees, though hardly anyone does this.

FIGURE 4.53. Razz's shagbark hickory syrup has become so popular that it is even being sold now in Wegmans—one of the best grocery chains on the East Coast. It's pictured here along with a recipe booklet from Michael Jaeb for shagbark hickory syrup. PHOTO BY NANCIE BATTAGLIA

It has been reported that shagbark hickory trees do yield sap in the spring, but I have never been successful at obtaining any. If you can get sap out of the trees, then I recommend making the syrup by using the hickory sap in place of water. And if you can't get any hickory sap, then perhaps you could make the shagbark hickory syrup with other tree saps (maple, birch, walnut) or with permeate from your RO. This could help with the marketing of your hickory syrup as an all-natural, tree-based syrup. If you really want to make it a gourmet item, use maple sugar instead of white sugar—or just boil down the toasted bark pieces in maple sap until you have shagbark hickory maple syrup!

CHAPTER 5

Gathering Sap: Cost-Effective Collection Techniques to Enhance Your Operation

In deciding how you are going to harvest sap, the first choice you have to make is whether to hang buckets (or bags) on individual trees or install tubing to bring sap to a larger tank. Your decision should depend largely on the number of taps you are putting out as well as your labor supply, topography, desire for profit, availability of materials, and aesthetic preferences. This chapter outlines some of the basic principles involved in gathering sap with buckets or tubing systems so that you can make a more informed decision on what method will be best for you.

There are very few large sugaring operations that still use buckets to gather sap. Many of the Amish and Mennonite sugarmakers in New York, Ohio, and elsewhere put out thousands of buckets (or bags) each year, but these folks are the exception to the rule. Bucket or bag collection is extremely time consuming and labor-intensive, and most sugarmakers do not have a large labor force to do the work. Furthermore, since you can double or triple your sap production by hooking up a vacuum pump to a tubing system, the vast majority of large sugarmakers use tubing to collect sap. For this reason, if you are trying to generate a substantial part of your income from maple production, you can't afford to not have vacuum tubing.

FIGURE 5.1. Properly designed tubing systems can be aesthetically pleasing and save a lot of time and effort when you're gathering sap. PHOTO COURTESY OF PETER SMALLIDGE

The topography of your sugarbush may also dictate whether you use tubing or buckets. If the hillsides are excessively steep, gathering with buckets is often impractical. It is hard enough climbing those hills to put the taps in, let along trying to carry buckets full of sap without spilling most of it on the ground. On the other extreme, sometimes the land will be so flat that it is hard to get any pitch to a tubing system. You could overcome this lack of slope by installing vacuum, though the sugarbush would need to be large enough (500 to 1,000 taps or more) to justify the expense of a vacuum pump. Without sufficient vacuum, these extremely flat areas may be better suited for bucket collection, especially for small producers.

You should give careful thought to your labor situation. If you happen to have a lot of young, enthusiastic people who want to help, then hanging buckets or bags makes a lot of sense. It gives them something very useful to do and allows them to be actively engaged in the process. On the other hand, if you don't have a good labor pool and your time is limited during sugaring season, tubing systems can save you a great deal of time and energy. You can install the tubing anytime during the year when your schedule allows; maintaining the tubing during the season requires only a fraction of the time of gathering buckets. Fall is one of the best times to install tubing, as it is easy to see throughout the woods, the temperatures are cool, there aren't any bugs to contend with, and it's easy to move around before the snow arrives.

Finally, you should also consider the cost and availability of certain materials. Do you have a readily accessible supply of food-grade buckets or pails that could be used to gather sap? Is there a local source of relatively inexpensive tubing in your area? If you are trying to save money, and don't care if your sap yield is diminished, you may be able to pick up some really cheap used tubing. Just remember that you get what you pay for. It will certainly reduce your sap yields and it's a pain to work with, but if you are willing to spend the time dealing with it and don't mind making less syrup, then you could consider buying used tubing. Your tubing costs (not including labor) could be as low as $1 per tap (or free) when salvaging used tubing. Personally I feel that buying used tubing actually costs you a lot of money in extra time and lost productivity, but if you aren't concerned about that, then it may be a viable alternative, especially for hobby producers. There are many options when it comes to materials and collection methods. The rest of this chapter covers some of the details of collecting sap with buckets or tubing systems.

Backyard Sap Collection

If you are tapping a few dozen or less trees for a backyard operation, the minimum materials you will need to gather sap include:

1. A drill bit of the appropriate size that will easily drill into wood.
2. The right size spouts for your drill bit (I recommend 5/16" for the drill bit and spout).
3. A hammer to gently tap the spouts into the hole.
4. Food-grade buckets or bags to collect the sap; these can hang directly from the spouts, or you can use plastic tubing to connect the spouts to a bucket on the ground.

Buckets

There are many different types of ways to collect sap from individual trees. The traditional galvanized sap buckets pictured in figure 5.4 are the most commonly used method. Many producers like the traditional look of these buckets hanging on trees, and this is often a key consideration for roadside applications where appearance is important. Another reason some producers prefer them is the belief that the zinc coating on the inside of the buckets helps to kill any bacteria or yeast that may be growing in the sap. New buckets sell for between $15 and $25 each whereas used buckets range from $2 to $10, so it's worth looking around for quality used ones. Many sugarmakers take good care of their equipment, so there are still plenty of high-quality used galvanized buckets available. You are also likely to find many old sap buckets in poor condition, so it's well worth your time to inspect the buckets before purchasing them. First of all, check to see if the seams are intact and that they hold water. Also make sure that the buckets are not painted, rusty, or otherwise in poor condition. Research in the 1990s found that some old sap buckets were the source of lead contamination in maple sap, so it is best to err on the side of caution when using old buckets. There are a few different kinds of plastic sap buckets on the market today. These work well, but don't have the traditional look of the galvanized sap bucket that many sugarmakers are aiming for. Aluminum sap buckets are also available and work well, though they typically do not hold as much sap and thus must be gathered more regularly. This isn't necessarily a bad thing, as it could motivate you to get out there and gather the sap more often than you otherwise would.

The most important consideration in collecting with buckets is to make sure you have a good road or trail system to gather and transport the sap. This is not an issue for trees located along roadsides, in parks, and in

FIGURE 5.2. Running short pieces of tubing to a 5-gallon bucket is another popular way to gather sap from maple (or birch) trees. PHOTO COURTESY OF FRANK FIEBER

FIGURE 5.3. A potential problem with bucket collection is the need to drive on forest roads during mud season. Tractors, ATVs, and even horses can do a lot of damage when the ground is saturated. In this sugarbush there was tremendous rutting and erosion where sap was being collected along the road, whereas the forest floor with tubing in the background was perfectly intact.

FIGURE 5.4. Nothing looks better than long rows of large sugar maples with the traditional galvanized sap buckets hanging on them in the spring. PHOTO BY NANCIE BATTAGLIA

backyards, but is crucial when considering a sugarbush setting. Sugaring season usually coincides with mud season every year, so it's extremely important to have good roads that you can drive a vehicle on or lead a team of horses through without causing excessive rutting or erosion. Depending on your soil condition and topography, building these trails could either be relatively easy or extremely expensive. Aside from the lower yields and higher labor costs, the cost of installing and maintaining good roads is one of the biggest impediments to collecting with buckets. If you have existing roads in your woods, you may be able to use them for gathering sap. However, just because you can drive on these roads when the soil is relatively dry or frozen does not mean it will withstand traffic during sugaring season, which is often the wettest, muddiest time of the year.

The Science of Sap Flow

Scientists have been studying the mechanism of sap flow in maples since at least the 1800s. Dozens of papers have been written on how and why sap flows, yet it still remains somewhat of a mystery. Trying to predict if and how much sap might flow in a given day can make a fool out of anyone. Sometimes we think we have the perfect conditions, but the sap barely runs. Other times the temperature regime seems as if we may not get much sap at all, yet our buckets and tanks are overflowing. On any given day, some trees may give several gallons of sap whereas others leave us with dry, empty buckets. Gary Backlund, the author of *Bigleaf Sugaring: Tapping the Western Maple*, likes to state that "we know more about teenagers than we do about maple trees." Despite all the vagaries in sap production, there are a few things that we have figured out over the years.

It is well known that alternating temperatures between freezing and thawing cause the sap to flow in maple (and walnut) trees. When the temperatures dip below freezing, ice crystals form within the air-filled fiber vessels, causing the air bubbles to become compressed. When the ice crystals are forming, the humidity in the fiber cells drops and moist air from adjacent vessels moves in. The pull of water from the vessels creates a suction that is transferred down the branches and trunk into the roots, thereby drawing in water from the ground (or through the taphole if it is connected to a tubing system). Water will continue to be drawn in until the tree is completely frozen. When temperatures rise above freezing, the ice crystals begin to thaw, allowing the compressed air bubbles to expand and force the sap back into the vessels. The expansion and movement of sap back into the vessels, coupled with gravity and osmotic forces, can lead to the development of very rapid sap flow at the taphole. Sap flow is usually greatest at the beginning of a run and tails off as the pressure gradient falls. The amount and rate of sap flow is influenced by a wide array of factors, including (but not limited to) the size of the tree, sugar concentration of the sap, duration and rate at which the tree froze, atmospheric pressure, soil moisture, size and cleanliness of the taphole, temperature during the thawing period, and so on. With so many variables, trying to predict how much sap you may get in a run can be very challenging. Whenever we think we know what kind of sap flow to expect, Mother Nature usually throws us a curveball that can make us look foolish.

FIGURE 5.5. Installing a pressure/vacuum gauge on an easily accessible tree is a great way to predict daily sap flows. However, even with this type of sophisticated machinery, predicting sap flow amounts can still be a difficult venture. PHOTO BY NANCIE BATTAGLIA

FIGURE 5.6. By the time the sap starts flowing in birch trees (at least in the eastern United States), the soil has usually started to dry out and driving on your forest roads to gather sap isn't as much of a concern. PHOTO COURTESY OF FRANK FIEBER

FIGURE 5.7. It is very important to only use food-grade containers when collecting sap. Even though they had been thoroughly cleaned, the sheet-rock buckets used in this sugarbush could contaminate the sap, and they detract from the image of pure maple syrup, especially when used along roadsides where they are clearly visible.

One of the advantages of collecting with buckets or bags is that you get to know each of your trees very well. When every tree is hooked up to a pipeline, it is impossible to know how much sap each one is producing or how sweet it is. However, every time you gather sap from buckets, you get to see how well each tree is producing. You could also periodically check the sugar content by simply dipping a refractometer into the bucket before you empty it. Knowing which trees are the best producers and which ones don't yield much sap will greatly help you when deciding which trees to cut and which ones to favor when thinning the sugarbush.

Bags

Although most people prefer to use buckets when collecting sap from individual trees, there are plenty of sugarmakers who prefer plastic bags. One of the main advantages of clear plastic bags is that you can see from a distance how much sap is in them. This makes it easy to know which trees you have to gather sap from and which bags have already been collected. They are also easy to empty out into your 5-gallon buckets for collecting. Just tip them on the spout and let them pour into your bucket. The main disadvantage is aesthetics—plastic bags don't have the same appeal as the traditional galvanized buckets and just don't look right hanging on the trees. If we could somehow get the look of the traditional galvanized bucket with the functionality and cost savings of a plastic bag, I would expect to see a lot more buckets or bags hanging on trees during sugaring season.

As a modern alternative to the original "sap sak," wine bags or soda bags are becoming increasingly popular among sugarmakers. I recently heard a story in which one dealer in Ohio purchased two tractor trailer loads (400,000 bags) just to supply the growing market demand for them. The clear plastic bags come with a built-in spout holder and are a quick and easy way to start collecting sap—you can have a brand-new, basically sterile collection system for approximately $1 per tap. They don't take up much room in your sugarhouse and are easy to handle and deal with. Whereas a few hundred buckets require a lot of storage space, handling, and care, the same number of plastic bags can be stored and moved around very easily. The bags are durable enough that they can be washed and reused yet inexpensive enough that they could be recycled at the end of the season, depending on the desires of the sugarmaker. One complaint I have heard is that squirrels and other wildlife may chew holes in them to get at the sweet sap. Thus, if you have an extensive squirrel population that is beyond your ability to control, then you may want to reconsider using bags.

FIGURE 5.8. A 75-cent 3-gallon wine bag hanging from a maple tree held up by a 25-cent spout. For this particular setup, I cut off the spout extension on the bag and used a spout adapter with wings to prevent the bag from blowing off. I also cut a slit on one side of the bag top to get the spout adapter inside and provide a means to empty the contents into a bucket.

Tubing Systems

Before you purchase tubing or install it in your woods, it is important to spend plenty of time planning and designing the system. Tubing systems are a large investment and usually last for 10 to 20 years before they need to be replaced. It makes sense to carefully plan and evaluate all options before you start purchasing materials or installing tubing. I highly recommend looking around at different sugarbushes, talking with equipment dealers, and doing as much reading and research as you can. Of course, the challenge then comes in processing all this information and deciding how to proceed. This chapter describes my preferred way of installing tubing, but this is certainly not the only way. Also, since the materials and practices are constantly evolving, it is important to keep up with the latest developments and technological breakthroughs. While the general principles will remain the same, I have no doubt that the practices I describe here will be modified and improved even further by the time you are reading this. By staying up to date on all the latest developments, you can ensure that your system is performing at its optimum level.

This chapter is written primarily for those who wish to install vacuum tubing systems, but much of the information is also relevant to gravity-based systems. The main differences between vacuum and gravity tubing systems have to do with the size of the mainlines and the number of taps on lateral lines. Vacuum tubing systems require more *and* larger mainlines to facilitate

FIGURE 5.9. Whenever you are transferring sap from the trees to your 5-gallon gathering buckets, I recommend pouring the sap through a paint filter that sits directly on top of your 5-gallon buckets. This will remove all the bark, lichen, bugs, et cetera, and make processing the sap later a much easier task. PHOTO BY NANCIE BATTAGLIA

the removal of both air and sap from the tubing system. If the pipe diameter is too small, it may fill up with sap without leaving any space for air to be removed from the system. Therefore, vacuum systems require larger-diameter pipes and may also necessitate "dry lines" that are installed primarily to facilitate air movement (more on this later).

With vacuum systems, you want to keep lateral lines as short and direct as possible with as few taps as possible; this allows for greater transfer of vacuum to the taphole and gets the sap into the mainline and down to your collection tanks sooner. Getting the sap out of the tubing lines quickly means less bacterial and yeast contamination of the sap. In turn, that allows you to make more syrup that tends to be lighter in color. On the other hand, with a gravity-based system you can install up to 25 taps per lateral line. With enough sap, a decent pitch, and no air leaks, the tubing lines can develop a natural vacuum that can improve sap yields. This is especially true with smaller-diameter 3⁄16" tubing. Tim Wilmot, a maple specialist at the University of Vermont, has been experimenting with this for several years now, and several sugarmakers have also tried it out. The positive results from these trials are discussed later in this chapter.

FIGURE 5.10. This vacuum pump was retrofitted to include an additional cooling fan that allows it to achieve higher vacuum levels and run more efficiently. It is worth doing your homework to find out which vacuum pump will work best for your situation.

Vacuum Pumps

In the 1960s and '70s, when tubing systems were just being developed in the maple industry, a great number of sugarmakers were also dairy farmers. Some got the idea of using the same vacuum pumps that help get milk out of their cows' udders to also get more sap out of their maple trees. The pumps maxed out at 15" Hg because the cows don't particularly like suction on their teats any higher than that. Once it became evident how helpful vacuum pumping was in boosting sap yields, sugarmakers started looking at other styles of pumps that are able to achieve much higher vacuum levels. There are now several models on the market that can achieve vacuum levels above 25". Research from UVM has found that each additional inch of vacuum at the taphole will result in 5–7 percent greater yield of sap, so if you are serious about getting the most sap out of your trees, installing a modern vacuum pump is a necessity.

Without a vacuum pump, sugarmakers are at the mercy of Mother Nature. With gravity-based systems, there must be enough temperature fluctuation to create positive pressure inside a maple tree, allowing sap to flow out of a taphole. By hooking a vacuum pump to a tubing system, however, you can lower the pressure within the tubing system and cause sap to flow from areas of higher pressure within the tree to those of lower pressure near the spout and tubing. Thus, even when temperature conditions haven't been ideal for sap flow, sugarmakers with high vacuum levels at the taphole are able to generate decent yields of sap. As long as temperatures remain above freezing, you can continue to gather sap without necessarily relying on a freeze–thaw cycle. Buying a good vacuum pump is probably the best insurance policy you can get for a changing and variable climate.

There are many types and sizes of vacuum pumps on the market today. As with most technology, the capacity has been increasing while the cost has been coming down. Purchasing a high-quality vacuum pump may be one of your best investments in your sugaring operation, so if you are going to splurge anywhere, do it on the vacuum pump. It is worth talking with many different manufacturers and dealers to determine which pump will work best for you. It's also worth noting that spending a lot of money on a high-powered vacuum pump with a lot of horsepower and CFM (cubic feet per minute) capacity won't necessarily get you a lot more sap. You also need to spend time in the woods fixing all your leaks to make your investment in the pump pay off. If you are diligent about fixing leaks, you may be able to buy a relatively smaller vacuum pump, since large pumps are only necessary when dealing with a *lot* of taps or a leaky sugarbush. I would rather save some money by getting a slightly smaller pump and spend more time out in the woods making sure there aren't any leaks in my tubing system.

Sap Releasers/Extractors

Without vacuum hooked up to the tubing lines, there is no need for any type of equipment to help the sap drain out of your tubing lines into a collection tank—just let gravity do its thing. However, for most situations in which a vacuum pump is involved, you will need to install a sap extractor (or releaser) to separate the air from the sap. The air continues to travel out through the vacuum pump while the sap either falls out through the bottom of the chamber into a tank (with a mechanical releaser) or is pumped into a collection tank (with an electrical releaser). The only vacuum pump that doesn't require a releaser is the Sap Puller, a diaphragm pump that can handle both air and sap. These are useful for sugarmakers who have 400 taps or less and want to get as much sap as possible without spending large sums of money.

Mechanical Releasers

When it comes to releasers, there are a wide variety of models to choose from. The one you select should be dependent on the number of taps flowing into it, your personal preference, and what is readily available. Mechanical releasers work by allowing a small amount of air into the system, which in turn allows sap to release into a holding tank. You should get an instruction and maintenance manual with your purchase; it

Does Vacuum Pumping Hurt Maple Trees?

Despite decades of evidence showing that vacuum tubing systems have little if any effect on tree health and growth, many people still think that vacuum harms maple trees. This has caused many producers to forgo the use of vacuum out of fear of harming the trees. To address this issue, the Proctor Maple Research Center has performed multiple tests to determine if applying vacuum does indeed harm trees. After analyzing the wounds caused by tapping and examining taphole closure rates, the UVM researchers found no significant differences among trees that had been tapped with gravity systems, low vacuum, or high vacuum. Furthermore, from a plant physiology perspective, there is no reason why high vacuum would cause damage. Unlike the cells of people and other animals, plant cells are encased in a rigid wall that protects them from the effects of vacuum. Vacuum simply allows more sap to be collected from a taphole without negatively impacting the structure or health of the tree. Thus, my recommendation (as well as that of Dr. Tim Perkins, who conducted a lot of this research) is to apply as high a vacuum as you can to generate the greatest yields of sap.

FIGURE 5.11. Mechanical releasers work well but require routine maintenance to ensure that they continue to function properly at all times. PHOTO COURTESY OF SIBYL QUAYLE

FIGURE 5.12. If you are handling large volumes of sap, consider using a diatomaceous earth or sand filter (pictured here).

FIGURE 5.13. It is important to filter your sap before processing to remove any impurities. Filters must be changed out regularly and kept clean to prevent you from inoculating relatively clean sap with bacteria and yeast as it passes through a dirty filter. Sock filters can work well, but require frequent washing and changing for optimum performance.

is *very* important to read it carefully and have a solid understanding of how your releaser works. There are several features that need periodic attention to ensure that it functions properly all season long.

Electric Releasers

If you have electric power readily available at your collection location, it may make sense to utilize an electric releaser. One of the advantages of these systems is that they allow you to pump sap higher than the height of the releaser, which may be necessary depending on your site (mechanical releasers can also be outfitted with a submersible pump to accomplish the same task). With an electric releaser, you can place the tubing lines lower to the ground, as the electric releaser doesn't need to be perched on top of a collection tank (unlike some mechanical releasers). This is especially important for relatively flat sugarbushes. Electric releasers also don't require letting air into the system to get the sap out. Pulses of air disrupt the vacuum system and briefly send sap backward up toward the trees. If you can avoid introducing air, that is always a plus.

There are two main types of electric releasers available on the market today. Some have the pump located external to the main collection tank, whereas newer models may include a submersible pump directly within the tank. I know many sugarmakers (myself included) who have struggled with getting electric releasers to work under high vacuum. If you are not an expert plumber, it is worth the money to have a professional do the plumbing, as even the slightest air leak will cause the pump to fail under high vacuum. The newer models with a submersible pump within the tank are more forgiving and should work flawlessly under high vacuum, though you should still be extremely careful when installing them.

With electric releasers, I recommend putting a coarse filter on the end of the pipe where sap flows in. This will help catch any small bits and pieces of plastic tubing, wood shavings from drilling the hole, pieces of bark, lichens, and anything else that may make its way into the tubing system besides sap. I have been amazed at what we have cleaned out of our paint

FIGURE 5.14. It is always satisfying to watch a steady stream of sap coming into your releaser. If the sap comes in spurts, you know there are vacuum leaks letting air into the tubing system and causing turbulence in the sap flow. PHOTO COURTESY OF THE NEW YORK STATE MAPLE PRODUCERS ASSOCIATION

FIGURE 5.15. Throughout the course of the end of the season, take the time to properly wash your releasers. When temperatures are warm, slime can quickly build up in places that are not being continuously flushed with new clean sap.

filters over the years! Putting on a filter is especially important for new installations, as shavings of mainline can cause serious problems in electric sap pumps. It is important to change out the filters regularly, and to only use them in the beginning of the season when there isn't a lot of bacteria and yeast in the sap. Otherwise the filters will not be as effective and they will wind up contaminating relatively clean sap as soon as it enters the releaser.

Sap Collection Tank Location

Before running any tubing lines, the first thing you need to decide is where your tanks are going to be placed and what type of sap extractor system you will use. Your tanks, vacuum pump, and releaser may be housed in a separate building that some people refer to as the saphouse (as opposed to a sugarhouse, where the sap is processed into syrup). The location and style of the releaser will determine where you start running your lines from and the height that mainlines will be placed at. This location is typically the lowest point of the sugarbush or property, but it doesn't have to be. If the lowest point is subject to severe flooding or would make it difficult to access and transport the sap, then an alternative location should be chosen. If the saphouse is not located at the lowest point, then your tubing lines may not reach all the tappable maple trees. For these trees, you could either direct them to their own smaller tank located near the lowest point, use sap ladders or "reverse slope" units to move the sap uphill to your releaser, gather them with buckets, or just leave them untapped. Ultimately your decision will depend on the resources you are willing to invest and your desire to gather additional sap.

FIGURE 5.16. As seen in this photo, sometimes the absolute lowest point isn't the best place to situate your collection tank, especially if that area is prone to seasonal flooding. You should also use food-grade collection tanks (stainless steel is best; plastic is acceptable) and make sure they are covered to prevent rain, wind, snow, and animals from getting in.

Planning the Mainlines

Once you have the location for your tank and releasers figured out, then you can start planning mainline locations. The best tools for this are a clinometer and some highly visible flagging tape. With two people working on this, one person can look through the clinometer and tell the other where to tie the flagging to meet the desired pitch. When putting up the flagging, I always tie the knots on the side of the tree where the mainline will be located. This helps prevent the support wire and tubing from winding up on the wrong side of a tree when it is being stretched through the woods. It is especially important where mainlines curve or change direction, as you never want a mainline to have tension against a tree trunk. As seen in figure 5.17, this can eventually cause the mainline to kink and/or the wire to grow into the tree trunk, especially if careful maintenance is not carried out on a regular basis. It is better to run mainlines between (rather than around) trees and then tie them back to a corner tree with wire and tubing, as seen in figure 5.18.

FIGURE 5.17. Mainline wire wrapped around a tree; this will cause the wire to grow into the tree and the mainline to become kinked.

FIGURE 5.18. A proper way to install mainline: Rather than the wire and tubing being wrapped around the tree, they have been tied back to the tree with 14-gauge wire. The tree is protected with a piece of old tubing. PHOTOGRAPH COURTESY OF MAINLINE DESIGN

There are two main ways of laying out mainlines; the method you choose will depend primarily on personal preferences and the topography of your sugarbush. The most common way is to simply run mainlines up the valleys and have the lateral lines feed into both sides of the mainlines in a herringbone pattern. If you have distinct valleys in your sugarbush, the herringbone pattern usually makes the most sense. Another method involves running the mainlines on a 3 to 5 percent grade along the contours of the hillside, with the lateral lines running downhill into one side of the mainline. Scenario 1 in figure 5.19 depicts a traditional herringbone pattern whereas Scenario 2 in figure 5.20 illustrates putting mainlines along the contour of a hillside. There are advantages and disadvantages to each design, so you need to carefully examine the topography of your sugarbush and figure out what makes the most sense for you.

The herringbone pattern featured in Scenario 1 makes sense when there are clearly delineated valleys on a hillside, especially if they are relatively deep. When the topography naturally lends itself to ridges and valleys, it is fairly intuitive where to locate the mainlines. Trying to run mainlines along the contour of a hillside presents serious issues when crossing over deep valleys. The mainlines wind up being extremely high over ground level, making installation and maintenance difficult. These situations may necessitate the use of ladders, and if you can avoid using a ladder to put up mainlines, your life will be a lot easier.

Although the herringbone pattern works well in many sugarbushes, it can also present problems in some

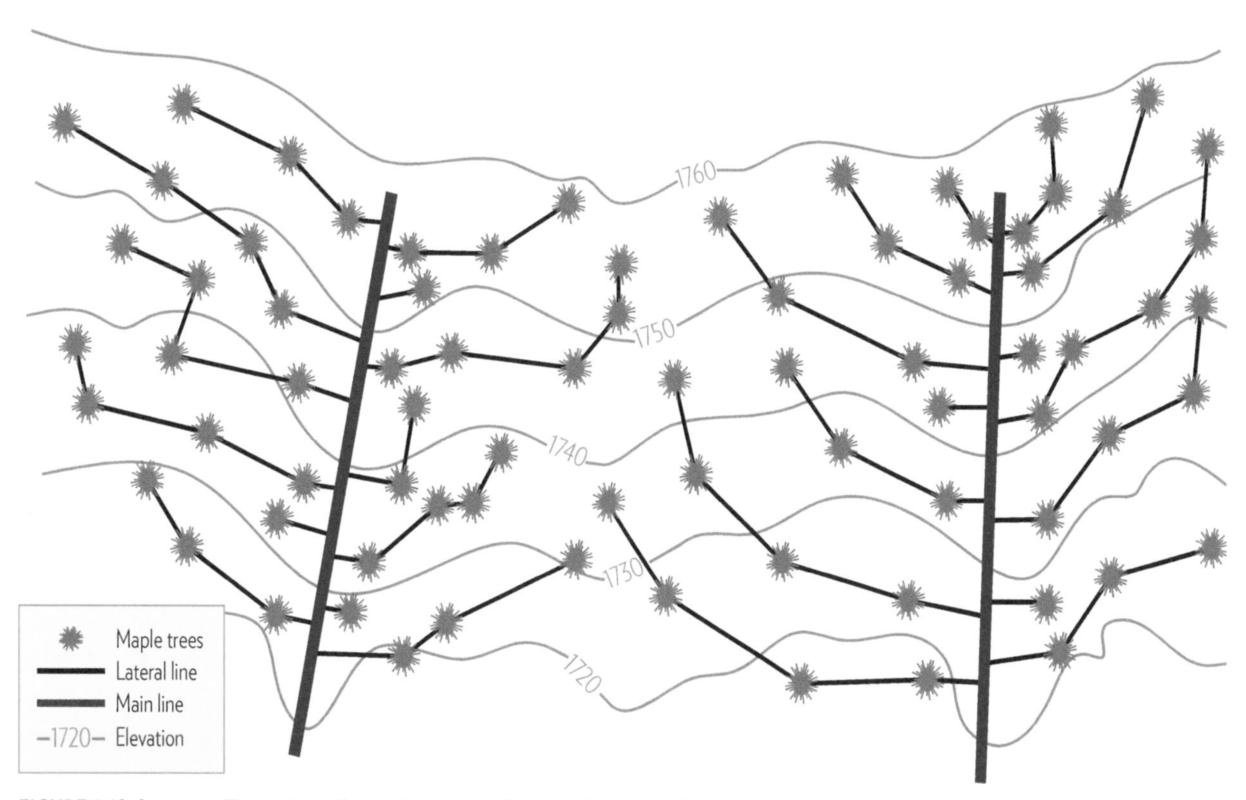

FIGURE 5.19. Scenario 1: The traditional herringbone tubing design works well on hillsides that have distinct valleys.

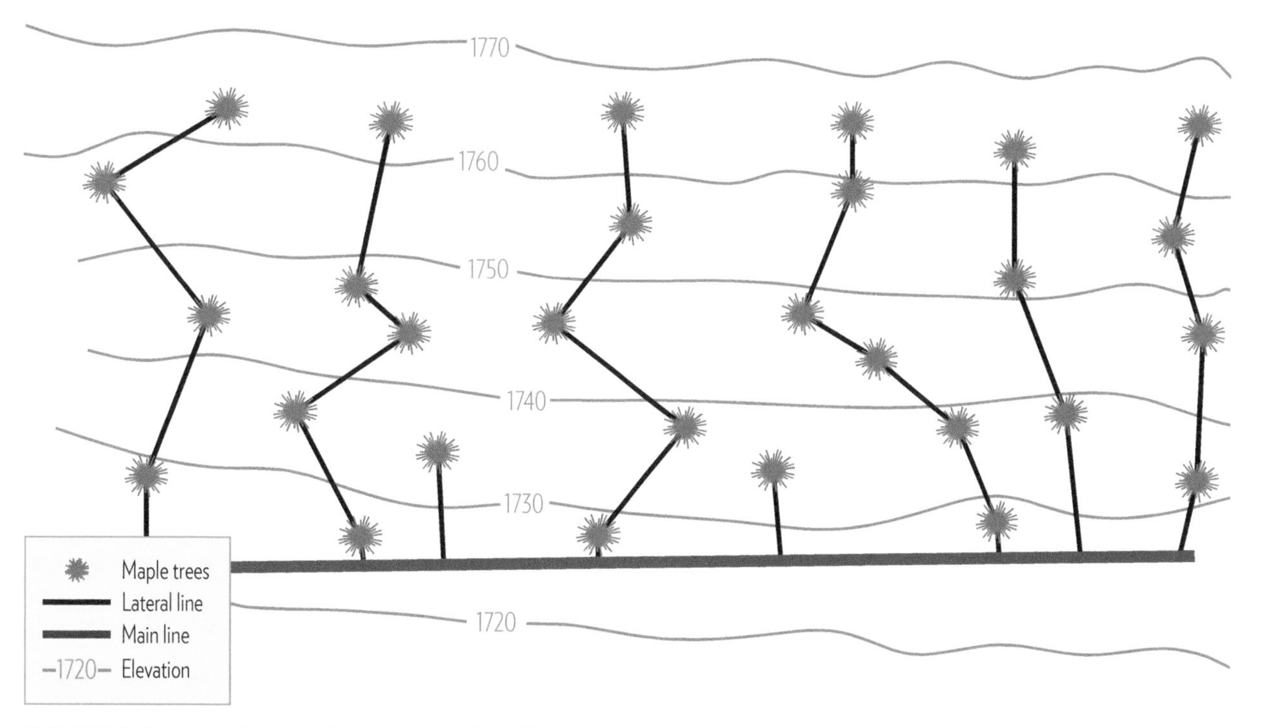

FIGURE 5.20. Scenario 2: A more modern method installs mainlines on a 3 to 5 percent slope along the contours of a hillside.

situations. It is better to have mainlines flow continuously at the same gentle slope the entire way into the releaser. Any abrupt changes in pitch can cause the sap to gurgle in the lines and interrupt the steady flow of sap and air out of the tubing system. When the mainlines that run up the valleys connect in to the conductor lines that run along the base of a hillside, there can be an abrupt transition in the pitch of the mainline. Furthermore, if you run a mainline up a hillside without any distinct ridges and valleys, it is possible that there will be little to no downward slope to the lateral lines. Thus, for hillsides that don't have much pitch variation, the design in Scenario 2 is usually a better fit.

It is worth noting that the design in Scenario 2 requires much more mainline than the design in Scenario 1 to keep lateral lines of equal length. Imagine that you wish to keep lateral lines a maximum of 100 feet long. In Scenario 1, the mainlines could be spaced 200 feet apart, and the 100-foot-long lateral lines could bisect the area between mainlines. However, the mainlines would have to be placed 100 feet apart in Scenario 2, since all of the lines must flow downhill to the closest mainline. Some people like being able to walk freely along the length of a mainline without having to duck underneath or climb over lateral lines. The design in Scenario 2 allows for this to a certain extent. However, all the mainlines running across the contour of the hillside must connect into a conductor line that runs up and down the hill, which usually prevents the free flow of ATVs and snowmobiles throughout the sugarbush.

FIGURE 5.21. This notebook contains valuable information that allows you to determine the optimum size of vacuum pump and mainlines for your particular tubing setup. PHOTO BY NANCIE BATTAGLIA

Distance Between Mainlines

How far apart you put your mainlines largely depends on whether you have a vacuum pump. With vacuum tubing, you want to keep your lateral lines short with as few taps as possible, which means keeping your mainlines relatively close together. Under Scenario 1, I would try to keep them roughly 120 feet apart (as the topography dictates), but with Scenario 2, I would keep them about 80 feet apart, depending on the density of the sugarbush and the general terrain. Without vacuum, I would probably install mainlines 300 to 400 feet apart, depending on the topography and natural drainage patterns.

Mainline Sizing

What size tubing to use for the mainline is a question that has long perplexed sugarmakers. You want to keep mainlines as small as possible, yet still be able to get a steady flow of air and sap out of the lines without any restrictions. Smaller-diameter tubing costs less, is easier to handle, and does not cause the sap to heat up as rapidly as larger-diameter tubing. However, mainline tubing that is too small won't be able to properly transfer vacuum from the pump to your trees. There is a balancing act involved in making sure that the tubing is large enough to do the job, but not so large that you are wasting money and causing unnecessary heating of your sap. Sugarmakers had used mainline tubing of 0.5" in the past, but nowadays hardly anyone uses tubing that is less than 0.75". Large sugarmakers with thousands of taps will often use 1" tubing for the smallest mainlines and 1.25" to 4" tubing for the conductor lines, depending on the number of taps and topography of the sugarbush.

Steve Childs has spent a lot of time trying to help sugarmakers determine the optimum mainline size for their particular situation. In the recently published *New York State Maple Tubing and Vacuum System Notebook*, you can find tables and charts that let you determine the number of taps that can be supported by a given vacuum pump on a certain slope. Rather than just deciding on a vacuum pump out of thin air, you can now figure out what will work best based on your particular situation. This can keep you from spending too much money on a vacuum pump that is larger than what you need—and it can also help you purchase a pump with a large enough capacity for the tubing system you have. Many sugarmakers have also used this to determine how many mainlines to put in their woods and what size they should be.

Installing Support Wire

After you have designed your mainlines and installed all the necessary flagging, the next step is to install the mainline support wire. The most common material used by sugarmakers for this purpose is 12.5-gauge high-tensile wire. If you can get the wire spooler to the end of the line, it is a lot easier to unspool it while heading downhill. However, this is often impractical, and you will usually wind up starting at the base of a hill that you can easily get to with the wire and spooler. I highly recommend having someone stay at the spooler to keep an eye on it, making sure it unravels properly. Hopefully everything goes smoothly and that person doesn't have to do anything, but occasionally problems will develop that make the person watching the wire spooler extremely valuable.

The main support wire needs to be anchored at both ends. One method involves setting an anchor bolt in to a tree and then attaching the wire to this bolt. Sugarmakers who prefer this method like the fact that it is relatively easy, it requires installing only one piece of hardware in a tree, and you easily can set the exact height of the mainline. However, since the bolt could potentially cause damage to the anchor tree, you should choose a sturdy yet otherwise non-valuable tree for this purpose. Furthermore, since the wire is anchored into

FIGURE 5.22. I recommend using a wire ratchet strainer to get the wire reasonably tight before attaching the mainline tubing to it. There should be more tubing surrounding the wire to provide additional protection for the maple tree. PHOTO BY NANCIE BATTAGLIA

a set point, the wire is more likely to break if a tree or large limb falls on it. An alternative method involves wrapping the wire around a tree and securing it back on itself, by either tying the wire in a knot (that won't come apart) or using a specialized piece of hardware (such as a Gripple or wire crimps). The advantage of this system is that you don't need to install any hardware in the trees and you can adjust the height of the mainline as necessary. When using this method, it is important to adequately protect the tree from the wire wrapped around it. Some people use blocks of wood, though these can become problematic over time as the tree grows into the wood. My preferred method is to wrap the wire in used lateral line tubing and then put this within an old section of mainline tubing. The plastic tubing can easily slide up and down the tree. You should adjust it every few years to ensure that the tree doesn't start growing into the wire.

No matter which system you use, I recommend putting in a ratchet strainer on one end of the wire, as seen in figure 5.22. This allows you to make the wire reasonably tight before attaching the tubing to the wire. When it becomes difficult to crank on the ratchet, you should stop. You want to leave a little bit of play in the wire and make it tighter by tying the wire back to nearby trees with a lighter gauge of wire. Most people use 12.5-gauge high-tensile wire for the mainline wire and 14-gauge wire for the side-tie wire. The theory behind this is that if a branch or tree falls on the mainline, the side ties will

FIGURE 5.23. The Rapi-Tube system is another way to install mainline stays tight without the use of a support wire. PHOTO TAKEN IN AN ALASKAN WHITE BIRCH "SUGARBUSH" COURTESY OF KEVIN SARGENT

FIGURE 5.24. Even without a specialized tool, you can easily spool out mainline tubing when there is a solid snowpack for the tubing to slide on. PHOTO COURTESY OF JEN KRETSER

just break or slide down and the entire mainline will simply fall to the ground without the main support wire breaking. Then you can simply cut the tree off the line, redo the side ties, and everything is back to normal.

Installing Mainlines

Once you have the wire up and reasonably tight, unroll the mainline directly underneath the wire. If you are going to be doing a lot of this, you may want to invest in a mainline spooler, as this will make the job relatively easy. If you are adept at welding and construction, you could also build one yourself. If there is snow on the ground, another option is having one person sit in the center of the roll and help unspool it while the other person pulls it out. The tubing easily slides on the snow, and the process usually goes smoothly. It is important to make sure you start pulling on the outside of the roll of tubing; otherwise it will get difficult very quickly. This rule also applies to the high-tensile wire and lateral line tubing. Once you have made the mistake once, you are unlikely to do so again!

Securing the mainline to the wire is a fairly low-skill, time-consuming job that is not a lot of fun, but it needs to be done right. You should first anchor the mainline to the end of the wire to make sure that it can't slide down the wire and become snaky during periods of warm weather. There are several ways to do this. The easiest and most expensive method is using the wire mesh grips commonly known as "Chinese fingers." I have also seen people slice a piece of mainline down the center, slide it over the mainline, and then affix a hose clamp over this assemblage. You can then use wire (and possibly a ratchet) to secure the mainline tubing to the anchor tree. Once this step is completed, then you start putting on wire ties every few feet to get the tubing off the ground and generally secure to the mainline. It is important to pull the mainline tight each time you do this to avoid dips, sags, and snakes in the line. Later on, after the lateral lines and saddle manifolds have been installed, you should go back and fill in the wire ties to make sure they are no more than 18" apart. Doing this will keep your mainline tight to the wire at all times.

Installing Lateral Lines

Lateral lines are used to connect the trees to the mainline. Almost all sugarmakers use 5⁄16"-diameter tubing for this, although some producers are now experimenting with 3⁄16"-diameter tubing for natural vacuum applications on steep hillsides. Once you have your wire installed and partially tensioned with the ratchet strainer, you can then start installing lateral lines. I usually tie the mainline tubing to the support wire before starting with lateral line installation, but you don't have to do this. By using the slide fitting seen in figure 5.25 (or a similar device), you can pull the lateral lines as tight as possible while also taking up slack in the wire. Simply hang the spool of tubing on the wire (figure 5.26), run the tubing out to the last tree on the line, and secure it with an end-line ring. As you walk back to the wire, pull the lateral line tight and place it at the desired height. When you get back to the wire, simultaneously pull the tubing and wire as close to each other as possible and secure the tubing with the slide fitting, making sure to leave at least 12" of tubing after the slide fitting to connect into the saddle manifold.

Even though I prefer to use the slide fittings, the most commonly used method is installing a hooked connector that slides on top of the wire (figure 5.27). This requires cutting and joining two pieces of tubing together with the barbs on the hooked connector. I prefer to use the slide fitting for several reasons. First of all, it's best to limit the number of barbed connections within a tubing system—each fitting creates a possible vacuum leak, is a place for bacteria to build, and puts more restriction on sap flow. The slide fitting is able to grip the tubing without any need to cut it, eliminating all these problems. Second, installing lateral lines is much quicker when using the slide fittings; you simply place the tubing inside the fitting, push the two pieces together, and pull tight. It doesn't require cutting the tubing, putting in the barbed fitting, and using the tubing tool to join the tubing together again. Using the slide fittings allows you to pull the lateral line tubing extremely tight and just cut the tubing 12" after the fitting, thereby reducing waste and eliminating the need

FIGURE 5.25. The slide fitting grips the lateral line and keeps tension on it while allowing a short piece of tubing without tension to connect the lateral line to the saddle manifold and mainline.

FIGURE 5.26. Tubing spoolers are the easiest way to run lateral line tubing–simply hook the spooler on top of your mainline and walk with the tubing to the trees you want to tap. PHOTO COURTESY OF BRIAN CHABOT

FIGURE 5.27. A hooked connector is often used instead of the slide fitting seen in figure 5.25. PHOTO BY NANCIE BATTAGLIA

FIGURE 5.28. Short lateral lines with only two or three taps on them facilitate better vacuum transfer to the taphole and make vacuum leak checking much easier.

to cut out short pieces of tubing when installing the droplines. Finally, the thick, wide hook on the slide fitting allows you to actually put tension on the mainline wire when pulling the lateral lines tight. You simply pull the lateral line tubing and mainline wires as close to each other as possible and hook on the slide fitting. This puts positive tension on both the lateral line and mainline, and the lateral lines effectively serve as "side ties" for the mainline.

When installing lateral lines, you want to keep them relatively straight and use tension on the trees to keep the lines from sagging. Lateral line tubing doesn't transfer vacuum as well as mainlines, so to achieve the greatest level of vacuum at the taphole, you want to keep lateral lines short and with as few taps as possible. One tap per line is ideal but impractical, so there is a saying... "Strive for 5, no more than 10." In our sugarbush, we average 2.6 taps per lateral line. Having such a low number of taps per line makes vacuum leak checking

FIGURE 5.29. Once you have your lateral lines run between trees, you can then install the droplines at each tree. A two-handed tubing tool makes gripping the tubing and installing the droplines very easy. PHOTO COURTESY OF BRIAN CHABOT

much easier. I'd much rather chase down a leak on a three-tap line that is 50 feet long than an eight-tap line stretching over several hundred feet. Since reworking our tubing system to add more mainlines and reduce the number of taps per lateral lines, our vacuum levels at the taphole have increased by 2" to 3". I attribute this primarily to the fact that it is now easier to find and fix leaks with shorter lateral lines.

End of Lateral Lines

There are many different fittings and methods for securing the end of the tubing at the last tree. My preferred method is to use a simple ring around the last tree and then cut in a one-way T just above the barbed fitting, as seen in figure 5.30. You can still use the ring to tension the line as needed, and the installation goes relatively quickly. A similar method utilizes a fitting that hooks on top of the lateral line after running around the tree. Both methods work well and are quick, easy, and inexpensive to install. You can also purchase an end-of-line fitting that includes a barbed outlet to place the dropline within the same fitting. I know some sugarmakers who like this system because it provides for a continuous flow of sap out of the tree and into the tubing system without any dead spots. However, you may encounter problems when using this fitting on very large trees, trees growing at an angle, and trees located on a relatively steep hillside. The main issue is that the sap must run uphill at some point to then get down the lateral line, breaking the main principle that sap should always run downhill. In these situations, sap tends to pool in the low spots, leading to excessive bacterial growth in warm weather and ice formation in cold weather. In places where the low spot is on the north side of the last tree, I have seen many instances where the ice never thawed even though the temperature may have gotten into the mid-30s on a clear, sunny day. Sap was running in the tree but it couldn't get into the tubing lines due to the frozen pocket of sap on the north side of the tree that never thawed out without direct sunlight hitting it.

For large, healthy trees above 18" that can support two taps, I have seen T or Y fittings used successfully. This takes more time to install, but it allows for a one-way T to be placed on either side of the tree, as seen in figure 5.31. Using a Y fitting makes for a smoother angle, but regular T's work just fine and you usually have extra of these fittings whereas the Y fittings are

Using Your Maple Tubing to Collect Birch Sap

If you happen to have a lot of birch trees mixed in with your maples, you may be able to utilize the same tubing system to gather the birch sap. At the end of the maple sugaring season, simply pull the maple spouts and wash the tubing to the greatest extent possible. (We once made the mistake of not washing, and the sap quickly became contaminated with the residual bacteria left over from maple season.) When you have all the maple taps out and you are confident that your tubing system is clean and drained of any remaining fluid, then you can tap in the birch spouts. Alternatively you can simply run all new tubing and mainlines to your birch trees, especially if they are concentrated in certain sections of your sugarbush. This would allow you to go straight in to birch sap collection without worrying about cleaning out the maple tubing and will result in higher quality sap. You can still use the same releasers, vacuum pumps, and tanks, as long as they are thoroughly cleaned and all the valves to your maple taps are closed. With only limited research at this point, we don't yet know the effect of vacuum tubing on sap yields from birch trees. I did not notice a huge difference in our yields between vacuum and gravity tubing, but it was apparent that the sap came in much cloudier when the vacuum was not on. It is likely that the vacuum was able to get the sap down to the collection tanks much faster so that there wasn't as much time for it to spoil in the lines. Until further research is available, if you already have vacuum for your maple tubing, I recommend also applying it to your birch taps.

FIGURE 5.30. My preferred method of installing the last dropline on a lateral line is to use an end-line ring and then place a one-way T in the proper position so that sap comes out of the dropline, flows into the one-way T, and then starts heading down the lateral line. In this photo, the right side of the T is plugged, so all sap has to flow to the left down the lateral line.

FIGURE 5.31. For large trees at the end of a lateral line that can support two taps, you can use a regular T to support the lateral line and then put in a pair of one-way T's for your droplines. PHOTO BY NANCIE BATTAGLIA

generally only used for this purpose. If you have extra T's lying around and don't know what to do with them, this would be a good use.

Installing Manifolds

Once all your lateral lines are in place, connect them to the mainlines by installing saddle manifolds. First drill a hole in the mainline several inches downhill from where the slide fitting connects the lateral line to the mainline wire. You can buy a specialized drill for this purpose, but I just put a good stopper on a regular drill bit and use a cordless drill. You want to make sure to drill slowly and not go through the other side of the mainline tubing! It is just as important to get all the shavings out of the hole. If you let any of the shavings fall in the mainline, these may cause problems farther down the line. We have had them get stuck within the impellers of sap pumps and clog up pre-filters housed within our sap meters. This has caused the sap pumps to fail, and the last thing you want to happen is have sap gushing in your releaser and no sap pumping out of it.

Once you have the hole drilled, it is important to properly install the manifold on the mainline tubing. The gasket needs to be properly placed in the hole, as vacuum leaks often occur when it is off-center. You must also make sure to properly secure the outside of the manifold. Some manifolds use plastic straps or ratchets; others rely on hose clamps or wire ties to fully secure the manifold to the tubing. Whatever manifolds you use, make sure you install them properly to avoid vacuum leaks.

Tapping

The process of drilling holes and inserting taps into the trees is one of the most crucial aspects of sugaring. It should be done slowly and deliberately by trained people who know what they are doing and who care about doing the best job they can. If sugaring is just a hobby and it doesn't really matter how much sap you get, then you can use volunteers to help tap. On the other hand,

FIGURE 5.32. Larger tapholes create longer and wider streaks of stained wood inside a maple tree that are no longer conducive to sap flow. The excessive staining is the main reason most sugarmakers have switched over to 5⁄16"-diameter health spouts. PHOTO COURTESY OF BRIAN CHABOT

FIGURE 5.33. Not only does this tree have *way* too many taps in it, but all the droplines have a J-shape where sap can pool before heading down the lateral line. This tree should only have two taps, and the tapholes should be placed high enough above the dropline to allow a continuous downward flow of sap from the taphole to the lateral line.

if you are relying on sugaring as a source of income, then it is extremely important to make sure that everyone who taps for you is experienced and fully understands all aspects of tapping. Although tapping may seem like a fairly simple practice, there is much more to it than just drilling a hole and putting in a spout.

Tapping Guidelines

To ensure the long-term sustainability of your sugaring operation, it is important to follow conservative tapping guidelines. If you use 5⁄16"-diameter spouts, start tapping trees when they are 10" in diameter, and don't add a second tap until they are 18", chances are you will be able to tap these trees indefinitely and always have enough clear white sapwood to maintain high yields. The issue with overtapping is not the possibility of killing the tree; rather, it is the fact that every hole you drill in a maple tree creates a stained column of wood that is no longer conducive to sap flow. If you drill too many holes in the tree and it's not growing fast enough to add new white sapwood for future tapping, eventually you will run into situations where it is hard to find a good spot to drill a hole. This is especially important in old sugarbushes where multiple 7⁄16" spouts have been put on the larger trees for several decades. As we have learned more about the effects of tapping wounds on internal staining and future sap yields, many sugarmakers with high-vacuum systems have switched to only putting in one 5⁄16" spout, no matter how large the tree is.

The trade-off you have to make when deciding how many spouts to put in a tree is your desire for additional sap today versus potential yields in the future. Since every hole we drill eliminates that portion of the stem from future sap flow, it is important to limit the number of tapholes placed in a given tree. For gravity-based systems, putting in a second spout can give you twice as much sap, but only if the tapholes are spaced far enough apart and were both drilled into clean sapwood. Thus, if you are on buckets or are using gravity-based tubing, putting a second spout on large trees above 18" dbh makes sense. On the other hand, in vacuum-based systems a second taphole will produce twice as much internal damage without necessarily giving you twice as much sap. The second taphole will be competing with your first for some of the same sap, so there isn't a direct linear relationship between the number of taps and amount of sap collected. If you have very large, healthy trees that are growing fast and can sustain two taps, then it makes sense to put two taps on large trees under vacuum. However, if your trees are no longer growing fast and exhibiting symptoms of decline, I recommend only one tap per tree, no matter how large it is.

When to Tap

One of the most difficult decisions you have to make from year to year in your sugaring operation is deciding when to tap. It's hard enough to predict the weather on a day-to-day basis, let alone over the course of several weeks. Sugarmakers typically start tapping in February when the five-day forecast calls for warm days above freezing. However, this doesn't always result in the optimum yield per taphole. Some years it may make sense to tap much earlier; other years the season may be delayed. Long-term trends have found that the sap flow season is moving earlier, so if you don't want to miss out on the early runs, it pays to be ready in January or February whenever sugaring weather arrives. Personally I would err on the side of tapping too early rather than too late, especially if you have vacuum tubing and are using new spouts every year.

I always recommend tapping just a few trees in January and February to determine what is going on with sap flow conditions. In relatively cold areas, even when the temperatures get above freezing in January and February, the amount of sap flow can be negligible. The trees are basically frozen, and it takes an extended period of warm temperatures to induce substantial sap flow. In warmer regions where the winter isn't as severe, optimum temperature fluctuations usually happen all winter and the trees may be producing a decent amount of sap in January and February. If you see this happening in your test trees, you'll want to tap the rest of your sugarbush to catch the early sap runs. However, it is also worth noting that the sap sugar content starts out low, builds throughout the course of the season, and then drops off toward the end. If the sap sugar concentration is still low (1 percent or less) during the early sap flow periods, you will probably want to hold off on tapping the rest of your trees until SSC gets up to viable levels (at least 1 percent and preferably closer to 2). This is especially true if you don't have a reverse osmosis unit to remove most of the water before finishing on the evaporator.

Even if the weather is appropriate for tapping, you have to first make sure that you are ready to start collecting and processing sap. There are many tasks that must be completed before you can start making syrup, and you don't want to tap your trees without being in a position to gather and process the sap into syrup. Another thing to consider is that if you start tapping in January or early February, you may encounter extremely low temperatures after you have already started using all your equipment. I have seen evaporators turn almost completely to ice, especially in the back pans, when temperatures drop well below freezing for an extended period. You'll want to make sure that all lines are properly drained, and you may want to drain the partially boiled sap in the back pan of your evaporator. For these reasons, a lot of small-scale sugarmakers will wait until the temperatures are warmer in late February or March to start their sugaring operations.

Which Spout to Use

There has been a tremendous amount of research and development over the years to develop the perfect

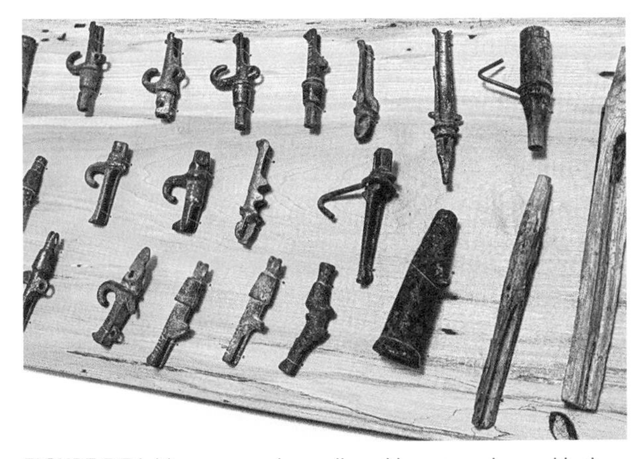

FIGURE 5.34. Many sugarmakers collect old spouts and assemble them in nice displays for educational purposes. This collection was developed by Joe Orefice, a forestry professor at Paul Smith's College. PHOTO BY NANCIE BATTAGLIA

FIGURE 5.35. Although these old rusty 7⁄16″ spouts may make for some nice vintage photographs, they belong on a display, not in your trees. It is worth the extra investment to buy new 5⁄16″-diameter spouts, preferably made out of stainless steel. PHOTO BY NANCIE BATTAGLIA

spout—one that will generate the greatest amount of sap flow. Figure 5.34 shows some of the many spouts that have been used over the past century for both tubing and bucket systems. I know many sugarmakers who love collecting these spouts and take great pride in finding a rare spout (or spile, as some people call them). It's worth noting that the best use for these old spouts is as a collectible! Several research projects in recent years have discovered that using a brand-new spout will result in much greater sap yields than old spouts that have been contaminated with bacteria and yeast. Thus, for the small price of a new spout each year, you can easily increase your sap yield by a substantial amount. The impact will vary each season based on the weather, vacuum level, and other factors, but you can almost guarantee to make your money back and then some. If somebody offered to give you $2 a few months from now if you gave them a quarter today, you would be crazy not to accept that offer. Indeed, very few investments can yield the type of return achieved with purchasing new spouts each year. As an alternative, some producers prefer to use stainless spouts—this is the only material that can be adequately cleaned and sterilized each year. If you are willing to make the larger initial investment in a stainless spout and take the time to properly sterilize it after every season, you can eliminate the need to purchase new spouts every year.

Dr. Timothy Perkins at the Proctor Maple Research Center in Vermont recently developed a check-valve spout in conjunction with Leader Evaporator Company that has become increasingly popular among sugarmakers. The reason it was developed was to prevent sap from getting sucked back up into the tree at the end of the day when the temperatures dip below freezing. When this happens, the tree develops a vacuum and up to a pint of sap can be sucked back in through the dropline. Part of the increase in sap yield is due to the fact that it is a new, sterile spout, while the other benefits are realized through the action of the check valve. Sap inside a maple tree is essentially sterile, yet once it enters the tubing system it becomes contaminated with bacteria and yeast—no matter how well you tried to clean it. By preventing the contaminated sap from going back into to the tree, the check-valve spout keeps the taphole cleaner. It is the bacteria and yeast that cause the taphole to "dry up" toward the end of the season, so sugarmakers who use check-valve spouts report getting more sap flow later in the season than those with regular spouts.

Check-valve spouts have been advertised to greatly boost sap yields, though not all sugarmakers get the same results. The type of impact you see depends on the following parameters:

- **Age and quality of tubing system.** If your system is relatively new and clean, check-valve spouts don't have as much impact as they do on older, more contaminated tubing systems.
- **Vacuum pump management.** If you keep your vacuum pump running whenever it is above freezing and don't turn it off until all the sap in your tubing system freezes, the impact of using a check-valve spout is significantly minimized. Under the constant pull of the vacuum, the sap won't be able to migrate back in to the taphole.
- **Type of releaser.** Since mechanical releasers allow air to enter the tubing system every time they dump, this action can cause sap to move backward into the tree, making check-valve spouts more useful. Electric releasers do not allow air into the system, so check-valves do not have as much of an impact in those types of tubing systems.
- **Date of tapping.** Using check-valve spouts makes sense for producers who want to tap in January or February to catch the early runs while still having viable tapholes in April to catch the late ones. Sugarmakers who tap later in the season will not experience as striking an effect from the check-valve spouts.

Check-valves aren't the only type of spout to gain a lot of interest and attention among sugarmakers. Clear polycarbonate spouts came out a few years ago and are now being carried by most maple equipment companies. Our research found little difference in yield between clear polycarbonate spouts and the black check-valve spout adapters. Leader has since come out with a clear polycarbonate check-valve spout that we are experimenting with this year. In 2013, researchers at UVM found that these spouts outperformed all others in their trials. One of the main advantages of clear polycarbonate spouts is that you can see through them, so it is easy to find leaky tapholes. They also seat very well in the taphole, so the chances of leaks developing are minimized. Finally, they are also relatively inexpensive, though I wouldn't necessarily base my choice of spout on the cost. For a relatively low investment, the potential returns from using a good spout far outweigh the cost of buying it. As seen in figure 5.36, a diverse array of spouts is available on the market and new ones are being developed almost every year, so be sure to keep up to date on the latest developments and choose the spout that you think will work best for your situation.

FIGURE 5.36. Some of the many spouts available from different maple supply companies in 2013. PHOTO BY NANCIE BATTAGLIA

Finally, I always recommend using 5⁄16" diameter spouts, especially if you have a vacuum tubing system. Research at Cornell, Proctor, and Centre ACER has found that 5⁄16"-diameter spouts will give you just as much sap as a 7⁄16" spout under vacuum, and slightly less with gravity-based systems. Even if you did get less sap with a 5⁄16" spout, the long-term benefits of drilling a smaller hole will eventually result in much greater yields than sticking with the 7⁄16" spouts. The only thing 7⁄16" spouts are good for is as a collectors' item. If you haven't already switched over to the smaller spouts, I strongly encourage you to do so.

How to Tap

As previously mentioned, how and where you drill the taphole and set the spout is extremely important to your overall yield of sap from a given tree. It is best to only use skilled, experienced people for this job. If you have folks who want to help and could do a good job—but they just don't have any experience yet—be sure to spend plenty of time showing them how to tap, and stay with them for at least the first 20 to 30 taps.

Before anyone ever drills a hole in a valuable maple tree, you should have them practice on low-value trees (such as beech and aspen) that you were planning on cutting down anyway.

Finding the Right Spot

The first step in tapping is to find a good spot to drill the hole. It doesn't matter how nice a hole you drill, what type of spout you use, or what level of vacuum you are pulling if you have drilled into a bad section of the tree. To get a decent amount of high-quality sap, you need to drill into clear, white sapwood. It is important to avoid previous tapholes and the associated stain columns as well as other defects and rotten areas on the trunk. Large seams and wounds are easy to identify and avoid, but it takes a trained eye to locate old tapholes. Figure 5.37 shows a tree trunk in our sugarbush. I have circled some of the old tapholes and left others unmarked—see if you can find the remaining tapholes. We try to stay at least 1" to the left and right and 12" up and down on the trunk from an old taphole.

Drilling the Hole

When I try to imagine the work involved in using a brace and bit, the drills that required a huge battery stored in a backpack, or gas-powered tappers, it makes me very appreciative of the cordless drill. Depending on the model you have and how old your batteries are, most sugarmakers can get at least 200 to 300 holes drilled on a single charge. They are lightweight, easy to use, and you can often find good deals on them. If you don't already have a good cordless drill, this is one of the best investments you can make in your sugaring operation.

FIGURE 5.37. Trees have that been previously tapped will show evidence of old tapholes. It is important to be able to recognize old tapholes to avoid drilling into stained columns of wood. See if you can find the previous tapholes that have not been circled.

FIGURE 5.38. This taphole was drilled at too much of an angle and in a very poor location, about 5" directly below the previous year's taphole. As you can see in the background, the spout was still in the tree as leaves were starting to develop. All your spouts should be pulled immediately after the sugaring season to give the trees maximum time to heal over tapholes. PHOTO COURTESY OF PETER SMALLIDGE

FIGURE 5.39. Always examine your shavings as you are pulling the drill out of the tree—they should look as white as (or even whiter than) the shavings seen here. PHOTO COURTESY OF PETER SMALLIDGE

FIGURE 5.40. Take as much time as you need to properly examine the tree before tapping. I tell my tapping crew that I want to see footprints around the entire tree and that they should be spending at least 30 seconds at each tree finding the best spot to drill. PHOTO COURTESY OF BRIAN CHABOT

When drilling the hole, some sugarmakers like to use the slow setting out of concern that high speed will cause the tissues to scorch or burn, thereby reducing sap yields. However, research performed by the USDA Forest Service back in the 1970s determined that drills operating at speeds as low as 120 rpms and as high as 6,600 rpms produced essentially the same amount of sap.[1] The main issue with drilling too fast is the possibility of making an oval taphole if you don't go straight in and straight out. Thus, it is essential that you have good footing and proper control of the drill to make sure your hole is perfectly round.

Some people advocate drilling the hole directly into the tree whereas others recommend drilling at a slight upward angle. The proponents of drilling straight in feel it is the only way to ensure a perfectly circular hole that will not have any air leaks around the spout. Those who recommend a slight upward angle (1 to 2 percent) believe that you can still get a good seal around the spout, and the slight angle allows any sap to drip out of the tree more easily. I can understand both points of view. I usually try to achieve a perfectly straight hole but always err on the side of making it at a slight upward angle whenever necessary. No matter how you drill the hole, be sure to use a relatively new, clean, sharp drill bit that is intended for drilling into maple trees. Old, rusty, and dirty drill bits immediately contaminate the taphole, so be sure to properly maintain your drill bits between seasons or simply purchase new ones every year.

When you are pulling the drill out of the tree, always examine the shavings to make sure that they are pure white. If you get brown or dark-colored shavings, you have drilled into a bad part of the tree. Your sap yield will be negligible, and any sap that does flow may have a yellow tinge to it and impart off-flavors to your syrup. Furthermore, drilling into a bad part of the tree will often result in a vacuum leak; without a steady flow of sap, the spout may just be sucking air into the tubing system.

One of the ways to avoid drilling into stained wood is by tapping at a shallower depth. Some producers drill 2" deep or even farther into the sapwood of the trees, and while this may be fine in new sugarbushes that have not been tapped previously, it is risky to do on trees that have been tapped for many years. Under vacuum, the taphole does not need to be as deep to achieve high yields of sap, so we try to keep our taphole depth at 1.5".

Setting the Spout

The final step is placing the spout in the tree. It takes some practice to figure out how hard to tap on the spout to get it nice and snug without overdoing things. Not tapping in hard enough can cause the spout to be too loose, creating a vacuum leak. On the other hand, tapping too hard can potentially cause the wood to split, which in turn leads to vacuum leaks, lost sap, and increased wounding at the taphole. This is especially true on older spout designs that had a heavy taper; most of the newer versions have a light taper that makes it very hard to split the wood. It is important to remember that the process is called "tapping" for a reason. Most sugarmakers use regular hammers to set the spouts, but you don't necessarily hammer the spouts in. Just a few gentle taps will usually do the trick until you hear a thumping sound. As soon as you can hear the difference, stop tapping on the spout.

Checking for Vacuum Leaks

If you are going to install any type of vacuum pump with your tubing system, it is very important to spend as much time as possible in your woods repairing vacuum leaks. Money invested in a vacuum pump only pays off if you maintain a high vacuum level at your tapholes. It's not good enough just to have a high reading on the vacuum gauge at the releaser—that only gives you a broad representation of how many leaks are out in the woods. What you need to do is get vacuum readings near the last taphole for each of your mainlines and make sure they are within 1" to 2" (or ideally the same level) as they are at your pump. Research from the University of Vermont suggests that your sap yield will increase by 5 to 7 percent for every additional inch of vacuum you can achieve at the taphole. Glenn Goodrich, a well-known maple producer and equipment dealer in Vermont, estimates that the time he is able to spend in the woods fixing leaks is worth well over $100 an hour. By fixing vacuum leaks in one section, not only will your sap yield increase on the lines that had leaks, but also your overall vacuum level for the entire sugarbush will rise. This is where you make your money in sugaring—spending time in the woods fixing leaks to get higher vacuum levels and more sap.

So What Causes Vacuum Leaks?

A vacuum leak is created when any part of the tubing system is damaged or not properly maintained and air is allowed to enter the system. Often the biggest culprits are red and gray squirrels, especially if you have a healthy squirrel population on your land. They love

FIGURE 5.41. Some sugarmakers purchase a lightweight tapping hammer to make sure that they don't tap too hard. However, ordinary, lightweight hammers can work just as well. Be sure to hold the hammer high on the shaft and gently tap on the spout until it's properly seated and you hear the thumping sound. PHOTO COURTESY OF BRIAN CHABOT

Squirrels: The Enemy of Vacuum Tubing

The easiest way to limit squirrel damage is to prevent the squirrels from wanting to take up residency in or near your sugarbush in the first place. Squirrels rely on coniferous species for food and shelter, so if you are worried about squirrels, I recommend cutting out all the conifers located in your sugarbush. Mast-producing species such as oaks, hickories, and beech also lure in squirrels in the fall to harvest the acorns and nuts, so you may want to cut these out as well. Of course you can always trap and/or shoot squirrels, but as with all issues, it is much better to get at the heart of the problem rather than treating the symptoms. I recommend letting the squirrels roam freely elsewhere and do everything you can to make your sugarbush less appealing for them.

to chew and gnaw on tubing and can wreak havoc in a tubing system in a short period of time. Squirrel chews usually occur around the droplines and spouts, since the critters are often running up and down the tree trunks where the droplines are placed. They may also occur on the spans of tubing connecting trees, especially if the tubing passes through some low-growing evergreens.

Vacuum leaks are commonly found around the spout and taphole. If spouts are not seated properly or if the taphole is not perfectly round, the spouts will not seal correctly against the taphole and air can leak in. This is one of the reasons why proper tapping is so critical. Vacuum leaks also develop when tubing is damaged by falling trees and limbs during severe weather events. Whenever it has been windy, it is a good idea to check your sugarbush for any major damage. Careful monitoring of your vacuum gauge can alert you to potential problems in the woods. Vacuum leaks are always developing throughout the course of the season, so you need to be vigilant and get out there to fix them as much as possible.

So How Do You Fix Leaks?

There is definitely a learning curve to finding and fixing vacuum leaks. It requires patience, persistence, determination—and a lot of hiking through the sugarbush—to maintain high vacuum levels in a tubing system. When the sap is flowing, you walk along the mainlines, stopping at every manifold where a lateral line enters the mainline. It is best to do this by yourself, and preferably on a calm day, so that your hearing is not impeded by other people's activities or the rustling of leaves in the wind. Small beech trees are notorious for sounding like a vacuum leak when the tan-colored leaves that hang on in the winter rub against one another on windy days.

You shouldn't count on hearing the vacuum leaks, though; rather, you have to know how to look for them. When it appears that the sap is not flowing at all and there are just a few small air bubbles that stay in the same place or move very slowly, then you can move on to the next line. Ideally it would look as if the sap isn't moving at all—even though it is. On the other hand, if there are lots of air bubbles and the sap appears to be rushing very fast toward the mainlines, that means there is an air leak somewhere up the line. Your job is then to walk up the line, stopping at each dropline, until you find the leak. If you get to a point on the line where the sap is no longer racing forward, it means you have already passed the problem. You then have to backtrack to look over all the possible locations where the leak could have occurred.

There are some vacuum leaks that are easier to hear than they are to see. When the saddle manifolds aren't placed properly on a mainline, you will hear a hissing sound at the manifold. It is important to drill a proper hole and set the manifold on securely to avoid vacuum leaks. Similarly, whenever mainline connections are not properly installed, air will find a way in around the fittings. To avoid leaks, use a torch to soften the plastic before putting the tubing on the fitting and tightening the hose clamps. There are systems out there that involve butt fusion of pipes in lieu of insert fittings. If you are able to utilize this type of system, it could eliminate a lot of potential leaks while also minimizing restrictions in sap and airflow. The butt fusion requires

FIGURE 5.42. When you are fixing vacuum leaks, you are always on the lookout for air bubbles in sap lines. If you have a lot of fast-moving bubbles, there is an air leak somewhere up the line that needs to be fixed. PHOTO BY NANCIE BATTAGLIA

FIGURE 5.43. One of the advantages of using clear spouts is that they allow you to see if there is a vacuum leak letting air in at the taphole. PHOTO BY NANCIE BATTAGLIA

specialized piping and tools, but it is worth exploring if you are going to be doing a lot of tubing installation.

Moving Sap Uphill

If we got to design the topography of our sugarbush, all the trees would be evenly spaced and occur on a constant 5 percent grade sloping downhill to our sugarhouse. In reality, the topography of sugarbushes is highly variable and can make accessing some of the maple trees a real challenge. Roads, property boundaries, streams, wet areas, and other topographic features can also complicate the design and installation of tubing systems. To address these challenges, you may want to move sap uphill with proper utilization of your vacuum pump. The two main ways of achieving this are with sap ladders and "reverse slope" systems. If you decide to use one of these systems, I recommend doing a lot of research and talking with other sugarmakers who have them to determine what will work best for you. You may decide that not putting one in is your best move. These types of systems require air to enter the tubing system to move the sap, thereby using up your CFM capacity and lowering the vacuum throughout your sugarbush. Therefore, they should only be used as a last resort when deemed absolutely necessary.

FIGURE 5.44. Sap ladders come in all different shapes and sizes, depending on the preference of the sugarmaker, the amount of sap that needs to be lifted, and the height gain that is necessary. Some sugarmakers prefer a star-fitting system as shown here, but other systems have also been developed that may be easier to install and maintain. PHOTO COURTESY OF BRIAN CHABOT

What to Do with Used Tubing

Most large sugarmakers put a lot of plastic tubing and fittings out in the woods to help gather sap. While this tubing could theoretically last a very long time (20-plus years), research has shown that our yields will drastically improve if we replace spouts, droplines, and other parts of the tubing system on a regular basis. Given all the plastic that you will be replacing over the years, having a viable market for recycling it is essential to the sustainability of syrup production. Lois Levitan has been spearheading the Recycling Agricultural Plastics Project (RAPP) at Cornell to help develop viable markets for used tubing and keep it out of the waste stream. She is helping to make the connections between producers who have to dispose of this "waste material" and manufacturers who can use the old tubing and fittings as a valuable raw material for various products. The most current and promising use is for a plastic plywood that is perfect for marine applications. Before you do anything with your old tubing, spouts, and fittings, be sure to check the RAPP website at http://environmentalrisk.cornell.edu/AgPlastics.

FIGURE 5.45. A sheet of plastic plywood created with old tubing at Northbrook Farms in Auburn, New York. PHOTO COURTESY OF LOIS LEVITAN

FIGURE 5.46. Rather than throwing your old tubing into a big, unsightly mess, you can wind it up into 3-foot-diameter coils to be taken to a tubing recycler.

Washing Tubing

There is a long-running debate about whether it makes economic sense to wash tubing, and if you do wash the tubing, how to go about it in a cost-effective manner. I know many sugarmakers who have given up on washing tubing because they haven't been able to do a thorough job and don't see enough benefits of doing so. Some people think that the best thing to do is simply close off the valves and let any sap remaining in the lines turn to vinegar. These sugarmakers then either dump out the first run of sap or sell the first syrup that comes off the evaporator in the bulk markets. Letting the sap turn to vinegar may actually help clean the lines—though I'm a bit skeptical about this, and there is no research to prove that it actually works. I also wouldn't want to dump the first runs (fresh sap is an expensive cleaner); nor would I want to produce questionable syrup that I'd be pawning off on someone else to deal with. One possibility would be to vacuum out the lines in the fall once the sap vinegar has done its thing all summer. Doing this in the fall, with the vacuum pump on, would also allow you to fix any vacuum leaks that developed over the

FIGURE 5.47. When tapping birch trees, the sap may continue to run until the leaves come out, so some people will put dowels in the tapholes to stop the sapflow and "prevent infection." However, it is better to simply pull the spouts and let the trees heal naturally. Although quite rare, yeast may grow near a taphole when sap continues to flow after the taps have been pulled, as seen on this yellow birch. However, the yeast colony disappeared within a couple weeks and the taphole began healing over that summer.

summer months. A possible downside to this strategy is that the sap won't turn into vinegar and clean the lines during the summer; rather it could simply serve as a breeding ground for bacteria and molds, which I'm sure we have all seen in tubing systems before.

Given everything we know about the effect of bacteria and yeast contamination on sap flow, it makes sense to devote some time and resources to keeping your tubing as clean as possible. Furthermore, since we are making a gourmet food product, it is important to keep our production practices and sanitary treatments in line with the gourmet image that we are selling to the public. If some people found out that you don't wash your tubing and they could see all the bacteria and mold colonies in the tubing, you might wind up losing a lot of customers. However, despite the importance of cleaning tubing, we don't yet have a standard, accepted practice in the industry to carry out this task. The most common systems include flushing the lines with air and water or sucking the lines dry under vacuum when pulling spouts. Some people put cleaning agents such as hydrogen peroxide, bleach, or acids in with the water, though this may not have much effect on overall tubing cleanliness. In Canada many producers are using denatured alcohol, though this is not yet legal in the United States, and the Vermont Agency of Agriculture recently advised producers against trying this method until further research has been done.

The University of Vermont Proctor Maple Research Center and Cornell Maple Program have recently received a research grant to explore the effects of various cleaning treatments on sap production and quality. Preliminary survey research from UVM has found a wide array of tubing cleaning practices being used by producers with no discernible effect on sap production or quality. Further research is necessary to come up with cost-effective recommendations that will ensure that producers are not wasting their time trying to clean the tubing. I wish this section could outline the best practices for cleaning tubing, but unfortunately we don't yet know what the best system is. At the very least, be sure to do no harm by ensuring that any chemical sanitizers you use do not wind up contaminating the sap and syrup. If you use any sort of chemical in cleaning, thoroughly rinse your entire system to make sure it is free of any chemical residues. Finally, I encourage you to stay current with the latest research findings and adopt the best management practices when they are eventually developed.

CHAPTER 6

From Sap to Syrup:
Processing Techniques That Will Save You Time, Money, Fuel, and Frustration

Making maple syrup should be a relaxing experience. If you have spent a long day at work, either at your main job or out in the woods fixing vacuum leaks and gathering sap, you don't want to be running around like crazy trying to make syrup in the evening. When everything is running smoothly, making syrup can be one of the most relaxing, enjoyable experiences there is. It can be very satisfying watching all your hard work in collecting sap being transformed into delicious syrup. On the other hand, when things are not going well, making syrup can be an extremely stressful venture, especially when pans are burning (or almost burning) and the syrup isn't turning out right. My goal in writing this chapter is to provide you with some of the basic principles that will help you turn your sap into syrup in a more efficient and effective manner. The idea is to make sugaring a relaxing, enjoyable, and profitable experience. Knowing how to properly use your equipment is essential to achieving this goal.

FIGURE 6.1. The product of a successful season: rows of sample jars proudly displayed in the windowsill for every barrel (or batch) of syrup you produced.
PHOTO BY NANCIE BATTAGLIA

Batch Processing Versus Continuous Processing

Many people producing syrup on a small scale use the batch method. This involves continually adding sap to a large container until you have run out of sap and everything has boiled down to the consistency of syrup. You wind up with one large batch of syrup that tends to be dark in color with a robust flavor since it has been boiled for so long. Once you have graduated past the backyard sugarmaker level and have purchased a modern evaporator, you will make the big step toward continuous processing. With a typical evaporator setup, there is an entry port for raw sap (or concentrate) in the back and a drawoff valve in the front for finished syrup. Ideally some sap is always coming in at the back and a much smaller amount of syrup is being drawn off in the front. The sap flows like a river through the different compartments of the evaporator, becoming more and more concentrated by the minute. Because the sap does not boil for as long, it tends to be lighter in color with a more delicate maple flavor.

Choosing an Evaporator

The style and size of evaporator you purchase depends on the amount of sap you plan to boil, what you can afford, and how much time you want to spend in the sugarhouse. I recommend getting quotes from all the evaporator companies and taking the time to fully listen to their sales pitches. The sales representative's job is to provide you with a compelling argument for why you should buy a certain evaporator; your job is to take in as much information as you can and then carefully weigh all your options. Although high-quality used evaporators are often hard to find, it is worth checking out the web and other publications for possible deals. If well taken care of, a used evaporator can be almost just as good as a new one, but you need to know what you are looking for and where to find it.

The *North American Maple Syrup Producers Manual* contains a great deal of information on the anatomy and configuration of different types of evaporators, so I have not included that detailed level of information here. New evaporators tend to be much more energy-efficient than older ones, so if you are interested in saving fuel, time, and money over the long term, it is worth the extra money to buy an energy-efficient evaporator today. If you don't have a lot of money to spend initially, you may want to just pick up a used evaporator until you can save enough to buy a newer one down the road.

One of the main decisions you'll need to make is what type of fuel to burn in your arch. The two most common fuel sources for an evaporator are wood and fuel oil, though I have also heard of evaporators that run on propane, natural gas, wood chips, pellets, coal, electricity, and steam. High-pressure steam evaporators are probably the best choice (in theory), since they are extremely energy-efficient and produce steady, even heat without scorching the syrup. However, they are not readily available and are somewhat difficult and more dangerous to operate. You should do some more research and seek out sugarmakers who boil with steam to see if that type of system will work for you. Personally, I am a strong believer in wood-fired evaporators though I also fully understand the benefits of oil (in full disclosure, we use an oil-fired evaporator at our facility).

The main benefit of using wood is that it is the natural, local source of fuel that we all have in our sugarbushes. As sugarmakers, we should be cutting out dead and dying trees as well as undesirable and interfering species on a regular basis. Burning this wood in your evaporator provides a great incentive to do the appropriate forest management that we all should be practicing. As you may expect, the best-managed sugarbushes are often found where sugarmakers burn wood in their evaporators. The other main advantage of using wood is that you don't have to pay the oil company to make syrup. Many sugarmakers would rather burn a renewable source of fuel that they can cut themselves rather than spend money on expensive fossil fuels.

If you cut and split the wood for your evaporator on your own land, it is important to realize that this fuel source is not free! There are real costs involved in buying and maintaining the necessary equipment to harvest firewood, and it also takes quite a bit of time to do the work. Furthermore, you could have sold that wood on the open market, so you also need to factor in the opportunity cost of burning the wood in your evaporator. To do this, you simply divide the amount of syrup produced from one cord of wood by the price that you could have sold that cord for on the open market. Typically this value will range anywhere from $2 all the way up to $20 for every gallon of syrup produced, depending on the efficiency of your evaporator and going rate for firewood in your region. One way to reduce the cost of wood is to use materials that don't have a lot of market value otherwise. Few people will purchase softwoods and inferior hardwoods such as aspen for their home heating systems. However, these species can provide hot, quick fires that are excellent for syrup making. When fully cured and properly loaded in the evaporator, slabwood and old pallets can also be excellent sources of fuel that otherwise have little value.

Another great reason to use wood is for the enjoyment and aesthetics of it. People love to see wood being loaded into the arch, and a wood-fired evaporator adds greatly to the ambiance and experience of sugaring. Visitors don't mind paying top dollar for maple syrup after spending time watching you load wood into the evaporator. Running a wood-fired evaporator can also be a very peaceful experience, especially if you have an older model without fans blowing air into the arch. To me nothing sounds better than a mostly quiet sugarhouse occasionally interrupted by the crackle of the fire and the bubbling sap being transformed into syrup. They may not be as efficient as the newer models that use forced air and gasification technology, but if you are making syrup on a small enough scale, the old ones work just fine. The added bonus of their inefficiency is that some of that heat escapes into the sugarhouse, keeping you warm on those cold, frosty nights.

Oil-fired evaporators make sense for large-scale producers who don't have the time or ability to cut firewood during the off-season or feed the evaporator with wood while trying to make syrup. The main advantage

FIGURE 6.2. One of the best ways to check out different evaporators is going to the open house events that all the major equipment companies host every spring after sugaring season ends. They will often boil water all day long to demonstrate how the evaporators run.

FIGURE 6.3. This is what all our woodsheds should look like—long pieces of firewood, mostly beech, finely split, and neatly stacked right next to the sugarhouse. Although this woodshed is covered on three sides, I recommend leaving the sides mostly open to allow more airflow, which will help your wood dry faster.

of oil is that it is easy—you just flip a switch to turn it on or off. This saves a lot of time, since you don't have to wait for the fire to heat up in the beginning of a boil or die down at the end of the night. Furthermore, oil-fired rigs are especially nice to have if things start to get out of control when you are boiling. With a flip of the switch you may be able to save a pan that would otherwise burn in a typical wood-fired arch. Our evaporator even contains a nozzle that allows us to turn the fire intensity up or down, which also greatly helps even things out during a difficult boil.

The disadvantages of firing with oil are the cost and source of the fuel. Burning oil is more expensive than wood, and the price of oil is only going up over time. If you are going to use an oil-fired evaporator, I recommend getting the most efficient oil burner you can afford and regularly servicing it to maintain high efficiency. Furthermore, it's hard to market your operation as sustainable when it is based on burning a finite, non-renewable resource that is a major source of global warming. Given the environmental damages caused by extracting and burning fossil fuels, you can't really feel good about burning oil, and there is a perception among many consumers that syrup is more environmentally friendly if it is produced using wood rather than oil. While I tend to agree with this sentiment, the reality is that larger operations that burn fuel oil tend to have a lower carbon footprint than smaller, hobby-sized operations that burn wood. To understand how this can be, consider the following...

Generally speaking, the smaller the evaporator, the less efficient it is at processing sap into syrup. Thus, larger operations with modern evaporators that feature specialized flue pans and heat recovery units (such as Steam-Away, piggyback, steam hoods with preheaters, and so forth) tend to emit less CO_2 per gallon of syrup produced, even if they are using fuel oil as opposed to wood. Furthermore, reverse osmosis is the most energy-efficient method of removing water from sap, and you need at least several hundred taps before it makes sense to purchase an RO. Since fuel oil is so expensive, large sugarmakers with oil-fired evaporators are the most likely to invest in powerful ROs. When you are able to remove 80 to 90 percent of the water before boiling, the type of fuel that you use to finish the concentrated sap in the evaporator doesn't matter as much. Whereas it may take 3 or 4 gallons of oil to produce 1 gallon of syrup without an RO, by concentrating the sap first, it is common to use 0.5 gallon of oil or less for every gallon of syrup produced. The most environmentally friendly operations are those that use ROs to remove most of the water and then boil the concentrated sap with one of the new wood-gasification evaporators.

I don't want to give the impression that small-scale sugaring operations are not environmentally friendly, but it is important to understand that larger-scale operations generally have more options to minimize energy usage and are therefore more sustainable from an economic and environmental perspective. Since energy consumption is one of the most pressing environmental issues of our time and will only become more important in the future, as sugarmakers we have an obligation to reduce our energy consumption to the greatest extent possible. The good news is that trying to reduce the amount of energy consumed per gallon of syrup produced often involves scaling up your operation, not down! Whereas I am a big proponent of small-scale agriculture when it comes to raising animals and growing many other crops, the situation is much different with maple sugaring. There aren't any negative externalities from large-scale sugaring operations, as the only waste products from making syrup are steam from the evaporator and pure water from the RO. We aren't confining maple trees to make syrup in a large-scale operation; nor do we have to spray pesticides or herbicides on our cropland (the sugarbush) when sugaring on a commercial scale. Because of the potential energy savings when using modern technology, sugaring is one of the few agricultural enterprises where the environmental impact tends to fall as the size of the operation grows.

Reverse Osmosis

The process of reverse osmosis, or RO, in which sap is pumped through a membrane to remove most of the water molecules, is a very efficient process that can save you a great deal of time and money. In fact, the maple industry could never have developed the way it has over the past 30 years without RO technology. Boiling sap to remove water is a fairly inefficient and time-consuming process that had previously limited the amount of sap an operation could process. Once ROs came along, sugarmakers were able to greatly increase the number of trees they tapped since the amount of time and fuel required to process the sap fell drastically. Without an RO to first concentrate the sap, it would take roughly 4 gallons of oil to produce 1 gallon of syrup on a typical evaporator. By concentrating the sap first, most sugarmakers who use ROs are now using less than 1 gallon of oil per gallon of syrup produced, and the most efficient producers use less than 0.5 gallon of oil per gallon of syrup. While the fuel savings are incredible, the time saved is equally as impressive. There used to be many stories of 24- or 48-hour continuous boils during a huge sap run, but those are now legends of the past, as anyone who uses an RO can reduce their boiling time by 75 to 85 percent. A lot of sugarmakers I know who have gotten an RO now have the problem that they don't think they boil enough!

FIGURE 6.4. A cross section of an RO membrane. Water molecules can make it through the small pores of the membrane, yet sugar molecules are too large to pass through.

Although you don't need to know every detail of how RO technology works to use one, it is important to have a general understanding of the principles involved. Traditionally, sugarmakers would simply let the sap freeze in buckets or tanks and then throw out the ice chunks, which were mostly pure water. The theory of removing water before boiling hasn't changed, only the technology and process of doing so. Basically, a high-pressure pump is forcing sap against a semi-permeable membrane at pressures of 250 to 500 pounds per square inch (psi). Water molecules are small enough to make it through the extremely small pore sizes (0.0001 micron for a typical membrane) and come out of the RO as "permeate." The sugar molecules, minerals, and biological organisms are too large to fit through the small pores of the membrane and come out of the RO as "concentrate." By turning up the pressure, more water is forced through the membrane and less water remains with the sugars and minerals, so the concentrate becomes even more concentrated.

There are two types of membranes that you can put in your RO machine. Most sugarmakers use an RO membrane with a pore size of 0.0001 micron. However, you can also put in nanofiltration membranes with a pore size that is 0.001 micron, 10 times the size of a typical RO membrane. The nanofiltration membranes allow some minerals and simple sugars to also pass through the membrane, and they can achieve higher flow rates at lower pressures. However, they are more prone to abrasion and clogging, are more difficult to clean, and have a lower life expectancy, so they aren't as popular among sugarmakers. With a typical RO membrane of 0.0001 micron pore size, the permeate that passes through the RO is essentially pure water—H_2O. You should

Maple Sap Permeate: From a Waste Product to a Gourmet Beverage

Since most permeate is pure H_2O and lacks any minerals, its flavor is much different from that of the water we typically drink. Despite its unusual flavor, permeate does have excellent marketing appeal since it is essentially pure water that comes out of the ground through a maple tree. Although I've heard many people discuss bottling it over the years, the first company that has successfully done so is Eau Matelo in Quebec. In 2012 siblings Elodie and Mathieu Fleury started selling permeate in attractive glass bottles at premium prices under the De L'Aubier brand, focusing primarily on high-end restaurants. With sharp engineering, clever marketing, and entrepreneurship, they have turned what nearly all sugarmakers see as a waste product into a successful business venture. Rather than creating absolutely pure water with their RO permeate, they utilize a membrane that allows some of the minerals to pass through while leaving the sugars behind to be turned into syrup. The permeate then goes through a patent-pending process to give it a distinct flavor and long shelf life. De L'Aubier has already been extremely successful, earning a best-water award in the Water Innovation Awards at the 9th Global Bottled Water Congress in Barcelona in 2012.

FIGURE 6.5. Eau Matelo has taken permeate—a "waste" product of the RO—and transformed it into gourmet water through innovative marketing and packaging. PHOTO COURTESY OF ELODIE FLEURY

include plenty of storage to keep as much permeate as you can for cleaning and rinsing your RO. Because it lacks any minerals, it is also the best product for cleaning the mineral deposits (sugar sand) that form on your evaporator pans. Many sugarmakers let their pans soak in permeate overnight while others continuously spray permeate over the pans with an automated pan washer.

It is very important to follow the manufacturer's recommendations and guidelines for proper use, care, and maintenance of your RO. If the machine you purchased didn't come with an owners' manual, track down a copy from the manufacturer. You should thoroughly study the manual and become familiar with all aspects of it before you start running sap through it. I also recommend getting some practice by running well water through your RO before putting any sap in it. Be sure that you never put any water in the RO that has been treated with chlorine or contains large levels of iron or heavy metals. Many owners' manuals come with a logbook allowing you to keep track of RO procedures, performance, and maintenance. Although most sugarmakers (including me) have no interest in keeping these kinds of records, doing so will help you understand how your machine is performing and what kind of maintenance might be required. It is one of those things that you may not enjoy doing, but it will help you get the most performance out of your machine, thereby saving you a great deal of time and money. One of the most important things you need to do on a regular basis is the PEP test to determine how your membranes are performing—check your manual for proper instructions on how to do this.

Setting up your RO requires careful thought and planning to ensure that you get the most out of your investment. Figure 6.6 shows one way of setting up your system. You certainly don't have to follow this method entirely, but there are some general principles to keep in mind.

In an ideal situation, your RO would be large enough to produce the same amount of concentrate per minute that the evaporator is processing. However, I've found that most sugarmakers crank their pressure valves down to the point that they are only yielding 1 or 2 gallons of concentrate per minute (and perhaps 6 to 10 gallons of permeate) to get as much water as possible out of the sap. Unless your evaporator only boils off 1 or 2 gallons of water per minute, you will need to run the RO for several hours to get a decent amount of concentrate before starting up the evaporator. Your raw sap feed tank should be large enough to keep the RO running for several hours when full. This way you could fill up your feed tanks, turn the RO on, and then leave to take care of other things while the RO is processing the raw sap. It helps to have the raw sap feed tank elevated above the intake of the RO to allow sap to flow freely and avoid air pockets in the line. Permeate tanks should also be elevated above the RO intake, and your concentrate tank needs to be high enough to let gravity feed the sap into your evaporator. When deciding on how large to make your concentrate tank, remember that bigger is always better, especially if you are concentrating sap ahead of a boil. You never want to run the risk of having your concentrate tank overflow, as each gallon that is lost is equal to about a pint (or possibly more) of syrup. It is also useful to be able to direct your concentrate output line back into your raw sap tank. This feature should be used when you are concentrating permeate before you start the rinse/cleaning cycle. It's important to push the permeate through to salvage the sugar that has become caked on the membranes, but since you

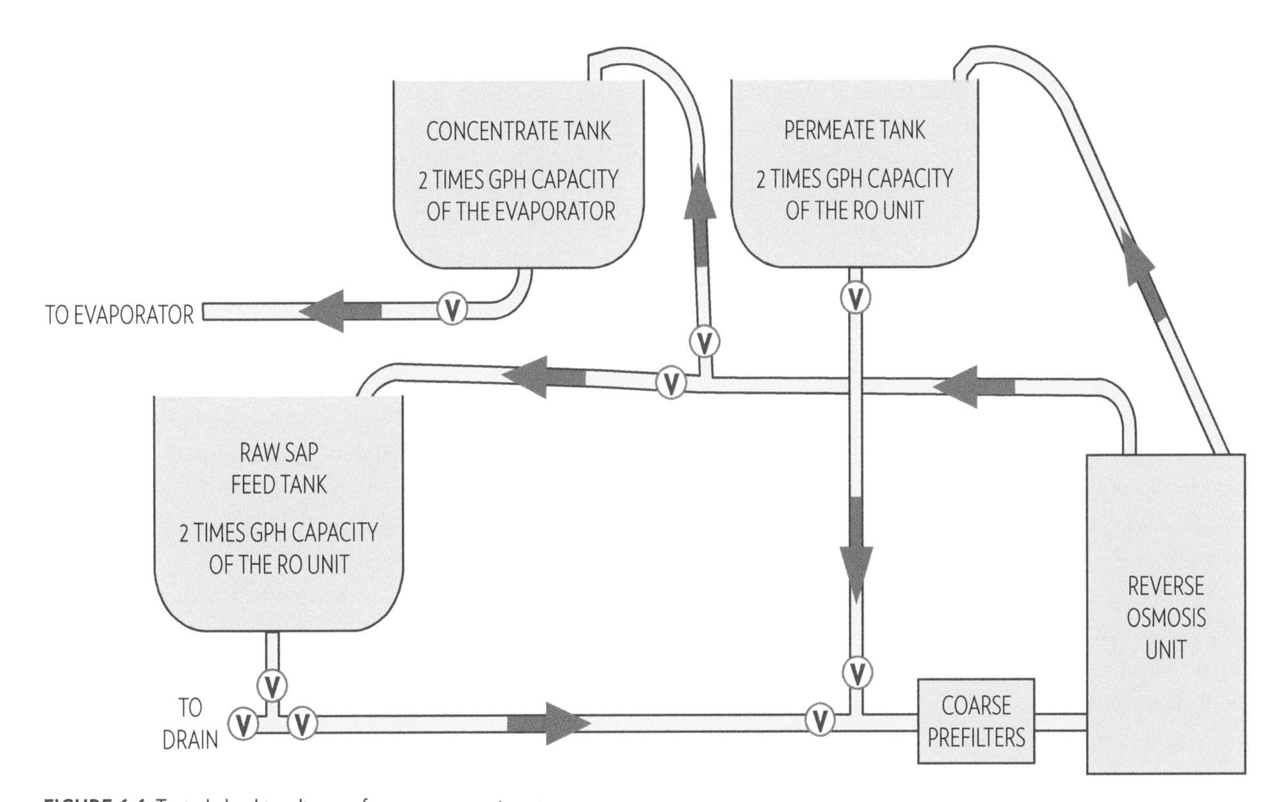

FIGURE 6.6. Typical plumbing diagram for reverse osmosis system

wind up making a lot of 2 percent (and lower) "concentrate," it is better to direct this back into your raw sap tank to be concentrated again. Additionally, when you first start up your machine, you may also want to put the first 10 minutes of concentrate back in the raw sap tank, as it takes a little while for the RO to achieve its peak concentration performance.

I have heard of some producers who choose not to use reverse osmosis because they feel it makes "techno-syrup." Some people complain that the syrup is too light and doesn't have enough flavor; others believe that the syrup comes out too dark. I've even heard some people state that ROs "take the nutrients out of the syrup." Since maple syrup is a luxury food in which the flavor is developed during sap processing, it is important to make sure that any new technology used to convert sap into syrup does not negatively affect the flavor and nutritional profiles. To address the concerns some people have with reverse osmosis, the Proctor Maple Research Center has conducted significant research to determine the impact of RO on the chemical composition of syrup, even under very high concentrations.

In 2008 researchers at Proctor conducted controlled experiments in which a single source of sap was concentrated to 15 percent with a standard reverse osmosis unit. Some of this concentrate was then watered down with the permeate to produce less sweet concentrate with brix levels of 12, 8, and 2 percent. All the sap was then boiled on four separate evaporators using the same fuel source and evaporation techniques. They took syrup samples from four different points in the season and examined the physiochemical properties (pH, conductivity, color, and density) and chemical composition (including minerals, carbohydrates, and volatile flavor and aroma compounds). The syrup produced from more concentrated sap had a lighter color, slightly higher pH, and less invert sugar and volatile flavor compounds. Although there were statistically significant differences in pH and invert sugar levels, these differences were very small and not of practical concern for sugarmakers or consumers. Researchers did find noticeable differences in volatile flavor compounds, but people were not able to detect any differences in flavor during the sensory evaluation trials. The only aspect with any real-world significance was the fact that higher concentrations led to lighter syrups, which sell at higher prices in the bulk syrup marketplace—not a bad thing!

Despite the overwhelming evidence that RO concentration does not negatively impact flavor, some people were still not satisfied and wanted Proctor to conduct similar tests with very high-brix concentrate. Large producers are increasingly concentrating sap to much higher levels, often at or above 20 brix, to reduce their fuel oil or wood consumption to the greatest extent possible. Thus, the next year they conducted similar trials, yet tested standard concentrate of 8 brix with a much sweeter concentrate of 22 brix. Once again, the results indicated that the only notable difference between high- and low-concentrated sap was the color of the finished syrup.

It is worth mentioning that all the research conducted at Proctor was done under appropriate conditions by skilled practitioners. In the real world, it is possible to make darker syrup with an RO if you let the concentrate sit around for a long time before boiling. Without proper maintenance of the membranes and regular changing of filters, the RO can also contaminate incoming sap with microorganisms that lead to darker syrups. Despite the potential pitfalls, most people who use ROs manage them effectively and make lighter syrup as a result. Since the concentrate doesn't have to boil for as long on the evaporator as raw sap would, there is less time for the sugars to caramelize and develop the characteristic maple flavor. For this reason syrup produced with an RO is more likely to take on the subtler and more delicate flavor often found in light and medium amber syrups. The impact of RO concentration on syrup flavor has as much to do with the operator as the machine itself.

Putting aside the effect of RO on syrup flavor (good or bad), ROs are simply the most cost-effective means of removing water from sap. Both the initial purchase price and fuel costs for processing sap are significantly less with reverse osmosis than with an evaporator. For a large investment of about $20,000, you could either buy an RO that processes 1,000 gallons of sap per hour or an evaporator that can boil off a couple hundred gallons of water per hour. On a smaller scale, you can get an RO that will process 200 gallons of sap per hour for roughly the same price of an evaporator that will do 40 gallons per hour. Not only can the RO process sap more quickly than an evaporator, the electricity cost to run the RO is usually significantly less than the fuel oil or firewood used in the evaporator (and if you think your firewood is free, see Choosing an Evaporator earlier in this chapter for an explanation of how you should value the firewood that you cut yourself). Finally, in addition to higher processing capability and lower fuel costs, the labor investment is significantly less since ROs basically run themselves. Whereas an evaporator needs constant attention and maintenance to run smoothly, your RO will continue to work for hours (while you are out doing something else) until it runs out of sap or needs to be rinsed or cleaned. Even if using an RO had a negative effect on syrup quality (it doesn't), I would still recommend them because all these other benefits are too large to ignore.

When ROs first came out in the 1960s, they were extremely large, cumbersome, and fairly difficult to operate. They were also much more expensive, and the price of fuel oil was a fraction of the price it is today. However, when oil prices started going up during the 1970s, many sugarmakers started looking into them as a time- and money-saving device. The technology has improved drastically over the years, and the cost has come down considerably. A parallel analysis can be made with computers. The old mainframes and desktop computers took up a lot of space, had limited capabilities, and were cost-prohibitive for most people. Now almost everyone has laptop computers and/or smartphones with incredible computing capability at the fraction of the cost. The ROs on the market today are compact, easy to use, and can process much more sap per hour at a fraction of the price of evaporating. I can sympathize with those who want to keep sugaring as traditional as possible, but remember that our ancestors adapted new technology when it made sense to do so. Hardly anyone would advocate using a typewriter when much better computers are available today at a fraction of the cost. Similarly, if you have at least 500 taps, value your time at anything, and want to reduce the environmental impact of your sugaring operation, buying an RO is probably the best investment you can make in your sugarhouse. If you don't already have one, I would strongly encourage you to look into it further.

Running Your Evaporator

Even if you have an RO, you are still going to need to get an evaporator to finish processing the concentrated sap into syrup. The maximum sugar content that producers are bringing their sap to with reverse osmosis is about 22 percent, so there is still a lot of boiling left to do before you get to syrup. Every evaporator behaves a little bit differently and will vary tremendously based on the sugar content of the sap you are boiling and the time of the season. You will not be an expert after reading this chapter, but I've tried to include the general concepts and principles that will make your sugaring experience an enjoyable and productive one. Believe it or not, it's actually easier to run a larger evaporator than a smaller one. With a large evaporator and efficient RO, you can have almost a steady stream of sap coming in and syrup coming out. On the other hand, smaller evaporators have long time intervals between drawoffs and there isn't a steady movement of sap through the evaporator, which can make drawing off syrup at the right consistency a challenge. With a lot of practice and patience, you will eventually become an expert in running your evaporator, no matter what size it is.

I've had the good fortune of listening to Glenn Goodrich and Brad Gillian speak about running evaporators on several occasions. They always offer sage advice from their experiences working with a lot of different evaporators and operators over the years. The take-home message I've come away with is that running an evaporator should be like driving a car. Just as a car runs best at a constant 55 mph cruising down the highway rather than in stop-and-go traffic in the city, its best to operate our evaporators with a steady temperature and constant flow of sap being transformed into syrup. The goal is to be able to set it on cruise control to the greatest extent possible. An evaporator has a certain sound and feel to it when things are going right, and you want to keep your eyes and ears open for any

Plumbing in Your Evaporator

There are a lot of things to like about the way this evaporator is plumbed together. Although you should always be able to see a sight level on your sap/concentrate feed tank, having clear PVC pipe in your feed line is a cheap insurance feature to make sure sap is coming into the evaporator. If you look closely, you can also see the PVC union and chains/pulley system that allow the plumbing to be easily disconnected and the steam hood raised for cleaning. An added bonus of this steam hood is the extra preheater used to heat permeate. The two blue pipes coming in and out of the hood are being recirculated through the preheater as a way of capturing the energy of the trapped steam to heat the permeate (similar to a regular preheater for sap). Nothing cleans an RO membrane as quickly or as effectively as hot permeate. In our sugarhouse, we have a thermostat regulator on the recirculation pump that shuts off when the permeate reaches 110°F, which is the maximum temperature for washing the RO.

FIGURE 6.7. Plumbing sap and permeate into an evaporator.

sudden changes to the norm. In particular, there are five aspects you should be constantly monitoring on your evaporator: foam, floats, drawoff valve, levels, and niter.

Foam

As sap boils into syrup, it can produce a tremendous amount of foam within your evaporator. Foam insulates the boiling sap and keeps the steam from escaping, so you want to minimize it as much as possible. Too much foam can cause sap to bubble over the sides of the evaporator and may also cause your pans to burn if all the boiling sap turns to foam. There is more foam created toward the end of the season when bacteria and yeast are more plentiful in the sap. Keeping everything as clean as possible will help reduce problems with foam, but you are always going to have to deal with it to some extent. Sugarmakers use a wide variety of defoamers ranging from the chemical defoamers sold by the maple equipment companies to butter, various oils, or even pork fat and bacon grease. Basically anything with fat molecules in it will do the trick, but in today's world we need to be particularly careful about what we use as a defoaming agent.

When choosing a defoamer, it is important to consider the final market for your product and everyone who may be consuming it. If you are making it entirely for your own use and for friends and family that you know well, then it may be okay to use dairy products or certain animal fats and oils as defoamers. However, if you will be selling any of your syrup or don't know exactly who will be consuming it, then you need to be very careful in your choice of defoamer. There are a lot of vegetarians and vegans who do not wish to consume animal products, so dairy and other animal fats should not be used unless you plan on clearly displaying this on the label. More important, there are a lot of people who are allergic to dairy and peanut products, and they could suffer serious consequences if exposed to them. People who purchase your product expect it to be pure maple syrup, not a mixture of maple sap and peanut oil, dairy products, or animal fat. Our goal as sugarmakers should be to keep the syrup as pure as possible. We use the Atmos 300K liquid defoamer because it only requires a small amount, works very well, and will not lead to any allergic reactions. The only downside is that it is not currently accepted for organic certification, so if your operation is certified organic, you'll need to use a mild-flavored vegetable oil such as safflower oil or sunflower oil. Vegetable oils don't work nearly as well, so you'll have to use a lot more to get the desired consistency. In fact, I know several sugarmakers who dropped their organic certification because they had really bad experiences using vegetable oils in their evaporator.

No matter what type of defoamer you choose to use, it is important to use it correctly. Defoamer should be applied in very small quantities and very often. For large evaporators, you can purchase automatic defoamers that will put in a small amount of defoamer every few minutes to get even distribution throughout the evaporator. For most sugarmakers, attention to detail and careful monitoring of your evaporator is all that is necessary. You never want to squirt defoamer into a pan; the key is to put in a couple of drops at regular intervals to prevent the foam from getting out of control. It is best to mimic an automatic defoamer to the greatest extent possible, adding a drop or two every few minutes for large evaporators and every 10 to 15 minutes in a small one. You should also avoid putting defoamer in the front syrup pans, especially the middle compartment. The only place you should apply defoamer in syrup pans is right by the drawoff valve—doing this will help to get finished syrup to flow out of the pan faster if you are having a hard time getting the syrup up to temperature there.

FIGURE 6.8. A lot of foam may look pretty cool to the untrained eye, yet having this much foam in your evaporator insulates the boiling sap and reduces your boiling efficiency. PHOTO BY NANCIE BATTAGLIA

Floats

Most evaporators come equipped with float mechanisms that allow sap to come into the evaporator and move from the back pans to the front pans (the exceptions are some small, hobby models and homemade evaporators). If you want to make syrup efficiently and not worry about burning your pans (who doesn't?), you must ensure that your floats are working properly. Mechanical floats occasionally can get plugged up and won't allow sap to flow through. If they get stuck in the completely open or closed positions, you will encounter serious problems. If they aren't letting sap through, there is a good chance that you will burn your pans if you don't catch it in time. On the other hand, if your float never closes and just continues to let sap into the evaporator, the level of sap will get too high, your boil rate will fall, and sap may eventually overflow the pans before you catch the problem. It's a good idea to take apart and clean your floats on a regular basis and keep an eye on them throughout the boil to make sure they are operating properly.

Drawoff

In an ideal world, you would set the drawoff valve on your evaporator to continuously allow a small amount of syrup to flow out of the finishing pan. Large sugarmakers who make 50 or more gallons of syrup per hour can do this, but the vast majority of us can only draw off

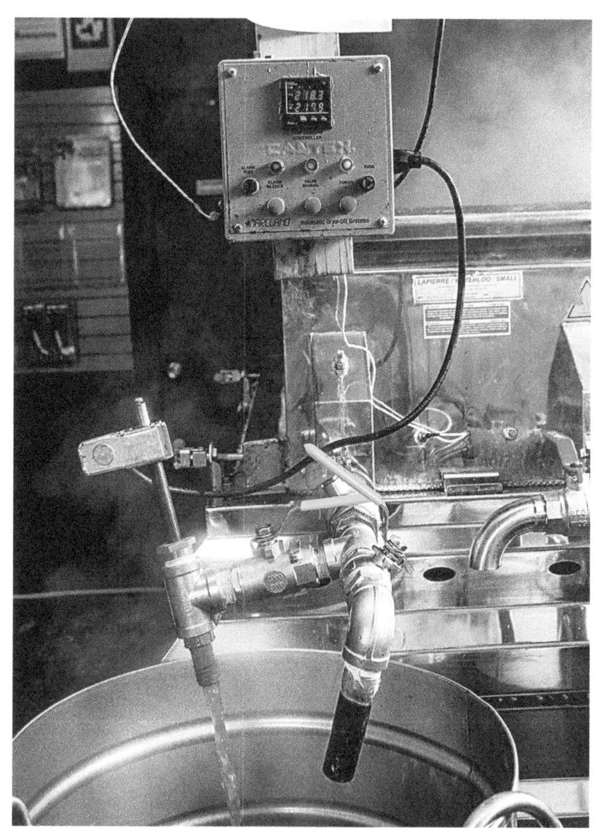

FIGURE 6.9. Although automatic drawoffs are very helpful tools, it's important to realize that the only automatic thing about them is that a valve opens when the boiling syrup in the very front pan reaches a given temperature. Use a hydrometer to determine what temperature to set the drawoff valve at. Also, adjust the output valve on a regular basis to ensure a steady flow of finished syrup. PHOTO BY NANCIE BATTAGLIA

FIGURE 6.10. You should continuously check the syrup coming off your evaporator with a hydrometer to make sure it meets appropriate density requirements. In this particular situation, the hydrometer reads 62.1 and the temperature is 146°F, so the actual brix content is 66.2. PHOTO COURTESY OF NANCIE BATTAGLIA

syrup every once in a while. Even if you can't get a steady flow, your goal should be to draw as often as you can with as small a batch as possible. Although it may feel great to have a huge batch of syrup come off the evaporator, this will simply result in a longer time interval until the next drawoff. Pulling off too much syrup at once can also lead to scorching of pans and producing syrup that is thicker than it has to be. One way to combat this issue is by throttling your drawoff valve to reduce the flow of syrup out of the evaporator. This will ensure more of a steady draw rather than spurts of syrup.

Many sugarmakers use an automatic drawoff unit that opens electronically whenever the temperature reaches the value you set it to. Although these work well, they are only as effective as the operator. You need to determine what temperature to set the unit at and constantly monitor it to make sure that the temperature is accurate. Often the temperature at which sap becomes syrup—7.1°F above the boiling point of water—will change throughout the course of the day based on weather patterns and atmospheric pressure. You must constantly check the syrup with a hydrometer to make sure your automatic drawoff unit is set to the right temperature, which may range anywhere from 215 to 220°F depending on your elevation and barometric pressure. Since each tenth of a degree results in a change of density of 0.25 brix, getting the right temperature is crucial to high-quality syrup production.

Properly Using Your Hydrometer

Most hydrometers sold by maple equipment companies are calibrated to provide the brix (sugar content) reading when the syrup is at either 60 or 68°F. However, syrup comes off the evaporator at nearly 220°F and cools to a lower temperature before we are able to filter and store it in its proper container. When you float your hydrometer in hot syrup, it will provide you with a lower-than-actual brix reading because the syrup is less dense at higher temperatures. Similarly, if you open a barrel of cold syrup that may be stored in your sugarhouse under cold conditions in the winter, the observed brix will be higher since the temperatures is below 60°F. To determine the actual brix content when dealing with syrup that is either above or below the temperature at which the hydrometer is calibrated, you must either add or subtract a given value. The formula for determining this value is as follows:

true density = observed brix + (0.047 × temp of syrup – temp at which hydrometer is calibrated)

Since very few sugarmakers would want to do this calculation, Edmund Grot developed table 6.1, which can be used to determine the actual brix content based on the observed hydrometer reading and temperature of the syrup. All you need to do is take your hydrometer and temperature reading at the same time and then add the corresponding value to your hydrometer reading. This will provide you with the actual brix of the syrup in question.

TABLE 6.1: Correction Value to Be Added to Observed Brix to Give the True Brix

Syrup	Hydrometer Calibration	
TEMP. (°F)	60°F	68°F
215	7.3	6.9
210	7.1	6.7
205	6.8	6.4
200	6.6	6.2
195	6.3	6.0
190	6.1	5.7
185	5.9	5.5
180	5.6	5.3
175	5.4	5.0
170	5.2	4.8
165	4.9	4.6
160	4.7	4.3
155	4.5	4.1
150	4.2	3.8
145	4.0	3.6
140	3.8	3.4
135	3.5	3.1
130	3.3	2.9
125	3.1	2.7
120	2.8	2.4
115	2.6	2.2
110	2.4	2
105	2.1	1.7
100	1.9	1.5
95	1.6	1.3
90	1.4	1
85	1.2	0.8
80	0.9	0.6
75	0.7	0.3
70	0.5	0.1
65	0.2	-0.1
60	0	-0.4
55	-0.2	-0.6
50	-0.5	-0.8
45	-0.7	-1.1
40	-0.9	-1.3
35	-1.2	-1.6
30	-1.4	-1.8

These figures are calculated from the formulation page 74 of the *Maple Syrup Producers Manual Agricultural Handbook* No. 134, July 1976.

Levels

I once heard Brad Gillian from Leader Evaporator say that there are two types of sugarmakers in this world—"those who have burned a syrup pan and liars." Another common saying is that "you aren't really a sugarmaker until you have burned your first pan." In reality I'm sure there are many sugarmakers who have never burned a pan, but given enough time, we all wind up doing it at some point. No matter how hot the fire is in the arch, as long as you have sap or syrup in the pans, you will be fine. However, as soon as something happens that leads to a shortage of sap in the evaporator, your pans will burn in a short period of time. The level of sap/syrup at which you operate has a big influence on the likelihood of this happening.

The lower the levels of sap in the pans, the faster the boil and the lighter the syrup you are able to produce. Thus, many sugarmakers boil with sap that is only 1" or less above the flues and/or in the syrup pan. However, those of us who are more cautious will have 2" or more sap in the pans. With higher levels, the syrup comes out a bit darker and takes a little while longer to process, but you may wind up saving significant time and money by not burning your pans. One thing to realize is that if you have a large evaporator or are boiling concentrated sap from an RO, you'll want to run deeper levels in your pans since your rate of syrup production will be so much faster.

You can purchase low-level alarms that remove much of the risk of burning your pans. The relatively low cost of these units can save you a ton of money in replacing or repairing burned flue pans. Some alarms just sound a buzzer, and while these are helpful, they won't fix the problem—they just alert you to do something about it. The most valuable and useful ones actually help to fix the problem! With oil-fired evaporators, you can get an alarm that will turn off the oil burners if the level gets too low for a certain period of time. We have one of these on our evaporator and it has saved our flue pans from burning on several occasions. Although you can't just turn off a wood-fired evaporator, the level alarm can also be wired to turn off the fans that control the

FIGURE 6.11. Since it is difficult to tell how much sap is actually in the syrup pans when it is boiling and full of foam, some evaporators come equipped with a sight level in the plumbing that connects the front pans together. PHOTO BY NANCIE BATTAGLIA

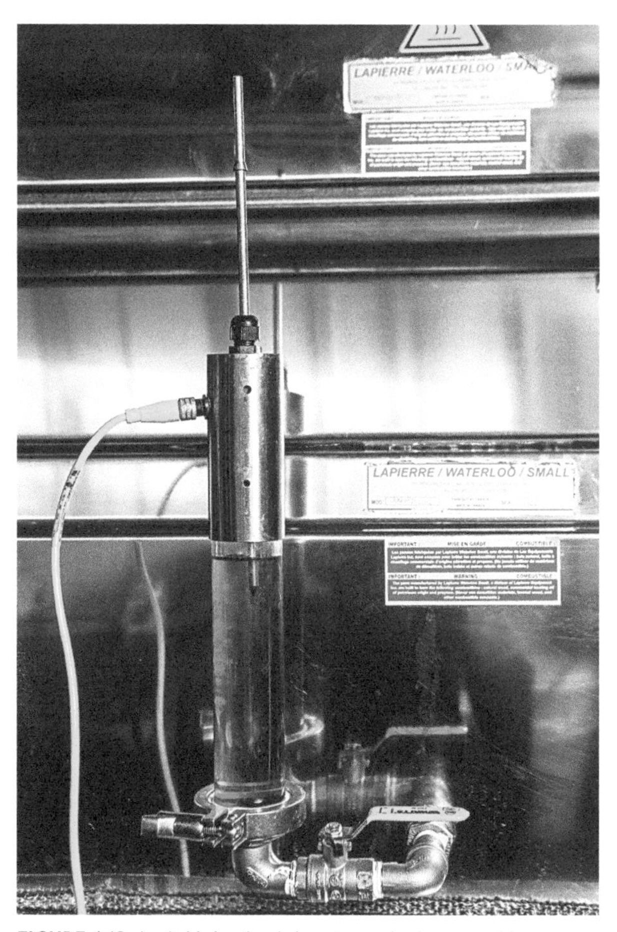

FIGURE 6.12. A reliable low-level alarm in your back pans could save you a great deal of time, money, and aggravation. PHOTO BY NANCIE BATTAGLIA

FIGURE 6.13. This wood-fired evaporator includes a feature that will start spraying cold permeate through the flue pan washer if the level in the back pans dips below the top of the flues. These spray nozzles are also used to clean any niter buildup on the flue pans throughout the course of the sugaring season.

intensity of the fire. Another option is to wire the alarm in order to open a relief valve that will start spraying the pans down with cold water (or sap, depending on how it is plumbed) if the levels get too low. Whether you have these types of backup systems or not, it is important to constantly monitor your levels to ensure consistent syrup production and to avoid burning your pans.

Niter

As we concentrate the sugars of maple sap into syrup, we also concentrate all the minerals found in the sap. Minerals that precipitate out of solution as the sap gets concentrated into syrup are called sugar sand, or niter.

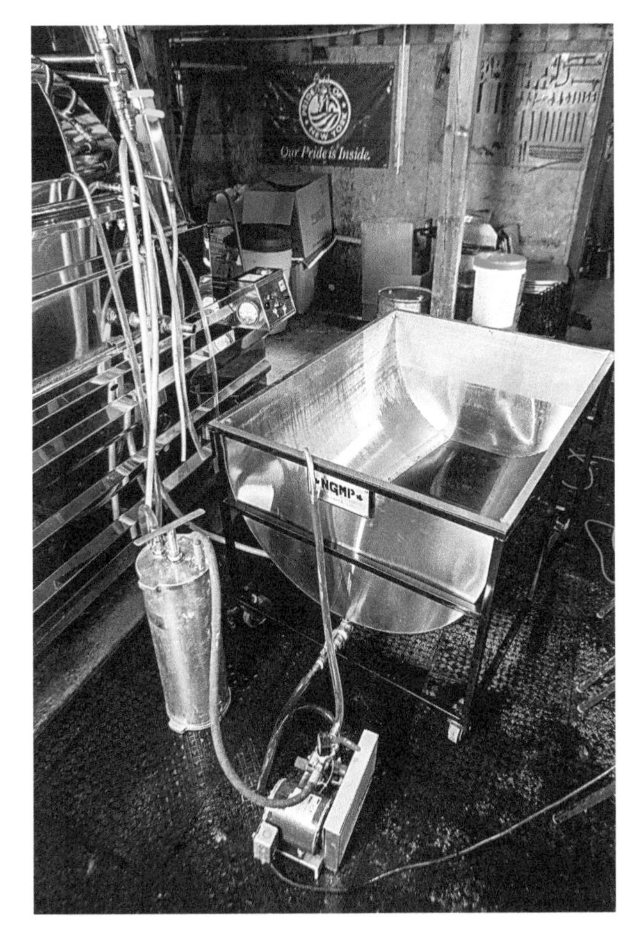

FIGURE 6.14. To wash the flue pans, all the partially boiled sap in the back pans is first drained off into a large tank and is then pumped through a canister filter on its way back into the cleaned flue pans. PHOTO BY NANCIE BATTAGLIA

FIGURE 6.15. A pan washer that continuously sprays permeate onto the surface of the pan, thereby removing the buildup of niter. PHOTO BY NANCIE BATTAGLIA

Air Injection in Evaporators

In recent years many producers have installed air injection (AI) systems in their evaporators. With AI, a series of stainless steel tubes is placed in the front syrup pans and/or the back flue pan in order to introduce air into the boiling sap. Anecdotally, using AI results in less sugar sand formation on evaporator pans and produces lighter colored syrups. Given the increasing popularity of these systems and their perceived (yet unproven benefits), the scientists at Proctor Maple Research Center set out to determine the impact of AI on syrup quality and composition. Using the same sap source and otherwise identical 3 × 10 evaporators, they equipped one of the evaporators with AI in the front and back pans. Over the course of the 2006 season, they tested the syrup coming out of each evaporator and reported on their results in the *International Sugar Journal*.[1] As expected, the syrup coming out of the AI evaporator was much lighter in color and did not contain as many volatile flavor compounds. This was the only major difference between the syrups produced in the AI and standard evaporators. They did notice that the AI syrup had more minerals suspended in the syrup (in particular potassium, magnesium, and manganese), yet the differences were minor and all syrups fell within the range reported for maple syrup. One of the surprising (and disappointing) findings was that the scale deposits on the evaporators pans were equal between both evaporators. This finding contradicts the belief among many sugarmakers that less niter builds up on pans when using AI. The take-home message is that if you want to make lighter syrup, then you should seriously consider using air injection in your front syrup pans and/or back flue pans. However, with prices for dark and light syrups converging in recent years, fewer sugarmakers are taking this extra step to produce lighter syrup. One word of caution is to ensure that the air you are injecting is clean, pure, and does not contain any noticeable odors. Thus, you need to locate your air intake in a proper location and conduct regular and proper maintenance on the air filter. It is also important to clean the piping after each boil to prevent sugar sand from plugging up the small holes used to blow air into the boiling sap.

FIGURE 6.16. If you run the evaporator too long with heavy sugar sand deposits, some of the mineral deposits are likely to burn on the surface of the pans and make cleaning difficult. PHOTO COURTESY OF PETER SMALLIDGE

FIGURE 6.17. It is best to have several backups available for replacing your syrup pans when they get a thick layer of niter. The sugarmakers at this operation take very good care of their syrup pans, treating them like patients at a hospital!

Making Birch Syrup

Making birch syrup is a lot like making maple syrup, but it's also a lot different! Both processes involve removing water from sap to wind up with syrup, but that is where the similarities end. Much of the difference in the flavor between birch and maple syrup can be attributed to the sugar composition of the sap. Whereas maple sap contains mostly sucrose, birch sap is almost entirely fructose and glucose. Sucrose does not caramelize until it reaches 280°F, yet fructose caramelizes at 160°F and glucose at 140°F. Caramelizing fructose and glucose under high heat is what creates the distinct maple flavor when processing maple sap. The more glucose and fructose in the maple sap, the stronger the flavor and the darker the color of the maple syrup. Since birch sap contains mostly glucose and fructose, these sugars caramelize for a much longer time period during the evaporation process. Thus, birch sap often develops a very strong flavor and is much darker in color.

By using reverse osmosis to remove at least 80 percent of the water before it is boiled, it's possible to make birch syrup with a much more delicate flavor and lighter color. Birch syrup that is first concentrated with reverse osmosis tastes *much* better than birch syrup made without this process. Because of the low sugar content of birch sap and the tendency for it to develop strong, bitter flavors when boiled for extensive periods, ROs are essential to produce high-quality birch syrup in a cost-efficient manner. I wouldn't even consider making birch syrup without an RO. Another strategy is simply processing the sap at lower temperatures to prevent the sugars from scorching; using hardwood instead of softwood for firewood can help make this possible. Some birch syrup producers just let the sap simmer over long periods of time, never letting it get to a full boil. Personally I wouldn't have the patience to do this, and all the syrup that I have seen processed this way still comes out very dark and strong tasting.

Because birch syrup is mostly fructose and glucose, it has a lower viscosity than maple syrup. To produce syrup that tastes as thick as maple syrup, some people will bring the sugar concentration to a higher brix. However, the extra time spent on the evaporator can lead to further darkening and possible scorching of the syrup, so you may be better off with a slightly runnier consistency. In fact, some producers in Canada only take their syrup to 60 brix and report good results. The acidity of the syrup supposedly helps preserve it from spoiling—and having the birch syrup a little more runny with a milder flavor is not a bad thing! Further research is necessary to determine the minimum concentration for birch syrup.

Excessive niter buildup can cause serious problems if not attended to on a regular basis. If you boil for too long without taking the time to remove the niter, it can form a thick layer that minimizes heat transfer to your boiling sap, or, worse yet, will burn on the pans and cause your syrup to become off-flavored. There are two main ways to deal with this issue. You can purchase an evaporator with Revolution pans that allow you to switch the point of entry and exit for sap and syrup in your front pans. By switching sides when the niter starts to build up near the drawoff valve, the incoming sap will have a lower concentration of minerals, thereby helping to remove some of the niter buildup from those parts of the pan. Another method involves switching the front pans with new ones whenever the niter starts to build up to significant levels. The more often you can switch your pans out, the easier it will be and the less likely you are to encounter problems. Often during big runs, we don't take the time to shut the evaporator down to switch pans out, but regular maintenance and cleaning makes our job much easier in the long run.

Filtering Syrup

In the process of making syrup, some of the minerals precipitate out as niter on the syrup pans while some remain suspended in the syrup and make it appear cloudy. If you did not filter the syrup, over time these minerals would eventually settle, leaving a layer of sediment on the bottom of the container. If you are only making syrup for yourself and are not planning on selling it, then filtering is not absolutely necessary. On the other hand, if you are selling your syrup on the open market and want to ensure that it's crystal clear, then properly filtering your syrup is a must. This is especially important if you package syrup in glass. Contrary to what some people may think, glass bottles do not make syrup cloudy, they only let people see what is inside. As Glenn Goodrich likes to state, packing syrup in glass is like taking a shower on Main Street—there is nothing hidden anymore!

Filtering syrup is best done while the syrup is very hot coming right off the evaporator. Cold (and even lukewarm) syrup does not filter well, so you are well advised to get the syrup filtered and stored away properly as soon as possible. There are a variety of methods for filtering syrup, with details for each below . . .

Gravity Filtering

Many small-scale sugarmakers use a cone filter with a felt liner and let gravity do the trick. This does a pretty good job, but the process is slow and the finished syrup won't be as clear as it could be with other methods. It is important to use clean filters that have been properly cleaned and stored. It is possible to ruin perfectly good syrup by putting it through a brand-new filter or an old, musty one. New filters must be rinsed thoroughly with hot water before they are used to filter syrup. Likewise, if you pull out the filter you used last year and it looks

FIGURE 6.18. Proper filtering is essential for packaging high-quality syrup in glass containers. PHOTO BY NANCIE BATTAGLIA

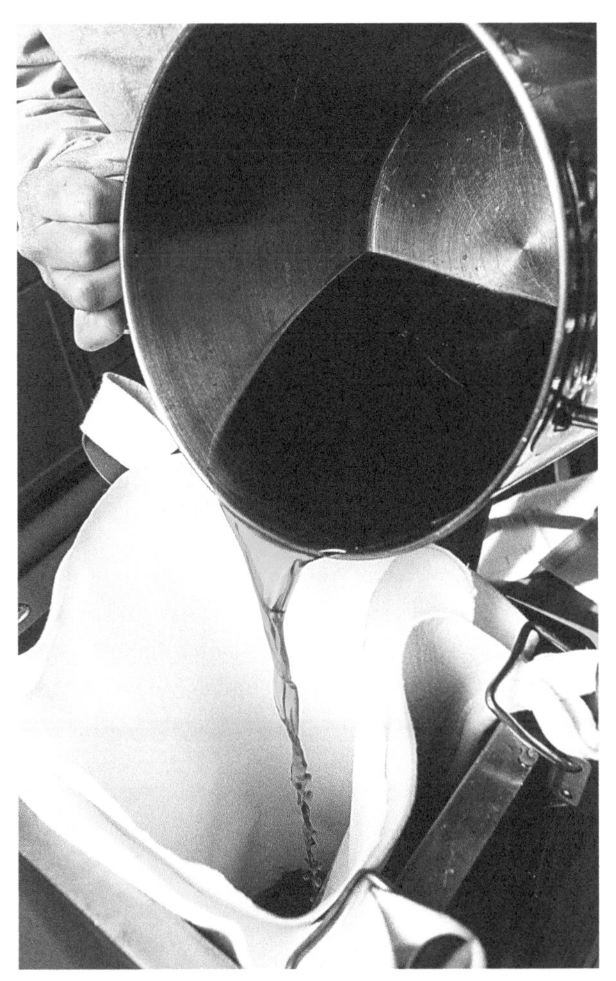

FIGURE 6.19. Cone filters are a popular way to filter syrup for small-scale sugarmakers. Although it won't come out looking as polished as syrup that has gone through a filter press, it is still perfectly acceptable for most people. PHOTO BY NANCIE BATTAGLIA

FIGURE 6.20. A filter press is the most common way of filtering syrup for large-scale sugarmakers.

FIGURE 6.21. You may want to run your syrup through a coarse screen before pumping it through a filter press or canister filter. This can catch large pieces of niter and extend the life of your filtering medium. PHOTO BY NANCIE BATTAGLIA

or smells questionable, you shouldn't use it. Gravity filtering may be fine for packaging syrup in plastic containers, but if you are putting it in glass, you'll probably need to use a filter press or comparable system.

Filter Press

This is the most common method of filtering syrup for large-scale sugarmakers. Filter presses come in a variety of sizes depending on the amount of syrup you need to filter and the speed at which you want to push it through. The Proctor Maple Research Center has developed an excellent guide for how to properly use a filter press, available as a free download at www.uvm.edu/~pmrc/filter_press_brochure.pdf. If you can get an older press made of cast iron, it will retain heat much better between batches of syrup than the newer ones that are usually made of aluminum. The type of diatomaceous earth (DE) that you use in the press is extremely important, as that is what does all the filtering. You need to ensure that the DE you purchase is food grade and meant for filtering syrup; different grades are meant for different applications, so be sure the DE that you buy is appropriate for maple syrup.

FIGURE 6.22. One of the main advantages of a canister filter is that it allows you to simply hose off the filter cloth whenever it gets plugged. PHOTO COURTESY OF ANDREA FARRELL

FIGURE 6.23. In this bottling setup, syrup is first heated with a steam kettle before it is refiltered as it is pumped into the water-jacketed bottling unit.

Canister Filter

Although our filter press worked well, I had heard from several sugarmakers that the canister filters they were using did just as good a job filtering the syrup and are easier to use. I decided to try out the sirofilter model from Lapierre and it worked so well that we sold our filter press. The operating cost of the sirofilter is less expensive over time, assuming we take good care of our filters. The 1- or 5-micron cloth filters can just be hosed off when they plug up with sugar sand and DE, whereas filter papers must be replaced when the filters get plugged. Finally, there is no risk of blowing a filter paper, and it's easy to switch out the canisters.

No matter what type of filter you use, it is important to recover as much syrup as you can before you break it down and clean it. We use hot condensate coming off our steam hood to run through the filter when we are done processing syrup. The diluted syrup that we collect is then poured back into the flue pans to be reprocessed into maple syrup. I think we save at least 10 to 15 gallons of syrup over the course of the season by taking the time to recover it from the filters. Doing this also makes washing the filter easier since there is less sticky syrup to clean up.

Bottling Your Syrup

Once you have finished producing and filtering your maple syrup, it is ready for bottling. There are several important things to consider when putting your syrup in consumer-sized containers:

- Always refilter the syrup before bottling after it has been stored in a barrel. This is especially important if you have purchased the syrup from another sugarmaker.
- Heat the syrup to 180°F (or 190°F if packaging in glass) and hold the temperature steady during the bottling process to kill off any bacteria, yeast, mold, et cetera.
- Double-check the grade of syrup as it is being bottled. Just because you labeled the barrel as medium amber when it was produced in March does not necessarily mean it is still medium amber after

being stored in the barrel for many months and then reheated for packaging purposes.

- Double-check the density of the syrup and adjust as necessary, either adding water if it is excessively high or letting some water evaporate if it is too low. Even though the legal minimum for most states is 66 percent sugar content, I recommend bottling syrup that is 67 percent. The extra 1 percent sugar content makes the syrup feel a lot thicker and is much more enjoyable to consume.
- Fill the bottles to the bottom of the neck with syrup, make sure you screw the cap on properly, and then set them on their side. Doing this will provide an induction seal on the cap while allowing you to check that the cap is on properly without any leakage.
- Keep detailed records on production batches. Label all bottles with the specific batch code and keep good records of what barrel(s) were used for the batch and how many bottles of different-sized containers were filled.

FIGURE 6.24. If you are a relatively small-scale producer but still want to produce crystal-clear syrup, there are hand pumps available at a lower cost that allows you to achieve nearly the same clarity in your syrup as large-scale filter presses.

Grading Your Syrup

After you have spent countless hours collecting sap, boiling it down into syrup, and packaging it in barrels or consumer-sized containers, there is still more work to be done. If you plan on selling the syrup, you must then label it according to the appropriate grading standards. Since maple syrup comes in a wide variety of colors and flavors, sugarmakers in different states and provinces have developed numerous ways of classifying it over the years. Because flavor is subjective using current assessment techniques, yet color can be objective and is more easily distinguished with light transmittance meters, all the grading systems have focused on the color of the syrup. Color is a good indicator of the intensity of the flavor, and these systems have worked reasonably well. However, the system in Canada differs from that in the United States, and many states and provinces have their own systems of classification, so consumers (and even producers) can get confused trying to understand it all.

Differences in grading standards have led to confusion and increased handling costs for buyers and processors of maple syrup. Since maple syrup is now a global commodity being shipped all over the world, the current grading system has been an obstacle to further development of these markets. Without one standard definition of pure maple syrup and the different grades, buying and selling syrup within and among states and provinces can be especially challenging. The differences in regulations have made it cumbersome to sell similar products in different states and provinces. For instance, the same syrup may need different labels to be stocked on store shelves in New York and Vermont. More important, international buyers would also like to know that the product they are purchasing is the same whether they are ordering it from someone in the United States or Canada. For these reasons and others, it has become apparent that a new standardized grading system is necessary to facilitate the expansion of both domestic and international markets.

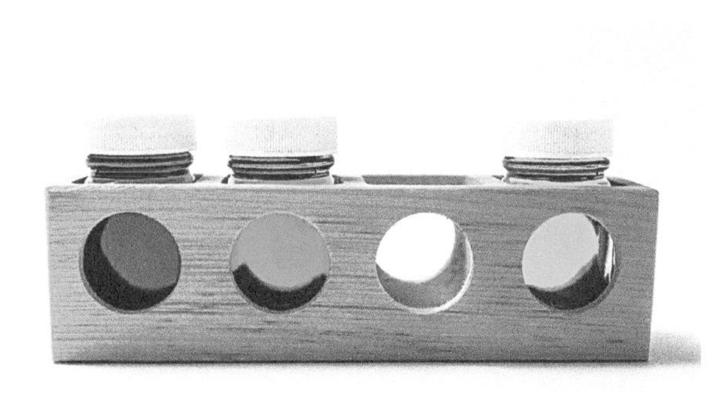

FIGURE 6.25. A new grading kit that has been proposed for a Standard International Grading System. Pictured from left to right are Dark, Amber, and Golden. By the time you are reading this, the proposed system may already be adopted into law. PHOTO BY NANCIE BATTAGLIA

Although it's hard for us to believe, many people who purchase syrup to put on their pancakes or waffles don't know there is a difference between pure and artificial syrups. Of those who do know, a large number are not aware that pure maple syrup comes in different colors, flavor intensities, and grades. And of the small fraction of people who know that there are different kinds of pure maple syrups, only a small percentage of them can actually explain the grading system that exists in their state or province. Having one standard grading system that is more easily understood will help to reduce any of this confusion, thereby marketing pure maple products to a larger audience. Furthermore, it's important to realize that nearly all the growth in pure maple syrup consumption is happening overseas. Anything we can do as an industry to make it easier to sell maple syrup in foreign countries will help all sugarmakers. Without the growth of international markets, we would currently have a huge surplus and bulk prices would be significantly lower than they are today. The new grading system will help foster development of additional markets and keep bulk syrup prices high for all sugarmakers.

Table 6.2 contains a summary of the current classification systems for the United States and Canada, and how they compare with the new IMSI grading standards. As you can see, there are broad similarities and key differences in the current grading systems. They all follow the same classification systems regarding the light transmittance of the syrup, but the names for the different grades can be very different and highly confusing to the consumer. For instance, the lightest grade in Vermont is called Fancy whereas other states and provinces call it Light Amber or No. 1 Extra Light. With a name like Fancy or No. 1, some consumers would tend to believe that this is the highest-quality, best-tasting syrup. Since it is also called Extra Light or Light Amber, other consumers may think that this has the least amount of calories and is the "diet version" of maple syrup. Under the new system, the lightest grade will be called Golden with a Delicate Taste. The new name should help clear up some confusion and actually help people understand that it has the most delicate maple flavor. I'm sure many Vermonters will still call it Fancy when talking about the lightest-colored syrup—old habits are hard to break! However, they will only be able to include the term *Fancy* as a marketing descriptor on their jugs along with the required color and flavor descriptor (Golden with a Delicate Taste) as per the new IMSI regulations. This has been a major point of contention among traditional Vermont sugarmakers, but the ability to still include *Fancy* on the label has allowed the grade changes to proceed without nearly as much controversy. I doubt it would have passed in Vermont if producers weren't also allowed to put *Fancy* on the label as a marketing descriptor.

Another major point of confusion is based on the darkest grades of syrup that can be sold to consumers. Some states have classifications labeling these as Grade B, Grade B for Reprocessing, Extra Dark Amber, Extra Dark for Cooking, or Commercial. Canada simply calls them No. 2 Amber or No. 3 Dark. These classifications are based on historical biases against dark syrup and the classification system does little to help sell this syrup. There is a small segment of society that understands that Grade B syrup is high quality, and these folks will go out of their way to buy Grade B syrup (many people mistakenly believe that Grade B syrup is healthier than Grade A). With the exception of the consumers who already know that they prefer the strong maple flavor of Grade B syrups, no one would purposefully buy Grade B syrup if Grade A was also available. For instance, have you ever gone to the supermarket looking for Grade B eggs?

TABLE 6.2: Comparison of Proposed Standard Maple Grades and Nomenclature with Existing Maple Grades and Nomenclature in Canada and the United States

Light Transmittance (Tc)	United States and All States Not Individually Listed	Vermont‡ and Ohio*	New Hampshire‡	New York	Maine	Canada and All Provinces	IMSI Proposed Option
100	US Grade A Light Amber (≥75.0%)	Vermont Fancy; Ohio Grade A Light Amber	Grade A Light Amber	Grade A Light Amber	Grade A Light Amber	Canada No. 1 Extra Light	Grade A Golden Delicate Taste (≥75% Tc)
	US Grade A Medium Amber (60.5–74.9%)	Grade A Medium Amber	Grade A Medium Amber	Grade A Medium Amber	Grade A Medium Amber	Canada No. 1 Extra Light	Grade A Amber Rich Taste (50–74.9% Tc)
	US Grade A Dark Amber (44.0–60.4%)	Grade A Dark Amber	Grade A Dark Amber	Grade A Dark Amber	Grade A Dark Amber	Canada No. 1 Medium	
	US Grade B for Reprocessing (27.0–43.9%)	Grade B	Grade B†	Extra Dark for Cooking or Grade B for Reprocessing	Grade A Extra Dark Amber	Canada No. 2 Amber	Grade A Dark Robust Taste (25–49.9% Tc)
0	US Grade B for Reprocessing (<27.0%)	Commercial Grade	Grade B†	Extra Dark for Cooking or Grade B for Reprocessing	Commercial Grade	Canada No. 3 Amber	Grade A Very Dark Strong Taste (<25.0% Tc)
Fails other grade requirements due to unacceptable density, color, or flavor	Substandard	Substandard	Substandard	Substandard	Substandard		Processing Grade Any color class Any off-flavored syrup

Source: *North American Maple Syrup Producers Manual*, 2006

* Grading is mandatory in Vermont but optional in Ohio. In Ohio, producers may sell syrup ungraded or graded with Ohio S or the USDA grading standards.

† Table Grade Syrup; may be packaged in consumer size containers.

‡ Vemont and New Hampshire require 66.9° brix mimimum density.

In the 1990s New York State tried to help the sales of Grade B by also allowing it to be called Extra Dark for Cooking. After all, if you are going to be cooking with maple syrup, and you don't want the other ingredients to mask your maple flavor, it makes sense to use the syrup with the strongest flavor. However, the *for Cooking* label makes many people think that you have to cook with it—or that it must be cooked down further before you can use it. This type of classification system weakens the market for dark, strong-flavored

Which Grade of Syrup Is Best?

Light-colored syrup has traditionally been considered the highest quality among the maple-syrup-producing community. This bias dates back to the times when it was more difficult to produce light syrup and only the best sugarmakers were capable of producing it. Back then darker syrups were much more prevalent and could contain serious off-flavors. On the other hand, light-colored syrups generally tasted good and were more similar in appearance to cane sugar when processed even further into granulated maple sugar. With new technology, it is now relatively easy to produce light syrup, which no longer commands such a high price premium in the bulk syrup markets. Furthermore, numerous research projects have found that consumers prefer dark syrups to lighter ones. In fact, during the taste trials performed for the new IMSI grading standards, Golden with a Delicate Taste (the former Light Amber or Fancy) was ranked last among the four choices presented to consumers. Although sugarmakers may pride themselves on making light syrup and prefer the flavor of it, it is important to realize what consumers like and market our syrup accordingly.

syrups that consumers actually prefer. In the taste trials performed by the IMSI as part of the new grading standards project, darker syrups were preferred to lighter syrups in nearly all focus groups. Extra-dark syrups are not just for cooking, but rather suitable for pancakes, waffles, and other typical uses.

In the new system, there will be no such thing as Grade B maple syrup. We will still be producing and selling dark-colored syrup with a strong flavor, but as long as the flavor is acceptable and it meets other requirements for density and clarity, it will be classified as Grade A Very Dark with a Robust Flavor. The main determining factor as to whether a dark syrup can be sold for Grade A or has to be used for commercial purposes is the flavor. Although this is currently a subjective criterion, Centre ACER is currently conducting research to come up with a scientific way of measuring this—stay tuned for more information. The new grading system could help sell more of the darker syrups that still taste great and are actually what the general public prefers.

While I fully support the idea of increasing markets for darker, more flavorful syrups, the only short-term concern I often hear is the loss of the Grade B designation for certain markets and locations. Grade B syrup has gained almost a cult-like following, as many people prefer the stronger flavors and seem to like the fact that they are knowledgeable enough about maple syrup to ask for the "inferior" grade. It appears that the people who tend to go against the grain are the ones who especially seek out the Grade B syrups. The recent spike in Grade B sales can also be attributed to the Master Cleanse Diet in which the only thing you consume for about 10 days is a tea made from Grade B maple syrup, lemon juice, cayenne pepper, and water. Stanley Burroughs, the inventor of this crash diet, claimed that the Grade B syrup contains more nutrients and is healthier than Grade A syrups (but, in reality, all maple syrup has the same basic nutritional qualities). Whether you think the Master Cleanse is a hoax or want to try it yourself, there is no denying that it has helped move a lot of the darker syrups in recent years. For those of you who have a well-established market for Grade B syrups, start the education process now and let your customers know that the classification system will be changing (or has already changed). Use this as a teachable moment to strengthen your existing ties with these customers by telling them that the Grade B syrup they know and love will still be available, just under a different name.

Attaching flavor descriptors to the grade names will give consumers more of the information that they actually care about. The color of the syrup means very little to most consumers. The vast majority of pure maple consumers just want to know that it's pure maple and not an imitation. If they have a preference on the grade of syrup, it's the flavor that they are most interested in—the color doesn't really matter. The new system won't completely eliminate the confusion surrounding maple syrup, and there are bound to be some difficulties in getting sugarmakers to switch to a new grading system. However, it is the right step forward to ensure the long-term viability of the maple syrup industry worldwide. There are a lot more potential pure maple customers than there are existing ones. I highly doubt the new grading system will cause any existing consumers to switch away from pure maple, whereas there is a strong possibility that it will open up new markets for pure maple. By the time you are reading this, hopefully your state or province will have already adopted the new IMSI regulations on maple syrup.

No matter what you call it, there are certain attributes of syrup that must be met before it can be sold to the general public. If you want to make syrup to give to family or friends, then you need not be concerned about the government regulations. However, if you will be selling it in an open marketplace, then you will have to abide by the following IMSI standards for Grade A maple syrup:

- Produced exclusively by the concentration of maple sap or by the solution or dilution of a maple product other than maple sap in potable water.
- Complies with appropriate federal and state/provincial regulations and policy directives.
- Minimum density of 66% brix and maximum density of 68.9% brix.
- Proper determination of grade and color class.
- Intensity of flavor (taste) normally associated with the color class.
- Free from objectionable odors and off-flavors.
- Free from turbidity.
- Traceable to batch (daily production).

One of the key new provisions in the grading standard is the requirement for producers to label each jug of syrup according to the batch it came from. This is nothing new for the large manufacturers that supply distributors and large grocery stores, but it will represent a whole new arena for most small-scale sugarmakers who bottle their own syrup. Although you may view it as yet another task when bottling syrup, this is a critically important job that every sugarmaker should already be doing. Maple syrup is a shelf-stable, safe food that hardly ever causes illness in people. That being said, you never know when someone may file a complaint, even if it is unwarranted. For instance, I often hear about people who think they have found pieces of glass in their syrup; what they actually found was crystallized sugar at the bottom of the jar. A more realistic and troubling issue is if you wind up bottling syrup that is below density or has an off-flavor. Being able to identify which batch that syrup is from and remove just those bottles from the shelves will save you a great deal of time and money.

Birch Syrup Regulations (or Lack Thereof)

As of 2013 there are no regulations for selling birch syrup in the United States or Canada. The Alaska Birch Syrupmakers' Association had previously requested that the Food and Drug Administration establish grading standards for the birch industry in the same way that the USDA regulates the maple industry. The FDA did not act on their request, likely because the birch syrup industry is small, with only several thousand gallons produced each year. In addition, the syrup producers themselves could not agree on what the standards should be, with questions remaining over the minimum brix content and how to label blended birch syrup. Even without formal regulations, some producers have taken it upon themselves to distinguish among the different colors and flavors of birch syrup. As the number of birch syrup producers continues to grow, eventually there may be a strong enough coalition to persuade the federal governments of the US and Canada to take action on developing grading standards for birch syrup.

Understanding and Avoiding Off-Flavored Syrup

Maple syrup is a luxury, gourmet food that customers pay a premium price for. They expect it to be 100 percent pure, all natural, and great tasting, but unfortunately not all the maple syrup produced will meet consumer expectations. It is important to understand how some common off-flavors develop in syrup so that you can avoid these pitfalls in your own operation. The following section is adapted from Henry Marckres's descriptions that appear in the *North American Maple Syrup Producers Manual*:

- **Chlorine.** If you wash your tubing with a chlorine and don't properly rinse it, a residue will remain in the tubing that can appear in the sap the following spring. Even small amounts of chlorine can become concentrated in the syrup as it boils down on the evaporator. Chlorine residues in the sap can also damage RO membranes.
- **Detergent.** Although nearly all the equipment in your sugaring operation can be cleaned with just hot water, some producers use soap and detergents that can impart an off-flavor in syrup.
- **Metallic.** If you package syrup in metal cans or store it long-term in galvanized drums, it can eventually take on a metallic taste.
- **Plastic.** A plastic taste can develop if you use non-food-grade plastic buckets to collect hot syrup off the evaporator, or package syrup in containers that are not rated for hot syrup. All the plastic syrup jugs on the market are BPA-free and food grade, so as long as you purchase them from a reputable company, you should be fine.
- **Filters.** Using brand-new filters or old filters that have not been stored properly can easily ruin a good batch of syrup. New filters need to be fully rinsed with hot water to avoid adding a chemical flavor to the syrup. Old filters need to washed with pure hot water and stored in a dry, odor-free environment to maintain their integrity.
- **Defoamer.** If you use too much of any defoamer, or even a little bit of a foul-tasting defoamer, this off-flavor could end up in the finished syrup.
- **Chemical.** ROs require chemical washing of membranes throughout the season to maintain peak performance. If you do not properly wash and rinse the RO following a chemical wash cycle, your next batch of syrup can have dangerous chemicals that not only make the syrup taste bad, but are hazardous to human health.
- **Ferment.** Maple syrup can start fermenting if the density of the syrup is not high enough or if perfectly good syrup is stored in a dirty container full of bacteria, yeast, and mold spores, especially when syrup is not properly hot-packed.
- **Burnt niter.** If you don't remove the niter buildup at regular intervals, it could burn on your pans and impart off-flavors to your syrup.
- **Earthy.** Tapping into dead and decaying parts of trees won't give you much sap, but anything you do collect out of those tapholes can give your syrup more of an earthy flavor than most people would prefer.
- **Metabolism.** This can occasionally occur during the course of the season due to changes in the metabolism of trees. Unfortunately there is nothing we can do as sugarmakers to prevent it. You just have to be aware of it and continuously taste your syrup to make sure this off-flavor isn't developing. Metabolism flavors are rare and have been described as woody, cardboard, peanut butter, popcorn, or chocolate.
- **Buddy.** Toward the end of the season, when the buds start to swell, the sap and resulting syrup take on a fruity, almost chocolaty flavor that has often been described as a Tootsie Roll. This syrup can be sold as commercial grade, but you never want to bottle it as table-grade syrup.

Hazardous Situations in the Sugarhouse

Vermont is well known for its high-quality maple syrup, but not all the syrup winds up meeting the standards

FIGURE 6.26. Henry Marckres testing a sample of pure maple syrup. PHOTO COURTESY OF BRIAN CHABOT

that people expect from pure Vermont maple. Henry Marckres runs the Consumer Protection Bureau for the Vermont Agency of Agriculture and is often called whenever there is a problem with pure maple syrup. Pure maple syrup should just be the fresh sap of maple trees processed down into sweet syrup, and anything used in the process of collecting the sap or transforming it into syrup should be food-grade materials. Unfortunately some people don't always think clearly when processing their sap into syrup, as seen in the following anecdotes that Henry shared with me about his experiences in dealing with substandard syrup over the years. Although most of the problems simply result in off-flavored syrup, sometimes the issues can make the syrup quite hazardous. In fact, Henry has been to the hospital on multiple occasions stemming from his job as Vermont's "syrup inspector in chief."

Non-Food-Grade Processing Materials

One of the times Henry wound up in the hospital was due to cyanide poisoning. A sugarmaker had purchased a steam hood for his evaporator, and the gasket material between the steam hood and flue pans was not made of food-grade materials. The gasket did not perform well in this hot, steamy environment, and it turned out that it was leaching cyanide into the boiling sap. Materials either do or don't have food-grade approval for a reason, so you must always examine a given product before you use it for sugaring.

Improper Use of Chemicals as Cleaning Agents

Once Henry had to taste some off-flavored syrup from a fairly large sugarmaker with a reverse osmosis system. The RO was starting to lose some of its performance capability, so the sugarmaker decided to use a lot of soap (sodium hydroxide) to clean the membranes. Afterward, being a frugal sugarmaker, he decided to save the wash water to put it in his evaporator to give it a "thorough cleaning." As if this wasn't bad enough, he destroyed the RO membranes by using too much soap, so he took someone's advice to flush them with muriatic acid, aka paint thinner. He then put the muriatic acid into the evaporator for some extra cleaning. It didn't take trying much of this sugarmaker's syrup to send Henry to the emergency room.

Be Careful of What's in Those Bottles

A customer once sent a bottle of syrup back to the sugarmaker and said it wouldn't pour out. He could tell there was syrup in there and a little bit would dribble out, but not at the rate that it should. Henry was called to examine the situation and noticed that there seemed to be some fibers and seeds in the syrup. After significant effort trying to get the syrup out of the bottle, eventually it all came through as a big glob. It turns out that a mouse had made a nice home with insulation in the bottle, and those weren't actually seeds that Henry had tasted . . . This story underscores the commonsense wisdom of keeping empty syrup containers in their original boxes stored away in a safe environment, and always thoroughly check the bottles before filling them.

Only Sap Should Be Cooked in the Evaporator

A sugarmaker was perplexed that his syrup had an odd, fishy flavor to it and called Henry to try to figure out what the problem was. They went through the usual series of questions—how was the sap collected, stored, processed, filtered, et cetera—but couldn't figure anything out. As the conversation was winding down, the sugarmaker told Henry that he often had people over to visit in the evenings. Recently, someone had discovered how easy it was to steam clams in the back pans of the evaporator, and they'd been enjoying these on a regular basis. I'm not sure why it took Henry's prodding to solve this mystery; common sense would tell us that anything we put in or near boiling sap will likely wind up in the finished syrup.

Be Careful How You Filter Syrup

As I've previously mentioned, filtering syrup is one of the most common ways to ruin a perfectly good batch of syrup. Henry has come across many situations of people using starched linen tablecloths and other odd materials to filter syrup. One sugarmaker didn't have time to wash a felt filter before putting syrup through it, so he just put the filter in the boiling sap in the back pan to wash it out. This certainly cleaned the filter. Unfortunately all the chemicals simply wound up in the boiling sap—and of course, the finished syrup.

CHAPTER 7

Registering and Certifying Your Sugaring Operation

Until recently there has been no registration or licensing requirements to produce maple syrup. Furthermore, even though the USDA's National Agricultural Statistics Service (NASS) attempts to track maple syrup production for 10 states, many sugarmakers fly under the radar of this government agency. If you are just producing on a hobby scale and don't sell any of your products, then registering your operation is not necessary and it really doesn't matter if the government knows what you are doing. However, if you make enough syrup to sell some on the open market, then you will need to follow the appropriate labeling and packaging requirements for your state or province. Accurately reporting your syrup production totals to NASS each year also benefits the entire maple industry. However, participation rates in the annual surveys and the five-year Census of Agriculture have historically been low, as many sugarmakers have no desire to let the government know what they are doing and certainly don't want anyone coming to inspect their facilities.

In reality, there are far more people producing syrup in the United States than the government reports indicate, and this lack of reporting has undermined the importance of maple in the eyes of the general public and our policy makers. If you are producing enough maple syrup to sell on the open market, then you should be reporting your activities to NASS. If you aren't already receiving a brief survey at the end of each sugaring season, then contact your regional office to make sure you are on their list. Rather than looking at the reporting as a hassle and intrusion on your business, consider it a free (or prepaid) service in which public servants keep track of output from year to year. The NASS data is an extremely valuable metric for how the industry has evolved over time, but it is only as good as the data that producers report.

The maple syrup industry will only be taken seriously by our elected officials if they have a full

FIGURE 7.1. The USDA National Agricultural Statistics Service does a great service for the maple industry by keeping track of production and pricing data over time. Statistics Canada performs a similar service for the Canadian provinces.

FIGURE 7.2. Every five years the Census of Agriculture provides a detailed outlook on all agricultural sectors in the United States.

understanding of the size and scope of production. For instance, when Senator Charles Schumer (D-NY) first learned about the development and growth potential of the maple industry, he introduced a bill in the Senate called the Maple TAP Act. This would provide funding for a competitive grant program aimed at developing the maple industry through research, education, and marketing initiatives. It took several years for this bill to go anywhere, but it finally made its way into the 2012 Farm Bill. The future of the bill is uncertain at this point, but it would have a much better chance of passage if policy makers fully understood the size of the maple industry and its potential for growth. In Canada, where maple syrup is big business, government agencies have been very friendly to the maple industry and have supported sugarmakers through a variety of mechanisms. If we would like the same type of support from our elected officials in the United States, the least we can do is report the full extent of our activities.

FIGURE 7.3. Senator Charles Schumer (D-NY) drinking some fresh maple syrup at Parker's Family Maple Farm in West Chazy, New York, as part of a press conference focusing on the Maple TAP Act.

FDA Registration

There has been great consternation and confusion surrounding the requirement of maple syrup producers to register their operation with the US Food and Drug Administration (FDA). In 2003, as a response to 9/11 and the threat of bioterrorism to our food supply, the FDA issued amendments to the food safety laws requiring facilities that manufacture, process, pack, or hold food for human or animal consumption in the United States to register with the FDA by December 12, 2003. This requirement did not make any waves in the maple industry at the time, and very few producers registered their operation over the past decade. More recently, there have been discussions about whether maple syrup will be excluded under the provisions of the Food Safety Modernization Act of 2011. During this deliberation, it became apparent that maple producers were supposed to have already registered with the FDA, so the North American Maple Syrup Council issued a statement to all producer organizations making them aware of the need to register by January 31, 2013. There is no cost to register, but it does take significant time and you will have to update your data every two years. Failing to register is a prohibited act under the FD&C (Federal Food, Drug, and Cosmetic) Act and could theoretically result in civil or criminal charges being pressed in a federal court. Although the likelihood of this happening is probably very low, simply taking the time to register will erase any chance of that occurring. I recommend visiting the FDA website[1] to learn more about this issue and to register your operation.

Vermont Sugar House Certification Program

If you produce syrup in Vermont, you now have the option of having your operation certified through the Vermont Agency of Agriculture, Food and Markets. Since 2010 a committee of sugarmakers, syrup buyers, and packaging companies—chaired by Tim Wilmot from UVM—has been working on a voluntary certification program that was unveiled for the 2013 season. Participating sugarmakers can pay a fee of $200 to have an inspector from the Agency of Agriculture come to their sugarhouse to examine it for food safety measures. The program was developed primarily at the request of large packers who handle a lot of bulk syrup. Even though pure maple syrup is generally considered a safe food, it can be contaminated in the production process, resulting in off-flavors at a minimum and possible health issues in a worst-case scenario. Grocery stores and other distributors want certification and traceability to make sure that the products they are selling meet a certain standard that we all expect from pure maple. The Vermont certification program is one way of getting ahead of the curve for producers who want to take the initiative to document the quality and integrity of their products.

Obtaining a passing grade shouldn't be very difficult since most of the requirements are commonsense measures that sugarmakers are doing anyway. The certification program includes a scoring sheet in which you are allocated a certain number of points for violating provisions in the different categories. For instance, failing to put a check-valve between your vacuum pump and releaser can cost you 3 points whereas using possible allergens in defoaming or failing to code your products can get you 10 points. If you score more than 10 points, or if your syrup tests over 250 ppb for lead, your operation will not pass.

I think this voluntary certification program is a great idea and encourage all sugarmakers to participate. If you live outside Vermont and can't get officially certified, you may still wish to download the scoring sheet and conduct your own inspection. You may discover things that you could do differently to ensure that your syrup always meets the highest standards for food safety. To learn more and to download the scoring sheet, visit http://vermontmaple.org/sugarmakers/join-vmsma/certification-program.

Organic Certification

Whether or not to get your syrup certified organic can be a tough decision. There are real costs and potential benefits to getting certified, so you should think critically about how much time and money it will take to become certified organic, and what benefits you will get out of it. You might think that all maple syrup would be considered organic. After all, it is simply pure sap collected from natural forests boiled down into sweet syrup. It's hard to get any more "organic" than that. If you already have a certified-organic farm, then it makes logical sense to include maple syrup in your overall certification. However, if you would have to go through the entire certification process just for your maple operation, then the choice isn't as easy. If you are a fairly small producer marketing all your syrup to rural people, then it is doubtful that getting certified organic would be very helpful from a marketing perspective. On the other hand, for larger producers who are selling bulk syrup, the premium price paid for organic syrup usually makes the certification process worthwhile. There is a limited quantity of organic syrup and high demand by food processors and other clients, so the larger buyers usually pay about 10 cents per pound extra for certified organic syrup. Furthermore, if you are direct-marketing your products into urban and suburban markets, the organic certification may help sell more syrup to people who generally prefer to purchase organic foods and don't understand how maple syrup is made.

Within the United States, all certified-organic maple syrup must follow the general National Organic Program Section 205.207 (b) Wild-Crop Harvesting Standard.

FIGURE 7.4.

FIGURE 7.5. There are many certifying organizations in the United States and Canada. MOFGA, NOFA-NY, and VOF certify the bulk of organic maple syrup production in the United States.

This states that "a wild crop must be harvested in a manner that ensures that such harvesting or gathering will not be destructive to the environment and will sustain the growth and production of the wild crop." Based on this notion, various agencies have developed their own sets of rules and guidelines for becoming certified in a particular state. Most of the guidelines make sense and merely dictate practices that we all should be doing anyway—certified organic or not. However, there are also some questionable guidelines that make you wonder who could have possibly come up with them. If you care about organic certification, I encourage you to become proactive with your state or regional certifying organization to make sure they have the best information available when determining the standards for organic maple syrup.

The Vermont organic certification program has the most extensive guidelines and puts a heavy emphasis on ecologically sound sugarbush management. They require a written management plan that focuses on maintaining species diversity and encouraging regeneration of new trees into the forest. This makes sense to me, as it falls in line with the spirit of organic agriculture—taking care of the land for long-term sustainability. Since diverse forests with a healthy mixture of many species and age classes are more resilient to pests, diseases, and natural disasters, it follows that sugarmakers producing organic maple syrup should follow guidelines ensuring that their sugarbush is as healthy as possible. I fully support all the guidelines that encourage long-term health and sustainability of the sugarbush. On the other hand, there are some guidelines have no impact on whether the sugarbush is well taken care of or if the finished product contains chemical residues (which is the main concern among organic food consumers). For instance, the Vermont guidelines mandate that you use droplines that are at least 24" long. While it is a good practice to make sure you have long enough droplines to be able to reach all parts of a tree trunk, you can certainly make "organic" syrup with a 20" dropline, especially if you are replacing droplines on a regular basis and moving them to different parts of the tree. I personally believe in and recommend longer droplines, but this shouldn't dictate whether syrup is considered organic. Another peculiar stipulation is the fact that if you use diatomaceous earth (which is not considered an organic material) as a filter aid, you can only label your syrup as organic, not 100 percent organic—even though DE never winds up in the finished syrup.

Some states put restrictions on the use of vacuum to protect tree health, yet these guidelines are not based on sound science or an understanding of the impact of vacuum on trees. For instance, in Maine you must keep vacuum levels below 20", whereas New York allows for the use of vacuum as long as "producers maintain as low a pressure as possible." This guideline confuses the terms *pressure* and *vacuum*, but it also misses the point of the purpose of vacuum. No scientific research has ever found that high vacuum levels have a negative impact on tree health or longevity. In fact, there is overwhelming evidence of sugarmakers using high vacuum tubing for decades, yet still having healthy, relatively fast-growing trees. If high levels of vacuum hurt or killed trees, we would certainly have seen evidence of it by now. Furthermore, due to the vagaries of seasonal weather patterns there are many years when producers can only achieve economical yields by utilizing high vacuum. It's important to keep in mind that one component of sustainability is economic viability; regulations should not place unnecessary burdens on sugarmakers who are trying to earn a living from their efforts.

Many certifying agencies will not allow trees to be marked with paint, as this is considered a synthetic substance. While I understand the desire to limit or negate any synthetic substances used in the production of maple syrup, this is another regulation that I tend

FIGURE 7.6. The small dots of paint sprayed on this tree to mark tapholes would not be allowed in organically certified sugaring operations. PHOTO BY NANCIE BATTAGLIA

FIGURE 7.7. Bottles of canola oil sitting on the shelf at a certified organic sugarmaker's operation in Quebec. The type of defoamer used is the only practical difference in many certified-organic sugarhouses.

to disagree with. At our sugarbush, we put a small dot of paint right below the taphole each year to mark the hole when removing the spout at the end of the season. Because the small tapholes heal over with new wood and it is not always apparent where previous years' tapholes were, this allows us to keep track of previous tapping and avoid drilling into stained wood in subsequent years. Perhaps a small dot of paint on the bark of maples could cause some harm to the trees, and if it did we would certainly stop this practice. However, I have yet to find any research supporting this theory or evidence of this in the field. On the other hand, using a small dot of paint below a taphole helps prevent drilling into stained wood in the future, which is the underlying reasoning behind having droplines of at least 24".

One of the only real differences in certified organic syrup is the type of defoamer used in the boiling process. Any type of fat molecules will stop and prevent foam from forming in your pans during boiling. Most sugarmakers use Atmos K defoamer, which is chemically synthesized, and therefore not currently allowed in organic certification guidelines. There is a potential that some certifying agencies will start allowing it, which would be a great step for certified organic maple producers, because defoamer alternatives are currently limited. Most certifying agencies only allow the use of certified organic canola oil or sunflower oil, yet these do not do nearly as good a job at keeping the foam down. Sugarmakers must use much more in their evaporator to keep the foam from bubbling over, leading to potential quality-control issues. I know several maple producers who dropped their certification because they wanted to produce pure maple syrup, not a mixture of maple syrup and canola oil!

Although some of the regulations are unnecessary (in my opinion), getting your syrup certified organic could be very helpful to your sugaring operation. One of the main benefits of organic certification is that it could force you to do beneficial things that you may not normally take the initiative to do. Having to keep the proper paperwork and go through inspections could also help you become more organized and deliberate in your practices. Most of us don't like record keeping, but if you are forced to do it through the certification process, you'll find that it provides very useful and interesting data. Note that if you sell less than $5,000 worth of organic products over the course of the year, then you don't need to get certified to call your syrup organic. However, if your gross sales are over $5,000 and you want to put the organic sticker on your syrup, then you should reach out to the certifying organization in your state to get the application process started.

Even if you have your syrup certified organic, I urge you not to use your certification to tout your maple syrup over that of other producers who haven't gotten certified. The reality is that most syrup would be considered organic, and we should never try to prop ourselves up at the expense of other sugarmakers. Rather, I recommend using the certified-organic label to compare your pure maple syrup with the artificial stuff. The organic label presents an excellent opportunity to inform your customers of the differences between pure maple syrups and the artificial pancake syrups that use high-fructose corn syrup as the main ingredient. Even if the HFCS came from organically grown corn, it's hardly an "organic" product in most people's view. It makes sense to use the organic label as a talking point to explain how pure maple syrup is, but please don't market your products in a way that would leave customers thinking that there is something wrong with maple syrup if it isn't certified organic. Always remember, other sugarmakers are not your competition—they are your collaborators. We're all in this together, certified organic or not, and our goal as an industry should be to strengthen the overall market share for pure maple products, not compete among ourselves for the tiny sliver we currently occupy. The certification process provides another way to broaden the market base for pure maple products, so you should explore whether it makes sense for your operation.

Certified Naturally Grown

Once the National Organic Program was implemented in 2002, farmers who wanted to call themselves organic had to become certified by a USDA-sanctioned agency. If you like the idea of organic certification but don't want to spend the time or money to get certified, you may want to consider the Certified Naturally Grown program. CNG is an alternative to the USDA's National Organic Program meant primarily for small farmers who sell their goods in local markets, not for large producers selling to wholesale accounts. The standards are basically the same, yet the application fee and reporting requirements are significantly less. Other farmers in the CNG program serve as the inspectors, which provides for excellent peer-to-peer learning opportunities. I know several maple producers who have gotten their operations certified through this program, and they have all been pleased with the process. Being able to add the attractive CNG sticker to the syrup bottles catches some people's eyes and provides a nice talking point about the pure and natural qualities of maple syrup. If you are interested in learning more, go to www.naturallygrown.org.

FIGURE 7.8.

Expanding Your Business Through Buying Sap and Leasing Taps

If you have tapped all the maple trees on your property, but you still want to make more syrup, what are your options? First of all, you could scrutinize your sap collection system to make sure that you are getting as much sap out of your trees as possible. This doesn't mean putting more taps on the same trees, but rather making sure your tubing system, annual maintenance, and spout selection adhere to the latest research and guidelines for optimizing sap yield (see chapter 5). You could also invest time in managing your forest to ensure that your trees are healthy and growing as vigorously as possible (chapter 13). Nevertheless, even if you have the best tubing system and the healthiest trees, you may want to make more syrup in your operation. This could mean buying more land (an expensive option) or working with your neighbors to lease trees for tapping or buy sap that others collect. This chapter focuses on these two collaborative strategies for expanding your operation.

Buying Sap

Before we discuss the details of buying sap, let's first consider the economic theory behind this practice, using maple syrup as an example. To understand why buying sap may be a worthwhile venture, you must first understand the difference between fixed and variable costs. Fixed costs include things like buildings, equipment, machinery, insurance, and so on. These do not vary with the amount of output (maple syrup) produced. As a sugarmaker, you must endure these costs whether you produce 100 gallons or 1,000 gallons of syrup in any given year. You will have already built your sugarhouse and bought an evaporator, reverse osmosis, filter press, and many other items necessary to produce high-quality syrup. You have already spent this money and the cost will remain the same whether you have a good sugaring season or a bad one. On the other hand, variable costs include items such as labor, utilities, and materials used in sap collection and processing. The more syrup you produce, the more time, fuel, and materials you must invest to produce that syrup. Variable costs tend to rise in direct proportion to the amount of a particular good or service (in this case maple syrup) produced. For instance, as you gather more sap, it will require more time and fuel to process that sap into syrup.

As long as the variable revenues (the amount of money generated by producing and selling a gallon of syrup) are greater than or equal to the variable costs of producing that gallon, then it makes financial sense for a sugarmaker to stay in business producing syrup. However, maple operations are only profitable if the sugarmaker produces enough syrup to cover both the variable *and* the fixed costs over time. Many sugarmakers easily exceed their variable costs with each gallon of syrup produced, but they do not make enough syrup to also cover their fixed costs. Buying sap is one way of expanding a sugaring operation so you can earn enough money to also cover your fixed costs.

While many of us consider sugaring a labor of love and accept limited (or no) profits from our efforts, some are able to earn a decent living producing and selling maple products. These tend to be the large producers who can gain economies of scale in their production and spread their fixed costs out over a greater number of taps. However, to become a sugarmaker on a fairly large scale (more than 2,000 taps), you must make significant investments to build a sugarhouse and buy an evaporator, reverse osmosis unit, collection tanks, filter press, and all the other items necessary to produce high-quality maple syrup in a cost-efficient manner. As previously stated, these are the fixed costs that you must endure each year no matter how many taps you put out or how good the sap runs. Thus, to make the most of your investments, it may make sense to purchase additional sap to spread your fixed costs out over a greater amount of syrup. As long as the variable revenues from processing the sap are greater than the additional costs of buying the sap, it makes financial sense to do so. Generally speaking, for large sugaring operations with ROs and efficient evaporators, the more sap you can put through your equipment, the more profitable your operation will be.

Should You Buy Sap?

Whether or not you should buy in extra sap to supplement your own sugaring operation depends on many factors. First of all, for new sugarmakers, you should go through at least one sugaring season making syrup from your own sap before agreeing to take on sap from others. Determine how much you like being in the sugarhouse making syrup as compared with being out in the woods collecting sap. These are two entirely different tasks that are geared toward distinctive personality types and skill sets.

Being out in the woods tapping trees, checking vacuum leaks, and gathering sap takes a lot of energy, athleticism, and endurance, especially if the terrain is tough and the snow is deep. It requires getting out in the woods not only on the nice warm, sunny days, but also on those windy, rainy 40°F days when the sap tends to flow readily. The conditions that make it enjoyable to be outside working in the woods don't always coincide with the weather that causes the trees to run sap. If you really enjoy being out in the woods, then perhaps you would be better off collecting and selling sap than buying it.

Processing sap in the sugarhouse can be a more relaxed activity that does not require as much physical labor, especially if you have an oil-fired evaporator and things are running smoothly. On the other hand, wood-fired evaporators require more work and keep sugarmakers constantly active filling the firebox and bringing in wood. Although boiling sap isn't always easy, it can be much less physically demanding than tapping trees and gathering sap. If you would rather be in the sugarhouse processing sap than out in the woods collecting it, then purchasing sap may be a good option for you.

Another good reason to buy sap is if you have a regular job during the day and don't have enough time to gather sap yourself during daylight hours. Or perhaps you have tapped all the trees on your own property and can't find any more trees to lease. If there are people nearby who are willing and able to gather sap for you to boil, and you have the capability to process it quickly and efficiently, then it could make sense for you to buy their sap or boil it on shares. Finally, if you have to buy in bulk syrup from other sugarmakers each year to fill your own market demand, then you could save money by simply buying in sap and producing more syrup yourself.

The Economics of Buying Sap

If you are considering buying sap to supplement your existing sugaring operation, you should first determine whether it would be profitable to do so. To help with this, I developed a fairly simple Microsoft Excel calculator that allows you to determine the amount of money your sugaring operation would make per hour (your hourly wage rate) for purchasing and processing additional sap. As with all the spreadsheets mentioned in this book, they can be downloaded from either the Chelsea Green Publishing website, www.chelseagreen.com/sapguide or the Cornell Maple Program website at http://maple.dnr.cornell.edu/sapbuying.htm.

To use this calculator, you must have a good grasp of your costs and revenues for processing sap. You need to know (or have a good estimate for) the amount of syrup you produce per hour, the bulk price of syrup, the total cost (including fuel, electricity, filters, depreciation on equipment, and so forth) to produce a gallon of syrup, and any cost incurred to store the additional syrup in barrels. You must also determine what percentage of the syrup (or bulk syrup revenues) you will retain and what percentage you will give to the sap seller. When you input values for these variables, the spreadsheet provides the hourly wage (the amount of profit a sugaring operation generates per hour) for processing sap according to the following formula:

Hourly Wage for Purchasing and Processing Sap=

gallons of syrup produced per hour ×
(bulk price of syrup ($/lb) × 11.14 lbs/gallon) –
(fuel cost + storage cost + filtering costs of sap
and syrup + depreciation on equipment
for every gallon of syrup produced) ×
percentage of bulk syrup revenues
retained by the person processing the sap

Let's consider how this works. Table 8.1 shows the amount of money a sugaring operation could make per hour under three different scenarios. Scenario 1 simulates a small hobby producer, Scenario 2 represents a medium-sized sugarmaker, and Scenario 3 could be a large-scale operation with energy-efficient equipment.

In performing this analysis, the only constant variable is the bulk price of syrup, as that price is set by the world market and we have little control over what price we get for our bulk syrup. However, the production of syrup per hour and fuel cost to produce a gallon of syrup will vary tremendously based on the size of the operation and equipment used to process the sap. Filtering costs for sap and syrup are negligible and do not vary much between sugaring operations, so we will not consider them here. You may also have to purchase additional barrels to store the syrup, but those costs are not factored in the simplified version. The Excel file contains another worksheet titled Advanced Hourly Wage Calculator that allows you to factor in these additional costs. It usually winds up costing between 5 and 20 cents per gallon of syrup produced to filter raw sap and the finished syrup. Since this represents less than 1 percent of the value of syrup, I do not include it in the Simple Hourly Wage Calculator. Purchasing a stainless-steel barrel usually costs between $5 and $6 per gallon of storage; that is a one-time fixed cost that is very noticeable the first year but goes away in future years. The main purpose of this exercise is to simply show the difference in profitability for buying sap based on different-sized sugaring operations. In reality, the actual earnings will be slightly less when you factor in these additional costs and even more so if you factor in depreciation on equipment.

Scenario 1. Hobby Sugarmaker with Small, Inefficient Equipment

As a hobbyist, it is difficult to make money boiling your own sap, let alone the sap that others gather and sell to you. That is why it's a hobby and not a business! For example, imagine that you can only produce 1 gallon of syrup per hour and the fuel cost to produce that gallon is $18. If you give the sap seller 40 percent of the revenue and bulk syrup prices are $2.75 per pound, you would only be making $7.58 an hour. Bear in mind that this also doesn't include storage costs, the variable costs for filtering sap and syrup, or depreciation on the equipment. When it is very time consuming and energy

TABLE 8.1: Hourly Wage Calculator for Processing Sap under Three Scenarios

	Scenario 1	Scenario 2	Scenario 3
	HOBBY SIZE	MEDIUM SIZE	LARGE SIZE
Gallons of syrup produced per hour	1	6	30
Total fuel/electricity cost per gallon of syrup produced ($)	$18.00	$10.00	$4.00
Percentage of additional syrup produced that is given to the sap seller	40%	50%	60%
Bulk price of maple syrup produced ($/lb)	$2.75	$2.75	$2.75
Hourly wage for producing syrup from purchased sap ($/hr)	$7.58	$61.89	$319.56

intensive to process sap, it is hard to offer someone even 40 percent of the revenues and still make decent money. As a hobbyist, even when you get to keep 100 percent of the syrup revenues by boiling your own sap, that only equates to $12.63 per hour (and doesn't take into account the sap collection costs). If you are making syrup on a small scale as a hobby, it is quite obvious that you aren't doing it to get rich. If you want to boil sap for other people, know that you are doing it as a favor and not as a business venture.

Scenario 2. Medium-Sized Sugarmaker with Older Equipment

If you have a medium-sized operation (about 2,000 taps) using a standard evaporator, it is possible to earn a decent profit by buying sap. If you make 6 gallons of syrup per hour and the fuel cost to produce a gallon of syrup is $10, you could give the sap seller 50 percent of the syrup revenue and still earn $61.89 per hour. Although you could greatly reduce your fuel costs and time spent boiling by using reverse osmosis, buying sap is profitable even without this technology. To be able to make roughly $60 an hour doing something you love while generating a tremendous product is certainly a worthwhile venture (in my opinion).

Scenario 3. Large Sugarmaker with Energy-Efficient Equipment

If you have made a large investment in an energy-efficient evaporator and reverse osmosis system, I recommend buying as much sap as you have time to process. With the right equipment, the marginal costs of buying sap

FIGURE 8.1. If you have invested in a large reverse osmosis unit such as this one, buying sap or boiling on shares can be a very profitable venture.

are usually much lower than the revenues gained from selling the additional syrup produced. Because you have invested so much money in the energy-efficient equipment, you can't afford *not* to buy sap. The expensive equipment you bought isn't generating any revenue when it sits idle; you need to use it as much as possible to get a good return on your investment. Imagine if you could make 30 gallons of syrup per hour and it only costs $4 in fuel for each gallon of syrup produced, which are common figures for large sugarmakers. Under these circumstances you could give the sap seller 60 percent of the bulk syrup revenue and still earn around $320 per hour by purchasing and processing additional sap. Even if you gave away 70 percent of the syrup revenue, you would still be making about $240 an hour. These figures overestimate the actual earnings since they do not account for additional factors such as depreciation on the evaporator and reverse osmosis unit. However, it is hard to imagine being able to make as much money in any other venture, and if you have already invested in the equipment to process your own sap, it certainly makes sense to use your expensive equipment to the greatest extent possible.

It is important to realize that these "hourly wages" are simply the value of the extra syrup produced minus the variable costs of processing that sap, determined on an hourly basis for the time spent processing. You will not always earn that much money processing your own sap because these figures do not factor in the fixed costs for processing sap or the costs of sap collection. However, when deciding whether or not to purchase *additional* sap to make greater use of your equipment, this type of analysis can be very helpful in determining your marginal revenues versus marginal costs. Maple production can be quite profitable *on the margin* for existing operations; it is the fixed costs that often drive down profitability. Purchasing additional sap can be one way to increase the overall profitability of a sugaring operation and help pay for the fixed costs of investments.

It is worth noting that the economic analyses presented for these three scenarios would not be appropriate if the sugarmakers made substantial renovations to their sugarhouse and/or purchased new equipment to buy and process more sap. If a large percentage of your processing capability is being used to process sap that you have purchased from others, then you really need to factor in the depreciation on equipment into your analysis. Under these circumstances, buying sap becomes much less profitable, so you will probably want to offer a lower percentage of syrup revenues to the sap sellers, especially for the first few years while you pay off your equipment and renovation costs. If you want to determine the depreciation costs on your evaporator and reverse osmosis units, I have included these features on the Advanced Hourly Wage Calculator within the Excel spreadsheet.

In doing these calculations, you may also be asking yourself . . . "I sell most of my syrup in retail-sized containers, so why should I be using bulk prices to estimate revenue?" The answer is quite simple. When syrup is first drawn off the evaporator and put in a barrel, it is basically a commodity item that can be bought or sold at roughly the same price. Although you earn greater revenues by selling syrup in retail-sized containers, you must also factor in the costs of time and materials to package the syrup and do the sales and marketing. Only focusing on the revenues without addressing the costs confuses the issue. Thus, when calculating the value of the syrup you produce, you should value it as either (1) the price you could sell it for in the barrel or (2) the price you would have to pay to purchase that syrup in a barrel. The second option is for producers who do not make enough of their own syrup to supply their customers and therefore must buy in syrup to satisfy their market demand. For instance, if you would normally have to pay $30 per gallon to buy bulk syrup, then every gallon of syrup you produce should be worth $30 to you. If purchasing raw sap will allow you to produce syrup for less than $30 per gallon (with all costs considered), then buying sap is a viable option.

Pricing Sap

Some people would like to have syrup in exchange for their sap whereas others are more interested in getting paid. For those who want cash for their sap, how do you know what to pay for it? The *Maple Syrup Digest* has traditionally published sap prices each year in a table that is based on the sugar concentration of the sap. The table comes with a few paragraphs offering a disclaimer that these are only suggested prices and will not work for all producers and all situations. In reality, there are many factors that can and should affect sap prices. The following paragraphs provide a detailed overview of how to calculate sap prices.

Since sap is the raw material (and only ingredient) necessary in producing maple syrup, the value of the sap should be directly tied to the value of syrup (using bulk syrup prices). If sap pricing is based *solely* on sugar content, it only accounts for how many gallons of sap are needed to produce a gallon of syrup. In actuality, pricing should also be based upon the quality of syrup produced from that sap, the bulk prices for the grade of syrup produced, and the distribution of syrup (or bulk syrup revenues) between the person who gathers the sap and the person who processes that sap into syrup. To account for these additional variables, I developed a sap-buying and pricing Excel file that can be downloaded from the Chelsea Green Publishing website, www.chelseagreen.com/sapguide. This is the same spreadsheet that includes the hourly wage calculator used in the previous analyses. Downloading the spreadsheet and experimenting with it will greatly improve your understanding of sap pricing and the variables involved.

Once you have downloaded the spreadsheet, the only variable you need to input a value for is the "Percentage of Bulk Syrup Price Provided to Sap Seller." This determines how to distribute the syrup (or syrup revenues) between the sap buyer and seller and should be based on a mutual agreement between the two parties. Using this value, the spreadsheet does the rest of the work calculating sap prices based on various combinations of sap sugar content and bulk syrup prices. The formula used to determine the price of a gallon of sap is as follows:

$$= ((1/(87.1/\text{ sap sugar content})) \times 11.1382) \times \\ \%\text{ distribution of syrup revenues} \times \text{bulk syrup price}$$

Table 8.2 provides a snapshot of when the sap seller receives 50 percent of the bulk price of syrup that is produced. This was chosen as the default value since it is the most commonly reported method for distributing syrup by people who boil sap "on shares." Once you download the spreadsheet, you can enter in whatever value works for your situation; the sap prices will change accordingly. If the sap seller gets more than 50 percent of the syrup revenues, the prices will go up. Likewise, if the sap seller gets less than 50 percent of the syrup revenues, then of course the sap prices would be less.

In the sap pricing table presented here, sap sugar concentrations range from 1 to 4 percent at 0.1 percent intervals while bulk syrup prices range from $1.40 to $4.00 per pound at 20-cent intervals. You can match up the sugar concentration and bulk syrup prices to figure out how much a gallon of sap is worth. The table provides the price per gallon of sap based on these two variables *and* whatever value is entered in for Percentage of Bulk Syrup Price Provided to Sap Seller. In this example, that value is set to 50 percent; the sap prices reflect this equal sharing of revenues throughout.

There are a few important observations to be made about the sap pricing table. Not surprisingly, the higher the sugar concentration of the sap, the more valuable the sap is. Likewise, the greater the price of bulk syrup, the more money the sap is worth. For example, consider a gallon of sap that contains 2 percent sugar. If bulk syrup was selling for $3.00 a pound, then the sap would be worth 38 cents per gallon. If bulk prices were $2.20 a pound, then the sap would only be worth 28 cents a gallon. While this is only 10 cents a gallon, that money adds up quickly, especially when thousands of gallons of sap are being purchased.

Since we have no control over bulk syrup prices, rather than locking in a set price for sap before the season starts, it may be better to just agree on how to split the revenues once the syrup is sold. Waiting until the end of the season to be paid can be difficult since many folks want to be paid whenever they deliver a load of sap. However, if the sap seller can wait for payment until after bulk syrup prices are set and/or the syrup is sold, then it could allow for a more equitable sharing of revenues. One possible compromise is to estimate what bulk prices will be before the season and then adjust payments (as needed) if the bulk prices wind up being significantly different from what you thought they

Table 8.2: Sap Pricing Table When Revenues are Split at 50%

Sap Sugar Content	Bulk Syrup Price ($/lb)													
	$1.40	$1.60	$1.80	$2.00	$2.20	$2.40	$2.60	$2.80	$3.00	$3.20	$3.40	$3.60	$3.80	$4.00
1	$0.09	$0.10	$0.12	$0.13	$0.14	$0.15	$0.17	$0.18	$0.19	$0.20	$0.22	$0.23	$0.24	$0.26
1.1	$0.10	$0.11	$0.13	$0.14	$0.15	$0.17	$0.18	$0.20	$0.21	$0.23	$0.24	$0.25	$0.27	$0.28
1.2	$0.11	$0.12	$0.14	$0.15	$0.17	$0.18	$0.20	$0.21	$0.23	$0.25	$0.26	$0.28	$0.29	$0.31
1.3	$0.12	$0.13	$0.15	$0.17	$0.18	$0.20	$0.22	$0.23	$0.25	$0.27	$0.28	$0.30	$0.32	$0.33
1.4	$0.13	$0.14	$0.16	$0.18	$0.20	$0.21	$0.23	$0.25	$0.27	$0.29	$0.30	$0.32	$0.34	$0.36
1.5	$0.13	$0.15	$0.17	$0.19	$0.21	$0.23	$0.25	$0.27	$0.29	$0.31	$0.33	$0.35	$0.36	$0.38
1.6	$0.14	$0.16	$0.18	$0.20	$0.23	$0.25	$0.27	$0.29	$0.31	$0.33	$0.35	$0.37	$0.39	$0.41
1.7	$0.15	$0.17	$0.20	$0.22	$0.24	$0.26	$0.28	$0.30	$0.33	$0.35	$0.37	$0.39	$0.41	$0.43
1.8	$0.16	$0.18	$0.21	$0.23	$0.25	$0.28	$0.30	$0.32	$0.35	$0.37	$0.39	$0.41	$0.44	$0.46
1.9	$0.17	$0.19	$0.22	$0.24	$0.27	$0.29	$0.32	$0.34	$0.36	$0.39	$0.41	$0.44	$0.46	$0.49
2	$0.18	$0.20	$0.23	$0.26	$0.28	$0.31	$0.33	$0.36	$0.38	$0.41	$0.43	$0.46	$0.49	$0.51
2.1	$0.19	$0.21	$0.24	$0.27	$0.30	$0.32	$0.35	$0.38	$0.40	$0.43	$0.46	$0.48	$0.51	$0.54
2.2	$0.20	$0.23	$0.25	$0.28	$0.31	$0.34	$0.37	$0.39	$0.42	$0.45	$0.48	$0.51	$0.53	$0.56
2.3	$0.21	$0.24	$0.26	$0.29	$0.32	$0.35	$0.38	$0.41	$0.44	$0.47	$0.50	$0.53	$0.56	$0.59
2.4	$0.21	$0.25	$0.28	$0.31	$0.34	$0.37	$0.40	$0.43	$0.46	$0.49	$0.52	$0.55	$0.58	$0.61
2.5	$0.22	$0.26	$0.29	$0.32	$0.35	$0.38	$0.42	$0.45	$0.48	$0.51	$0.54	$0.58	$0.61	$0.64
2.6	$0.23	$0.27	$0.30	$0.33	$0.37	$0.40	$0.43	$0.47	$0.50	$0.53	$0.57	$0.60	$0.63	$0.66
2.7	$0.24	$0.28	$0.31	$0.35	$0.38	$0.41	$0.45	$0.48	$0.52	$0.55	$0.59	$0.62	$0.66	$0.69
2.8	$0.25	$0.29	$0.32	$0.36	$0.39	$0.43	$0.47	$0.50	$0.54	$0.57	$0.61	$0.64	$0.68	$0.72
2.9	$0.26	$0.30	$0.33	$0.37	$0.41	$0.45	$0.48	$0.52	$0.56	$0.59	$0.63	$0.67	$0.70	$0.74
3	$0.27	$0.31	$0.35	$0.38	$0.42	$0.46	$0.50	$0.54	$0.58	$0.61	$0.65	$0.69	$0.73	$0.77
3.1	$0.28	$0.32	$0.36	$0.40	$0.44	$0.48	$0.52	$0.55	$0.59	$0.63	$0.67	$0.71	$0.75	$0.79
3.2	$0.29	$0.33	$0.37	$0.41	$0.45	$0.49	$0.53	$0.57	$0.61	$0.65	$0.70	$0.74	$0.78	$0.82
3.3	$0.30	$0.34	$0.38	$0.42	$0.46	$0.51	$0.55	$0.59	$0.63	$0.68	$0.72	$0.76	$0.80	$0.84
3.4	$0.30	$0.35	$0.39	$0.43	$0.48	$0.52	$0.57	$0.61	$0.65	$0.70	$0.74	$0.78	$0.83	$0.87
3.5	$0.31	$0.36	$0.40	$0.45	$0.49	$0.54	$0.58	$0.63	$0.67	$0.72	$0.76	$0.81	$0.85	$0.90
3.6	$0.32	$0.37	$0.41	$0.46	$0.51	$0.55	$0.60	$0.64	$0.69	$0.74	$0.78	$0.83	$0.87	$0.92
3.7	$0.33	$0.38	$0.43	$0.47	$0.52	$0.57	$0.62	$0.66	$0.71	$0.76	$0.80	$0.85	$0.90	$0.95
3.8	$0.34	$0.39	$0.44	$0.49	$0.53	$0.58	$0.63	$0.68	$0.73	$0.78	$0.83	$0.87	$0.92	$0.97
3.9	$0.35	$0.40	$0.45	$0.50	$0.55	$0.60	$0.65	$0.70	$0.75	$0.80	$0.85	$0.90	$0.95	$1.00
4	$0.36	$0.41	$0.46	$0.51	$0.56	$0.61	$0.66	$0.72	$0.77	$0.82	$0.87	$0.92	$0.97	$1.02

Sap pricing table when the percentage of bulk syrup revenues provided to sap seller is 50 percent. Values are provided in the table as $/gallon of sap.

The Rule of 86?

Many sugarmakers have heard of the "Rule of 86" in which you divide the sugar concentration of the sap in to 86 as a means of determining the amount of sap it requires to produce a gallon of syrup. For example, when you have 2 percent sugar in your sap, it requires 86/2 = 43 gallons of sap to make a gallon of syrup. However, the Rule of 86 was developed many years ago in order to calculate what it would take to make syrup at 65.5° brix. In today's world, the legal minimum for syrup is 66° brix and many people bring their syrup to 67° brix because the extra sugar concentration makes the syrup taste much better (66.9° brix is also the legal minimum in Vermont). In order to bring sap to a higher sugar concentration, more water must be boiled off and the Rule of 86 no longer applies. For syrup being produced at 66° brix, you should use the Rule of 87.1 and for syrup produced at 67° brix, use the Rule of 88.3. I don't expect the Rule of 86 to go away anytime soon—it is a fixture of the sugaring world and provides a good enough approximation. However, the spreadsheets described in this book utilize the Rule of 87.1 in order to provide a more accurate representation based on 66° brix syrup.

would be. This protects both parties from fluctuations in the bulk syrup market, though it can add a layer of complexity and possible confusion.

The sap prices that have been historically posted in the *Maple Digest* are weighted to pay proportionately more for sweet sap and less for low-sugar-content sap. Since it takes less time and fuel to process sap with a higher sugar content, and sweeter sap tends to make higher-quality syrup, sap sellers are compensated proportionately more for sweeter sap. For instance, sap that is closer to 1 percent brix range may only provide 40 percent of the syrup revenues to the sap seller, whereas 3 percent brix sap may provide the sap seller with 60 percent. While this practice makes a lot of sense for small producers, it is not *as* necessary for large producers with energy-efficient ROs and evaporators. The fuel and time costs to process sap are much lower when you have efficient equipment, so you don't *need* to offer a relatively lower price for the less sweet sap than you do for high-sugar sap. On the other hand, for small producers who spend a lot of time and fuel processing sap, this type of stratified distribution makes a lot of sense.

To be able to change the revenue distribution based on sap sugar content (SSC), I created an advanced sap price table (found in the same Excel spreadsheet) that allows you to add this feature. You can set the target SSC value for your "normal" distribution of syrup revenues and then specify the percentage change in revenue distribution for a given change in SSC. If you have relatively high fuel and time costs for processing sap, you may want to tie revenue distribution closely with SSC. On the other hand, if you can process sap quickly and cost-effectively, you may not need to change the revenue distribution very much based on sap sugar content. This advanced calculator allows you to tailor the revenue distribution according to your particular situation. If you would like to use this feature when pricing sap, click on the worksheet titled Advanced % Calculator from the sap-buying Excel file. If you don't use this feature, all of the syrup revenues will be allocated at the one percentage you specify. However, if you would like to pay less for low-sugar sap and are willing to pay more for higher-sugar sap, then you should definitely use the advanced calculator.

So how do you decide what an equitable distribution of revenues from purchased sap should be? There is no easy, one-size-fits-all answer to that question, and I generally try to avoid giving direct advice on this. The default value is set at 50 percent only because I've found that producers typically give the sap buyers 50 percent of the syrup that the sap would produce or 50 percent of the bulk price of that syrup. However, it doesn't always have to work that way and I strongly recommend having an open discussion on how to split syrup revenues before any sap is bought or sold. The two major factors that should determine the percentage distribution are:

1. Whether the sap is delivered to the sugarhouse or needs to be picked up.
2. The processing capability of the sugarmaker.

If the sap seller delivers the sap, then he or she should receive a higher percentage than if it needs to be picked up. Furthermore, if the sugarmaker has efficient equipment and low processing costs, then he or she will be able to offer a higher percentage distribution and still make a decent profit. I developed the hourly wage calculator to provide a method for sugarmakers to figure out what percentage of syrup revenues they could offer a sap seller and still earn enough profits to make it worth their while. The take-home message is that you want to agree on something where both parties are happy and feel that the distribution of revenues is equitable; otherwise the business relationship is not likely to last.

It may be helpful to consider an example of how this all works. Imagine that John delivers 10,000 gallons of sap to Paul over the course of a sugaring season. With that extra 10,000 gallons of sap, Paul is able to make 250 additional gallons of syrup. The question Paul and John need to decide is how much of that 250 gallons goes to John and how much stays with Paul. Under a typical arrangement where the syrup is divided evenly, one of two things would happen:

1. John and Paul would each keep 125 gallons of the syrup.
2. Paul would keep all the syrup and pay John whatever the value of 125 gallons of syrup is when sold in the bulk markets.

What option they decide depends on whether John would prefer to have syrup, cash, or some combination of the two. It is possible that John may want to receive 10 gallons of syrup for his personal use and be paid for the other 115 gallons. There are certainly many different ways of handling compensation, and Paul and John need to figure out a reasonable and fair way to distribute the fruits of their joint efforts. While a 50–50 split is common, does it actually yield equitable results in all situations? If Paul was a relatively small producer with a 3 × 10 evaporator and did not use reverse osmosis, it would be very difficult (or impossible) to give 50 percent of the syrup income to John and still yield a decent profit for his efforts. On the other hand, if Paul has a large evaporator and reverse osmosis, then he could process John's sap quickly and efficiently. Paul may be able to distribute even more than 50 percent of the syrup revenues to John and still make a substantial profit. The more syrup or money that John receives, the happier he will be. John will have a greater incentive to gather and deliver sap each year, and as word travels around the community more people will want to start selling sap to Paul. Remember, without any people like John gathering and selling sap, it would be impossible for Paul to buy sap. When the person selling sap and the person buying and processing it can both enjoy themselves and earn a reasonable return by collaborating, it's a win–win for everyone involved.

Practical Considerations When Buying Sap

Buying sap can be a wise decision for some sugarmakers, but as with all things in life, the devil is in the details. There are many things to keep in mind and common pitfalls to avoid. Keeping accurate records is essential to ensure a fair distribution of syrup or syrup revenues. To make this easier, I developed a spreadsheet that allows sugarmakers to easily keep track of sap volumes, syrup production, and payments throughout the course of the sugaring season. On the sap-buying Excel file, go to the tab titled Blank Worksheet, and follow the instructions on what variables to provide input data for; the spreadsheet will keep track of payments (in cash or syrup) for each load of sap delivered.

Table 8.3 provides an example of a typical sap-selling arrangement. As you can see, the sap purchaser (Phil) kept track of the sap sugar content, the total volume of sap delivered, and the grade of syrup produced from

Table 8.3: Example of a Sap-Buying Spreadsheet

Sap Buying Records				
Sap Buyer: Phil	Sap Collector: Doug	% of Syrup Price Provided to Sap Collector		50%
BULK SYRUP PRICES	*** ENTER IN EXPECTED PRICES FOR ALL GRADES OF SYRUP**			
Light Amber	$2.80	$/lb	LA	
Medium Amber	$2.70	$/lb	MA	
Dark Amber	$2.60	$/lb	DA	
Extra Dark for cooking (Grade B)	$2.40	$/lb	XD	
Commercial	$2.00	$/lb	CO	

USER Controlled Variables*				**Results‡**					
Date	Gallons of sap delivered	Sap sugar Content	Bulk price of syrup Produced†	Gallons of sap needed to produce a gallon of syrup	Total gallons of syrup produced from the sap	Gallons of syrup provided to the sap seller	OR	Total price of the delivered sap	Price per gallon of the delivered sap
1-Mar	200	1.8	$2.70	48.4	4.1	2.1		$62.15	$0.31
3-Mar	365	1.9	$2.00	45.8	8.0	4.0		$88.68	$0.24
4-Mar	378	2.1	$2.80	41.5	9.1	4.6		$142.11	$0.38
7-Mar	470	2.1	$2.00	41.5	11.3	5.7		$126.22	$0.27
10-Mar	423	2.3	$2.80	37.9	11.2	5.6		$174.18	$0.41
12-Mar	437	2.3	$2.70	37.9	11.5	5.8		$173.52	$0.40
13-Mar	489	2.2	$2.70	39.6	12.4	6.2		$185.72	$0.38
15-Mar	467	2.2	$2.70	39.6	11.8	5.9		$177.37	$0.38
19-Mar	451	2.1	$2.70	41.5	10.9	5.4		$163.50	$0.36
22-Mar	304	2.2	$2.60	39.6	7.7	3.8		$111.18	$0.37
23-Mar	458	2.2	$2.60	39.6	11.6	5.8		$167.51	$0.37
24-Mar	490	2.1	$2.60	41.5	11.8	5.9		$171.06	$0.35
26-Mar	308	2	$2.60	43.6	7.1	3.5		$102.40	$0.33
27-Mar	398	2	$2.40	43.6	9.1	4.6		$122.15	$0.31
30-Mar	451	2	$2.40	43.6	10.4	5.2		$138.42	$0.31
1-Apr	407	1.9	$2.40	45.8	8.9	4.4		$118.67	$0.29
4-Apr	469	1.7	$2.00	51.2	9.2	4.6		$101.96	$0.22
SUMS AND AVERAGES	**6,965**	**2.1**	**$2.51**	**42.5**	**165.9**	**83.0**		**$2,326.79**	**$0.33**

* The user must enter in values for each of the columns highlighted in yellow.

† If you type in the code of the grade of syrup produced, this column will reference the prices listed above for that grade. You need to put an equals sign before the code. For instance, by typing in =LA the cell automatically references the price entered in above for Light Amber bulk syrup.

‡ The final 5 columns are the results of what is entered in the first 4 columns. Users should not change any values in this section.

the sap each day that sap was delivered. The spreadsheet does the rest of the work, calculating the gallons of syrup or the cash payment that is owed to the sap seller, Doug, for each load of sap delivered. The final row sums the daily entries to provide a total amount of syrup or cash payments owed to Doug.

Note that you only have to enter values for the yellow-highlighted cells; the spreadsheet calculates the outputs in the white cells. Before entering any values for sap that is delivered, you must first enter the percentage of the syrup or syrup revenues that you will provide the sap seller and the expected prices for different grades of bulk syrup. Although we usually don't know what the exact prices for bulk syrup will be until after the season has ended in Quebec, you may have a general idea of what they will be. The spreadsheet allows you to simply keep track of the grade of syrup you are producing—whatever value you have entered for the predicted bulk syrup prices will dictate what the sap payments will be. When the syrup is actually sold and/or prices are set, you can change the predicted prices to actual prices (if necessary), and the spreadsheet will automatically update.

All the calculations necessary for buying and selling sap require accurate measurements of sap sugar content and volume. If you decide to purchase sap, you should have at least one and preferably two reliable ways of measuring the sugar content and volume of sap as it is picked up or delivered to the sugarhouse. Refractometers and sap hydrometers should be tested annually and throughout the course of the sugaring season to make sure they are providing accurate readings. Whenever sap is transferred to a sugarhouse, it should flow through a totaling water meter to get a precise measure-

FIGURE 8.2. Analog refractometers provide one of the quickest, easiest-to-read, and least expensive ways to measure sap sugar concentration. PHOTO BY NANCIE BATTAGLIA

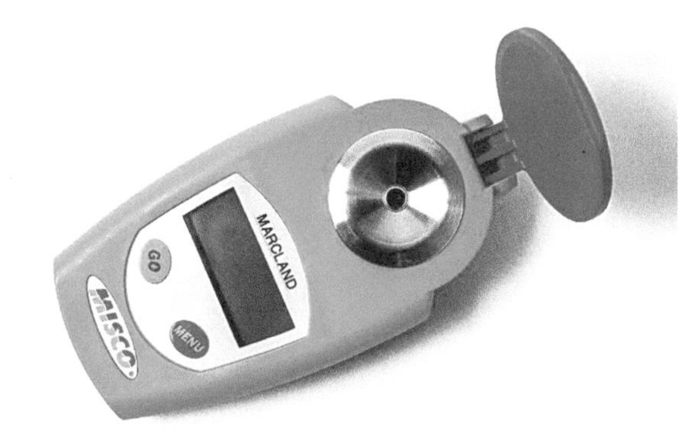

FIGURE 8.3. There are several digital refractometers on the market that are easy to use and can provide accurate readings; while they are more expensive than analog versions, they are worth the investment if you will be using them a lot. PHOTO BY NANCIE BATTAGLIA

FIGURE 8.4. Sap hydrometers are an inexpensive way to measure SSC, but they must be used carefully, periodically tested to ensure accuracy, and compensate for the temperature of the sap since most units are calibrated to 38°F. PHOTO BY NANCIE BATTAGLIA

Table 8.4: Sap Hydrometer Conversion Chart

Sap Temperature	Correction Factor
32–46°F	Keep same reading
47–56°F	add 0.1
57–61°F	add 0.2
62–64°F	add 0.3
65–68°F	add 0.4
69–72°F	add 0.5
73–75°F	add 0.6
Greater than 75°F	Stop sugaring!

If you use a sap hydrometer to measure sugar content, be sure to check the temperature of sap, especially later in the season when it is warm out, and add the appropriate value.

FIGURE 8.5. Whenever sap is being transferred, it should flow through a totaling water meter that keeps track of the total number of gallons. Although some tanks have marks indicating the number of gallons, using that method can only provide a rough estimate of the number of gallons. Metering the sap is essential to get an accurate measurement. PHOTO BY NANCIE BATTAGLIA

ment of the volume delivered. Having accurate measuring tools and keeping good records should help minimize any disputes between sap sellers and purchasers.

It is important to consider the cost of sap and syrup storage when you're buying sap or boiling it for someone else. If you need to purchase additional sap storage tanks, then you may want to reduce the percentage distribution to the sap seller. Similarly, if you have to buy extra barrels to store the additional syrup that you will produce, then the cost per gallon for storage should be accounted for. If the person providing you sap would like to receive bottled syrup (pints, quarts, and so forth) instead of cash payments, then the cost of the bottles and the time spent bottling should be factored into any agreements. There are several ways these costs can be accounted for:

1. The total cost of bottling can be subtracted from the syrup or money owed to the sap seller.
2. The sap seller can purchase the bottles and do the actual bottling at the sugarhouse.
3. The percentage of syrup given to the sap seller can be reduced by a mutually agreed-upon figure (for instance, reduce the percentage from 50 percent to 40 percent).

Finally, it is usually a good idea to develop a written contract between the sap buyer and seller should any questions arise throughout the course of the season. This contract should stipulate the terms of delivery, quality-control mechanisms for the sap, payment schedules and rates, and any other issues that could arise between a sap seller and buyer. Dealing with potential issues before they appear is a wise move and could save you lots of headaches down the road.

Possible Pitfalls When Buying Sap

Although buying sap may have positive benefits and increase your profitability, it is not without difficulties. Maple producers are often at the mercy of Mother Nature during sugaring season, and relying on others to bring you sap reduces your control even further. People will usually deliver sap when it is convenient for them, not when it is convenient for you. When you are having a big run and your tanks are almost full (or overflowing), chances are the folks who are selling you sap are also having a huge run and looking to drop off sap. You will have to increase your sap storage capacity to deal with these situations, which costs valuable space and money.

Furthermore, if your business model relies on others gathering and selling sap to you, this makes you vulnerable to the whims and desires of other people in any given year. Some years the people who normally gather and sell sap to you may not do it for a variety of reasons.

The snow might be too deep, they may be busy with more gainful employment, there could be a family issue that consumes their free time, and so on. To remedy these situations, I have heard of folks offering bonuses to people who continue to deliver sap year after year. For instance, one sugarmaker offers new people selling sap 50 percent of the syrup produced from their sap in the first year. If they come back the second year, they get 56 percent, and if they sell sap two years in a row they get 62 percent of the syrup produced from their sap. This gives the sap sellers an incentive to gather sap every year. The sugarmaker has had great loyalty and retention over the years by doing this.

Another issue can arise if someone delivers inferior-quality sap that could taint your high-quality sap when it is blended in the mix. This type of situation is common toward the end of the season when the temperatures are very warm and the sap is not flowing well. It may take a few days to collect enough sap to make a full load, and people don't want to bother delivering a small amount of sap. This is when the sap spoils more readily, however, and processing it ASAP becomes critical to producing high-quality syrup. Having clear expectations and consequences for not following the written agreement will help eliminate or minimize these types of issues. Occasionally you may have to reject a load of poor-quality sap; other times you may just pay for it according to the commercial grade syrup prices. It is best to avoid these situations altogether, so setting clear expectations that sap is delivered daily whenever it is running will help reduce or eliminate these types of issues.

Rather than having to penalize people for delivering poor-quality sap, you could take a proactive approach and teach them how to produce and maintain high-quality sap. We know that new equipment that is cleaned and maintained well will yield much more and higher-quality sap than older equipment that is not cleaned or maintained properly. By spending time with the sap sellers inspecting their operations and offering advice and guidance, you can ensure that they achieve the highest yields possible of high-quality sap. Not only will this help their bottom line, it will also help yours.

Finally, every industry has its bad eggs, so it is important to validate the character and trustworthiness of people you are buying sap from. Checking and measuring every load of sap delivered yourself can be time consuming, and often when someone shows up and wants to drop off sap, you may be busy doing something else. Being able to trust people to accurately measure the total volume and sugar content of the sap is important to ensure that you are not being cheated. A sugarmaker once told me about the time when he was buying sap delivered from someone in 5-gallon buckets. Rather than testing the sugar content of every bucket, the sugarmaker only tested the first couple of buckets and used that average to estimate all the buckets. He eventually became suspicious and decided to check some of the buckets in the back of the load—it turned out many of them were just water! Needless to say, that was the last load of sap he got from this person. This story is not meant to discourage you from buying sap; most people who gather and sell it are fair and honest. I just wanted to point out what can happen in a worst-case scenario so that you can try to avoid such a situation.

Leasing Taps

Buying sap can be a good strategy to grow your business if you can efficiently process sap quickly but don't have the ability or desire to collect more sap yourself. On the other hand, if you would like to tap more trees yourself and have already gotten to all the accessible maples on your own land, you may consider leasing options. Leasing another person's sugarbush is a low-cost alternative to purchasing forestland that allows you to add more taps without making a significant capital investment. Leasing also increases the pool of potential sugarbushes to choose from, as there are many more woodlots that could be leased for tapping than are available for sale. Rarely does a woodlot with easily accessible maples come up for sale in close proximity to your sugarhouse. However, there are often many potential sugarbushes within a 5- to 10-mile radius of your operation that are not being used. The owners may be reluctant to sell you the land, or you may not have enough money available to purchase it outright, but developing a contract allowing you to tap the trees can be a win–win for you and the landowner.

Winding Up in Court for Tapping Trees

I recently came across a newspaper article in which a family was sued for tapping trees on their neighbors' property. They had a handshake agreement in place for 10 years and things were fine, but as often occurs, there was a misunderstanding that led to the court case. Apparently the owners of the maple trees were elderly and never ventured into their woods. According to the lawsuit, they claimed that they thought that only 5 or 6 trees were being tapped whereas the sugarmaker had been putting out 258 taps a year over the past decade. When a friend of their neighbor alerted them to the fact that so many trees were being tapped, they hired a lawyer to sue in civil court for $20,000 to compensate for the lost revenue over the years and the "discomfort and annoyance" brought upon by the activities. The lawyer for the plaintiffs argued that . . . "who in their right mind would trade 258 quarts [of maple sugar] a year for four quarts of maple sugar in return?" Fortunately for the sugarmakers, the judge sided in their favor, and they were not required to pay the landowner any more money.

To me as an outside observer, the plaintiff's case was ridiculous and thankfully the judge also saw it that way. A sugarmaker cannot afford to give away a gallon of syrup for the rights to tap 5 or 6 trees, nor $20,000 to tap a couple hundred trees for 10 years. However, landowners usually receive more than a gallon of syrup for letting someone put out 258 taps on their property. Given that the average lease rate is about 50 cents per tap, the property owners would have received about $129 annually in a typical situation. Since they would prefer to have syrup instead, it would have made sense for the sugarmaker to offer them 2 or 3 gallons of syrup a year as adequate compensation. Due to the misunderstanding and lack of contract, the sugarmakers had to spend a great deal of time and money defending themselves in court, and they lost access to those trees as a result. I'm sure that if they could have seen into the future, they would have done things a little differently from the start. Although most leasing situations never wind up causing this much hassle, it is a good idea to develop a clear contract and make sure everyone is on the same page from the beginning.

One of the biggest questions facing sugarmakers is how to compensate landowners for the rights to tap their trees. Although this should go without saying, be sure to ask permission before tapping any trees and always offer fair compensation for this privilege. A story was recently picked up by the AP highlighting the growing trend of people tapping trees on other people's land in Maine without having ever asked for permission to do so. The article describes the people as "sap thieves" and explains how the tapholes devalue the commercial value of the sawlogs. Obviously this story isn't good for the maple industry; under no circumstances should you ever tap trees on another person's land without first asking permission.

Although the payment offered to landowners can vary tremendously depending on individual situations and regions, maple producers usually give some type of cash payment and/or syrup in exchange for the tapping rights. The most common lease payment that I hear of is 50 cents per tap per year, though there are certainly exceptions to this "rule." When only a small number of trees are involved or the trees are being leased from friends or family, the payments tend to be very low or nonexistent. On the other hand, if the woodlot being leased is large, the trees are high quality, and there are several sugarmakers competing to acquire a tapping lease, the price can go to $1 a tap or higher.

Many landowners prefer to receive syrup instead of cash for the rights to tap the trees on their property, especially if they don't need the money to begin with. They like the fact that the syrup "comes from their trees" and that they can use it as gifts for friends, family, and business associates. To determine how much syrup to provide a landowner, you should first figure out what you would be willing to pay per tap in cash. Based on this figure and the number of taps you are installing, you can figure out the total amount of money you would have to pay each year and correlate that to the value of the syrup. For example, imagine if Fred taps 200 trees on Joann's property and would be willing to pay 50 cents per tap for a total payment of $100. However, Joann would rather have syrup instead of being paid the $100 lease fee. If Joann would normally have to buy syrup at $50 a gallon, then Fred can simply provide her with 2 gallons of syrup instead of the $100 lease fee.

Contract Provisions

Although very few people actually do this, I highly recommend developing a written contract when entering into lease agreements. I have heard of many instances where misunderstandings and poor communication have led to tapping leases falling apart. The last thing you want to do is spend a lot of time and money installing a tubing system on someone else's property only to have to remove it after a short time. Without a solid contract allowing you to tap a property for a predetermined period, several things could happen:

1. The owners could sell the property to new owners who don't want you tapping anymore.
2. The owners could decide to have the trees harvested because they want to receive more money immediately.
3. The owners might not have realized that they were going to have to listen to a noisy vacuum pump and that the tubing would be up year-round, so they want it taken away.

These are only three of the many possible unforeseen events that could cause a lease agreement to turn sour. Discussing all the issues up front and putting the guidelines and expectations on paper will help ensure that there are no surprises in the future. Keeping this in

FIGURE 8.6. The tubing in this sugarbush was installed to purposefully bypass the large, veneer-quality sugar maple on the right, as per the lease agreement. PHOTO BY NANCIE BATTAGLIA

mind, the following items should all be included in any type of lease agreement:

Lease Payments

The contract should clearly indicate the amount of money (or syrup) that will be given to the landowner for each tap that is placed on an annual basis. It should also include a date (or dates) by which it should be paid and if there are penalties for late payments. Sometimes a down payment is made before tapping begins with the remainder given at the end of the season.

Tapping Guidelines

Include clear language on the number of taps that can be installed in a tree based on its size. For instance, you may just want to adhere to the "conservative tapping guidelines" that suggest one tap for a tree 10" to 17" and two taps on trees that are 18" or larger. This section can also identify the size of the spout that must be used—for example, 5⁄16" versus 7⁄16".

FIGURE 8.7. Lease contracts should clearly indicate a date by which taps should be removed, preferably before the growing season so that the tree can heal over the taphole wound. Landowners should not see taps still in the trees by the time summer arrives.

Allowable Trees

Some landowners may specify that veneer-quality trees should not be tapped to maintain their high commercial value; these trees (or any other trees that a landowner does not want tapped) should be clearly marked before any tubing is installed or taps are set.

Date of Spout Removal

Taps should be removed before leaves appear so that the trees can heal over the taphole wound; some contracts will include a fine (say, 50 cents per tap) for each spout that is not removed by a certain date.

Use of Chemicals

Landowners may require written notification of any chemicals that will be used on the land as part of the sap collection operation. If a landowner does not want certain chemicals used in tubing cleaning or for other purposes, this should be clearly delineated beforehand.

Tree Cutting/Removal

If you would like permission to cut trees, there should be clear guidelines about how tree selection and permission is granted. Contracts may stipulate that all trees to be cut must first be marked and inspected by the landowner. Exceptions can be made for hazardous trees or those that have fallen across roads or tubing lines.

Performance Bond

A landowner may request that the producer provide a sum of money to be held in escrow should the producer cause any damages to structures, roads, trees, et cetera.

Road Crossings

Tubing systems often wind up crossing established roads and trails within a sugarbush. Some contracts stipulate that roads stay open year-round by placing mainlines in culverts under the roads or locating them high enough so that vehicles can pass underneath. If the tubing system is designed such that mainlines cut off vehicular access on established roads, the contract may provide dates at which the tubing lines can be connected and when they must be disconnected (these dates should coincide with the beginning and end of sugaring season).

FIGURE 8.8. This mainline crosses underneath a road through a culvert, ensuring that the sap collection never interferes with vehicles.

FIGURE 8.9. This sap ladder is used to bring sap over the road so that vehicles can get through during sugaring season. This type of road crossing is necessary when the topography does not allow the mainline to cross under a culvert.

FIGURE 8.10. This mainline is attached to a ratchet and union so that it can be easily removed during the off-season. Use this type of system for roads that do not need access during sugaring season but should be open to vehicles during the rest of the year.

Insurance and Liability

To protect the landowner, most contracts provide a clause mandating that the person leasing the sugarbush provide a certificate of general liability insurance of $1 to $2 million naming the landowner as an additional insured. They may also stipulate that the person leasing the sugarbush carry workers' compensation insurance for their employees.

Ownership Change

Most contracts contain a provision that if the ownership of the property changes hands, then the provisions of the lease agreement are transferred to the new owner.

As you can see, there are many provisions that should be addressed between a maple producer and landowner before entering a lease agreement. A sample lease agreement is available on the Cornell Maple Program website. I recommend downloading the Word document and modifying it to suit the needs of your particular situation.

Finding More Sap

If after reading this section you would like to look for more trees to tap or more sap to buy, there are several strategies that you could use to grow your business in this manner. The rest of this chapter discusses options for reaching out to others in your community. I encourage you to employ some of these strategies or come up with your own creative ways to get the word out that you would like more trees to tap and/or more sap to purchase.

Snail Mail

One option is sending a letter to woodland owners in nearby towns asking them if they have maple trees on their property that they would be willing to let you tap. The letter should introduce yourself, describe the nature of your sugaring operation, and explain why you are looking for more trees to tap. It should clearly outline what's in it for them—annual cash payment and/or syrup right from their property, property tax breaks (in New York, Wisconsin, Minnesota, and possibly other states), someone to look after their woodlot if they are an absentee landowner, access roads put on the property (if desired), and so on. Most people will be skeptical when receiving an unsolicited request, so you have to be as persuasive as possible without coming across as too pushy.

With printing, paper, and postage, you should figure on spending approximately 60 cents in material costs for each letter sent. Depending on your municipality, getting the contact information and mailing addresses for these landowners can be time consuming and difficult. If you are computer-savvy and happen to live in a place with tax maps online, modern technology can make finding potential sugarbushes and their landowners much easier than it used to be. By using an aerial photo as the background layer of web-based GIS (Geographic Information System) tax maps, you can tell what percentage of a parcel is forested and whether it contains mostly hardwood or softwood trees. This can help you focus on just the properties that appear most promising, which is important given the fact that sending letters can be so time consuming and expensive. Although 60 cents is not a lot of money for a single letter, based on the sheer number of landowners in any given town, these costs can add up quickly.

To see how this works, consider the following example of a town in Essex County, New York. This county has all its tax parcels online; you can easily go to the website to find detailed information on any parcel.

FIGURE 8.11. Some counties, such as Essex County in New York, have all their tax maps online through a user-friendly GIS system. The identify button shows contact info and other pertinent information for each parcel.

If you were to simply send a letter to all landowners, you'd waste a lot of time and money on landowners who do not have any trees to tap. As you can see in Figure 8.11, there are many parcels of land along this road, yet most of them do not contain any maple trees. With infrared aerial photography, the grayish brown colors indicate deciduous trees (not necessarily maples), whereas the evergreen stands appear as dark red. Farm fields are clearly visible in an either green or pinkish hue, depending on the state of tillage and time of year the photo was taken. All waterways show up as black, as clearly seen with the pond and stream in this example.

I have inserted numbers on many of the parcels to describe how to read the map and utilize the GIS mapping system. Parcels 4, 7, 8, 12, 13, and 14 are dominated by evergreens and probably do not have many (if any) maple trees growing on them. Parcel 2 is mostly farmland with a homesite and frontage on the pond. Parcel 10 may contain a small number of maple trees, but they are located far off the main road and it is probably not worth the effort to try tapping them. The most promising parcels are 5 and 6. Parcel 5 has two potential sugarbushes on either side of the farm field. Furthermore, the stand of hardwoods in the southwesterly portion of parcel 5 backs up to an even larger deciduous forest without road frontage on parcel 6. If it turns out that many of these hardwoods are maples and you could convince both landowners to lease their land, this would provide an excellent tapping opportunity.

If your county does not have the tax maps in an online database, you can still cruise around online looking at aerial photos. It will certainly be harder to find out who owns the nice sugarbushes, but looking at the aerial photos online will provide you with a good idea of where the best possibilities exist. Yahoo Maps, Google Maps, Mapquest, and other sites have excellent aerial photography that shows where the potential

FIGURE 8.12. The parts of a hillside dominated by maple are easy to spot during "leaf-peeping" season. This photo was taken in the Adirondacks in early October.

FIGURE 8.13. Approximately every four years, sugar maples will produce a bumper seed crop, which starts with a huge display of flowers before the leaves of other trees appear. This feature makes identifying potential sugarbushes from afar very easy. PHOTO COURTESY OF BRIAN CHABOT

FIGURE 8.14. An up-close view of the yellowish green flowers that make identifying sugar maples in the spring easy from a distance. PHOTO COURTESY OF BRIAN CHABOT

sugarbushes are. You can only see so much from a road, whereas the aerial photos allow you to see everywhere.

Finally, if you are not computer-savvy, don't have access to the Internet, or would rather search in real life, you can always use your eyes! The best time to look for potential sugarbushes is the autumn or spring, as different species put out and drop their leaves at different times. During seed years for sugar maple, the yellowish green flowers open up before the leaves, often the first signs of spring in an otherwise grayish brown hillside. In the fall, the yellows and reds of sugar maples start to appear while oaks are still green, allowing you to distinguish between maple or oak-dominated hillsides from afar. Consequently, there is usually a two week period in which all the maple leaves have fallen and the oaks are still hanging on.

Put a Sign on Your Property

If your sugarhouse is on a well-traveled road, you could post a sign out by the road advertising that you are looking for more trees to tap and/or sap to purchase. Keep the sign simple and to the point—the lettering needs to be large enough so that someone can easily read it when driving past at 45 miles an hour. You just need to have enough information to pique their interest. I recommend putting your phone number so people can easily contact you for more information.

Craigslist

Posting an ad on Craigslist is free and will reach a large and growing segment of the population that doesn't read newspapers. Because the space is free, you can go into more detail on the specifics of what you are looking for. To reach people about leasing, you may want to list the towns you would like to work in, the lease payment you would offer the landowner, and other benefits of leasing such as agricultural assessment, maple syrup from their own property, someone to look after their woods, and so on. For buying sap, you could include information about the percentage of the syrup the sap seller would receive and/or the typical prices paid per gallon of sap delivered. A good advertisement may read as follows . . . "Make fresh maple syrup from your property with a lot less time

FIGURE 8.15. A simple sign by the road is a great way to let people know you are interested in buying sap. You may also want to include your phone number so that people can easily get in touch with you. PHOTO COURTESY OF PETER GREGG/MAPLE NEWS

and money invested—you tap the trees, bring us your sap, and we give you ½ of the syrup that the sap would produce." Framing it this way plays to the desires of many people who want to tap trees and make "their own" syrup, but do not have the time or money to invest in a full-scale sugaring operation.

Newspaper Advertisement

Not everyone has Internet access, and there are many rural landowners who are more likely to read a newspaper than surf the web. If you are willing to invest in a newspaper advertisement, this will help reach the people who are not online, though it could cost you a decent amount of money to post an ad large enough to catch people's attention. I would only do this after exhausting all other options.

Free Press Coverage

Rather than placing a newspaper advertisement, one of the best ways to get your message out to the community is through a well-written press release. The public is generally interested in maple syrup, and reporters are usually looking for news stories about maple sugaring and how the season is going. Once you have started leasing a property or buying sap from at least one person, notify your local media about this connection. It could lead to a great story about how you are working with others to help grow your business and that you are looking for more partnerships. Another great angle would be if you helped save a property from logging and/or development by tapping the trees for syrup production. Not only are these types of media stories free, they will yield much better success than any paid advertisement.

Stickers on Your Sap Buckets

If you are looking to hang buckets, there are probably more trees available along roadsides than you have time to collect. It's just a matter of driving, biking, jogging, or walking around your neighborhood and town to see where the trees are. Then you must approach each landowner and ask permission to tap their trees. I suggest developing a simple permission form; no need for a lengthy contract. By hanging buckets in prominent locations and placing attractive and descriptive stickers on them, you will get lots of free advertising. Several people who have seen our buckets have contacted me to buy our maple syrup, gather sap themselves for us to boil, and offer trees on their own property for us to tap.

FIGURE 8.16. We affixed stickers to all the sap buckets in Lake Placid as a means of bringing awareness and visibility to the maple syrup project at the Shipman Youth Center. PHOTO COURTESY OF SHARON BURSTEIN

Word of Mouth

This can be the easiest and most fruitful method to grow your business. Simply spread the word that you are looking for woodland to lease or more sap to boil, and let your friends, family, and neighbors do the legwork. Taking advantage of the existing social networks is a great strategy to make these types of connections.

Sell Maple Sugaring Equipment

Becoming a dealer in maple sugaring equipment, in particular materials used to collect sap, will yield multiple benefits. The number of maple sugaring equipment suppliers is limited or nonexistent in many parts of the country, so this will provide you with great exposure to the other current and potential maple producers in your region. You will become one of the "go-to" contacts for folks who are interesting in getting into syrup production. The more connected you are, the more likely you are to hear about sap-purchasing and sugarbush-leasing opportunities. Furthermore, having the necessary equipment available for sale could provide the impetus for more people to start collecting sap (that they may sell to you). Finally, selling equipment can also yield some supplemental income, as equipment dealers are able to purchase supplies at wholesale prices and mark items up for retail sale. The four main companies that have dealer networks in the United States and Canada are CDL, Dominion & Grimm, Lapierre, and Leader. A number of smaller companies also have dealer representatives.

Final Thoughts: The Central Evaporator Plant

The idea of gathering the sap from many places to be processed at one central facility is not a new one! The first maple syrup producers manual published in 1958 contained an entire section on the benefits of a central evaporator plant, encouraging maple producers to join forces to process their sap jointly at one efficient location.[1] In 1962 Jerome Pasto and Reed Taylor from Penn State published a lengthy bulletin on the economics of the central evaporator plant.[2] They outlined all the costs and revenues associated with purchasing sap from a large number of farmers and boiling it at one facility. Comparing three different-sized plants, they found that larger factories had lower processing costs and were therefore more profitable. This should not come as a surprise to anyone with a basic understanding of economics and the efficiencies of scale. Central evaporation facilities made sense back in the 1960s, and the differences in processing costs have only become greater over time.

Modern ROs and evaporators are able to process sap for a fraction of the cost of smaller evaporators that many small-scale sugarmakers utilize. As long as the amount of fuel used to transfer the sap from the sugarbush to a central processing facility is significantly less than the additional fuel that would be used to process the sap with less efficient equipment, then it makes sense economically and environmentally to do so. The only downside is that it can take away from the romantic ideals of sugaring as a small, quaint, rural homestead experience. In theory, central evaporation plants are more like factories than a traditional sugarhouse. Although they may be much more efficient, the experience of making syrup in such a facility is vastly different from boiling sap in the old sugarshack in the woods. You can certainly develop a large-scale sugaring operation with efficient equipment that still has the look and feel of a traditional sugarhouse; you just need to decide what model works best for you.

Although there are certainly great opportunities to develop large central sugarhouses in the structure of a co-op, these types of arrangements are difficult to manage and prone to failure. However, I strongly believe that entrepreneurial sugarmakers could (and should) develop their own central evaporation facilities by employing the main strategies discussed in this chapter. By paying a high enough price for sap to make it worthwhile for people to collect it, you can encourage nearby landowners to forgo producing syrup and just sell you their sap. Similarly, by providing other landowners with a high enough lease payment, you will be able to entice more people to let you tap their woodland. There are tremendous economies of scale in maple production, and by getting others to join in on your sugaring operation through buying their sap or leasing their woodland, you will be able to develop a very profitable enterprise while simultaneously benefiting others in your community. These types of collaborative arrangements can be a win–win for everyone involved. It just takes the initiative to reach out to others rather than trying to do everything yourself on your own property.

CHAPTER 9

Community Sugaring:
Growing Your Operation with Community Involvement

If you like to spend time by yourself out in the woods, maple sugaring can be the perfect activity to get away from it all. The only sounds you may hear are the ping of sap dripping in an empty bucket, the crackle of the fire in the evaporator, or the bubbling of sap as it transforms into syrup. This can be one of the most relaxing activities, especially if you enjoy quiet time by yourself in a peaceful setting. On the other hand, maple sugaring can also be a very social experience that brings all sorts of people together around a common activity. Sugarhouses often serve as the social nexus of a family or community during that time of the year, and making syrup serves as the perfect excuse to get together. If you have more of an extroverted personality, then you may want to structure your sugaring operation to get others engaged and involved in the process. Whereas chapter 8 deals with buying sap and leasing taps, this chapter explores additional strategies and ideas for involving the community in your sugaring operation. If the first scenario sounds ideal to you, then chances are the rest of this chapter probably won't be so appealing. However, if you enjoy meeting and working with new people, and educating others about the maple industry, connecting your sugaring operation with your local community could be one of the best decisions you make.

This chapter highlights several sugaring operations that have strong ties with their local community. While there are certainly more community-based sugaring operations than the ones outlined below, these were selected to provide a broad overview of what is possible. With the exception of the operation in Lake Placid, I discovered the others through my research for this book. For several years I've had Google alerts sent to my email for "maple" "maple syrup" and "maple trees," so when articles appear about various things having to do with maple sugaring, I get notifications sent to my inbox on a daily basis (I also get a lot of unrelated news that just happens to have *maple* in the headline—a fire on Maple Avenue, and scores from Toronto Maple Leafs games). After sifting through all the relevant stories, it is quite obvious that the media is drawn to community-based sugaring operations. Although reporters will do stories on regular sugarhouses, the ones that involve their community—and especially children—get the most media attention.

As you read about the different operations in this chapter, think about how you could incorporate similar ideas into your own operation. I'm sure many of the programs described here would not be applicable to your sugaring business, but there are bound to be some programs and ideas that could work well for you. You'll need to assess your own strengths and weaknesses to determine what opportunities make the most sense for your operation.

Ambler Farm, Wilton, Connecticut

In 1999 the Town of Wilton purchased a historic 22-acre farm that operates under the direction of a nonprofit group called the Friends of Ambler Farm. Their mission is "to celebrate the community's agrarian roots through active learning programs, sustainable agriculture, responsible land stewardship, and historic preservation." In 2006 Ambler Farm hired Kevin Meehan on a part-time basis to live on and manage the farm and conduct educational activities. After a year spent researching maple syrup production and visiting other local sugarmakers, Kevin made a proposal to the board to start a maple sugaring program at the farm. Since he already had a full-time job as a schoolteacher, the time that he could devote to sugaring was limited, so he decided to get the community engaged as much as possible. By having volunteers do much of the sap collecting, Kevin was able to limit his own hours while simultaneously educating others about sugaring. He started an adopt-a-tree program in which families pay $65 for the opportunity to come out to the farm at least twice a week during sugaring season to collect sap from "their own bucket." Families get to choose their favorite tree and are responsible for gathering all the sap that drips into it over the course of the season. At the end they receive an 8-ounce bottle of syrup that they helped produce. The program has grown from 58 families in the first year to now being maxed out at 100 families with a waiting list—the only restriction to getting more people involved is that there just aren't any more tappable trees located at the farm. In fact, other families who aren't official members still come out on Sunday afternoon to help gather sap from any buckets that haven't been emptied by 3 pm.

In addition to the adopt-a-tree program at the farm, Kevin has branched out his tapping operations to include 12 different locations, including all the schools in the town. He usually grabs a few volunteers and drives around collecting sap with a 250-gallon tank in the back of his pickup truck—aka the "sap shuttle." He helped start a maple syrup club at the middle school, and those students help tap and collect sap from 55 trees on the school grounds. One of his best locations is a local private school, where he has 40 taps in large, beautiful sugar maples. As a "lease payment," the school receives 2 gallons of syrup, which they use in their dining hall for special events. There are also a lot of enthusiastic kids who love maple syrup—they aren't old enough to help yet, but they do get to watch the activities as future sugarmakers in training. All the buckets on the most visible trees throughout town have AMBLER FARM painted prominently on them, providing a lot of free publicity.

FIGURE 9.1. Kevin Meehan of Ambler Farm explains the tapping and gathering process to families who signed up for the adopt-a-tree program. PHOTO COURTESY OF EVE DONOVAN

Ambler Farm's sugaring operation has grown to the point where the original 2 × 6 evaporator cannot handle all the sap being collected, so they have recently purchased an RO to reduce their boiling time and the amount of firewood they need to cut. Ambler Farm also wanted to reduce its carbon footprint and knew that the most energy-efficient and environmentally friendly way to process sap is through an RO. Another benefit of the RO is that they can adjust their concentration rates based on how much sap they have to process. When the sap flow has been abundant, the RO will run steadily and remove about 75 percent of the water before boiling. However, if sap flow has been minimal and they want people to be able to see the evaporator running, they can simply bypass the RO and boil raw sap. Ambler Farm usually boils sap on the weekends or when there are school groups or others visiting during the week. Occasionally they boil on weekdays when there is a lot of sap flow, but that is the exception and not the norm. Kevin's daughter Kate helps run the evaporator along with many other farm apprentices from local schools (5th through 11th graders). These students like to spend their weekends hanging out in the sugarhouse, and Ambler Farm offers a productive, safe, and enjoyable place for the teenagers to get together.

The Ambler Farm model has been extremely successful. From the adopt-a tree program and sales of maple syrup ($15 for an 8-ounce bottle), Ambler Farm was able to cover all the initial start-up costs in the first year. It usually takes several years before a sugaring operation is able to become profitable, yet due to the nature of their program, they achieved this goal right away. In addition, one of the families was so thrilled with the program that they made a substantial donation that helped pay for another venture into beekeeping. The year 2011 was a great one for sugaring in the Northeast and the best season ever at Ambler Farm—they produced 950 bottles and were completely sold out by August. As part of their agreement with the town, Ambler Farm is only allowed to sell products grown and/or made on the farm. If they could buy in bulk syrup to repackage, their sales would be significantly greater.

Kevin feels that engaging with the community members is the most critical aspect of their sugaring operation. He wouldn't have such great participation and repeat members if people just went out there to

FIGURE 9.2. Collecting buckets with the maple syrup club at a local elementary school. PHOTO COURTESY OF ASHLEY KINEON

collect the sap and then left. Those who sign up for the program want to be a part of the process and feel connected to the people in charge and the other members. Kevin makes it a point to always reach out and talk with the members whenever he gets a chance. The reason Ambler Farm's program is so successful is that Kevin strives to create a sense of community that makes people want to stay involved. Kevin explains that "the farm is the draw and collecting sap is a good excuse for getting people out there—what keeps folks coming back is the connection to other families and the people running the program." People could easily go to the supermarket and get a lot more maple syrup for a lot less money and effort. What Ambler Farm is selling is not maple syrup, but the experiences that tie people to the land and one another. There are many rural parts of the Northeast where the Ambler Farm model wouldn't work, but there are also thousands of towns and communities that would be thrilled to have this type of program. Give careful thought to your own sugaring operation and nearby communities to determine if an adopt-a-tree program (or a variation on this theme) is right for you.

Groundwork Somerville, Somerville, Massachusetts

Every winter, through a partnership with the Somerville Community Growing Center, Somerville Public Schools, and Tufts University, Groundwork Somerville makes maple syrup in an urban environment just outside Boston. The Maple Syrup Project existed for several years before Groundwork Somerville took over. It had always been managed by interns, students, and volunteers, yet there was little continuity of staff over the years and no organizational framework to ensure it kept going. In 2009 Groundwork Somerville took the reins under the direction of their Gardens Coordinator at the time, Tai Dinnan. In addition to actually collecting sap and producing syrup, Groundwork Somerville staff and volunteers also teach a four-week arts and science curriculum based on maple syrup production to second graders at several public schools and libraries. The final event is a two-day sugaring festival held at the Somerville Community Growing Center, where all the sap is boiled down into syrup.

The Groundwork Somerville sugaring operation is certainly not large-scale. Depending on the weather, the sugaring season usually kicks off in late January with a tree-tapping event on Tufts University campus, where nearly all the trees used for tapping are located. The tapping is scheduled for a warm day, as it is always nice for people to actually see the sap running when they drill a hole in the tree. They save all the sap they collect from 20 taps over the course of the season to use for the Boil Down Festival in early March. The sap is stored in 5-gallon buckets that are kept in large freezers and coolers throughout various Somerville schools and businesses. They have a 2.5 × 3.5 wood-fired evaporator that was fabricated by metal shop students at Somerville High. All the syrup produced is put in small glass bottles and either given as thank-you gifts to community partners or sold at several farmers' markets. The 1.7-ounce bottles sell for $6, so it is definitely a premium product!

Tai states that the most profound impact of their maple sugaring project is the educational value it provides for schoolchildren and members of the community. Most people have never thought about trees providing food, especially in an urban environment, so it's an eye-opening experience for those who take part in the sugaring process. Groundwork Somerville recruits and trains many volunteers each year to be maple syrup ambassadors in the local schools. Working in teams of two, the volunteers visit up to 20 classrooms of 2nd graders in Somerville. They provide four sessions that are each an hour long and deal with various aspects of syrup production. The two-day Boil Down Festival also serves as the capstone activity, drawing in roughly 1,000 people each year. Three hundred students from local schools cap off their monthlong educational program with a visit to the festival on Friday, and it is opened up to the general public on Saturday. There are many events for the students and public to participate in, including a maple Jeopardy challenge and songs about maple sugaring.

FIGURE 9.3. Groundwork Somerville staff and community members tap maple trees on the Tufts University campus as part of their urban Maple Syrup Project. PHOTO COURTESY OF TAI DINNAN

In 2013 Groundwork Somerville expanded its reach by partnering with NOFA Mass to offer maple syrup programming to adults. Given the rising interest in urban agriculture and homesteading, there are many community members curious about continuing this old New England tradition in their own neighborhoods. As with many sugaring operations, the Groundwork Somerville community maple project continues to grow and expand in new and exciting directions.

Shipman Youth Center Maple Syrup Project, Lake Placid, New York

In 2008 I was asked to become a board member for the Shipman Youth Center (SYC) in Lake Placid. I

FIGURE 9.4. All the sap is refrigerated or frozen and saved for the two-day Boil Down Festival at the end of the collection season. PHOTO COURTESY OF TAI DINNAN

FIGURE 9.5. Dmitry Feld, Board President of the Shipman Youth Center, helps Lake Placid H.S. students Casey DiNicola (left) and Chris Kordziel (right) gather sap by the village beachhouse. PHOTO BY NANCIE BATTAGLIA

FIGURE 9.6. A mobile sugarhouse was originally parked at the Golden Arrow Resort on Main Street and served as a great marketing tool for pure maple. Here Anthony Kordziel and Jon Fremante unload 5 gallon buckets of sap in to the holding tank. PHOTO COURTESY OF SHARON BURSTEIN

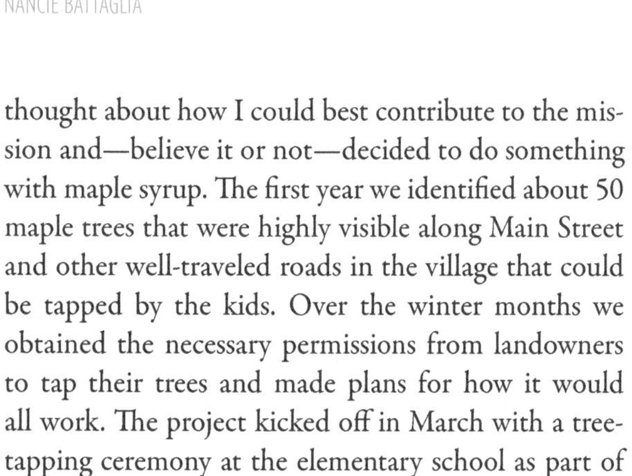

thought about how I could best contribute to the mission and—believe it or not—decided to do something with maple syrup. The first year we identified about 50 maple trees that were highly visible along Main Street and other well-traveled roads in the village that could be tapped by the kids. Over the winter months we obtained the necessary permissions from landowners to tap their trees and made plans for how it would all work. The project kicked off in March with a tree-tapping ceremony at the elementary school as part of a pancake breakfast. Jamie Rogers, a local sugarmaker who was mayor at the time, led the ceremony and read a proclamation declaring Lake Placid a "maple city."

For the first two years all the sap was collected by the SYC and transported to the Cornell sugarhouse just a couple of miles from the village center. The kids, accompanied by a staff member, unloaded the sap into our large collection tanks on a regular basis. We kept track of how much sap the students collected and gave them back the equivalent amount of syrup. At the end of the season, we hosted a community-wide pancake breakfast at the youth center, featuring the freshly made syrup that the kids were instrumental in helping to produce.

Although the project was moderately successful in the first two years, I knew it could be much bigger. For the third year, one of the board members who has his own construction business built a mobile sugarhouse on a trailer frame. Instead of having to transport the sap out to the Cornell sugarhouse, the students could simply collect the buckets along Main Street and boil the sap right in the village. After we had many difficulties in finding an appropriate place to park the sugarhouse, the Golden Arrow Resort offered part of their parking lot in front of their hotel on Main Street for the six-week time frame. Both residents of and visitors to Lake Placid were surprised and impressed to see an actual sugarhouse making pure maple syrup right in the village. We sold a wide variety of maple products out of the sugarhouse, made a lot of sugar-on-snow, and generated additional funds for the SYC.

The spin-offs from this project have been remarkable. Seven other families in Lake Placid have started to collect sap from their trees and bring the sap to us in exchange for some syrup back. Two private schools (National Sports Academy and Mountain Lake Academy) also decided to tap trees on or near their campuses to have us process it into syrup. Upon hearing of this project and seeing the kids in action, several residents decided to take up syrup production as a hobby themselves. It has also helped our sales at the sugarhouse. Many residents and tourists have told me they said they saw the buckets hanging on trees in town and knew it must be sugaring season, so they came out to see the process and purchase some fresh syrup.

The project got a big boost in 2010 when a local couple, Randy and Sibyl Quayle, found out about the

FIGURE 9.7. Randy and Sibyl Quayle of Red Fox Maple decided to start a maple operation on their property to support the Shipman Youth Center. PHOTO COURTESY OF DAN FOXX

SYC project and decided to get involved. They offered to tap their 15-acre woodlot and have the proceeds benefit the youth center. That fall, a team of volunteers from the SYC, Cornell, and the Quayles installed approximately 800 taps on vacuum. All the sap is hauled to the Cornell sugarhouse to be processed into syrup. The SYC received a few thousand dollars from the syrup proceeds for each of the first two years, and now that the Quayles have recovered their infrastructure investment, all the profits go to support the youth center. The generosity of the Quayles has sparked interest among other landowners in doing the same—in fact, the youth center now has more offers for land to tap than the labor and time to gather the sap.

The greatest benefit from this project has been the impact it has had on the kids who have gotten involved. One student in particular, Anthony Kordziel, liked sugaring so much that he decided to become a "maple sugar apprentice" with me for his senior project at the high school. He put in more hours than any other student ever has on a senior project, and it certainly showed. During his final presentation, I watched two teachers with tears in their eyes listening to Anthony talk about his accomplishments over the course of the spring semester. They commented about how rewarding and encouraging it is to see a young person become so passionate and enthusiastic about something. Anthony is now studying forestry and natural

Generations Restaurant

When the mobile sugarhouse was moved to the nearby Wild Center in Tupper Lake, the Shipman Youth Center was looking for another partner in their efforts. Boiling the sap on Main Street provided a lot more exposure and was more convenient than always bringing the sap out to our sugarhouse on the outskirts of town. They found a willing partner in David Hunt, the executive chef for Generations Restaurant, which is a part of the Golden Arrow Resort. They have a prime location in the center of town with a covered porch about 10 feet away from all the foot traffic on Main Street. Chef Hunt purchased a small 2 × 4 evaporator from Next Generation Maple Products in Syracuse and spent many weekends boiling sap and educating tourists about the sugaring process. The Shipman Youth Center now sells their syrup to Generations, and the evaporator on Main Street draws in a lot of extra business for the restaurant when it is in operation.

FIGURE 9.8. Anthony Kordziel and Chef Dave Hunt from Generations make sugar-on-snow by pouring hot syrup on a cup of crushed ice. PHOTO COURTESY OF JACK LADUKE

resources at the State University of New York College of Environmental Science & Forestry in Syracuse. His goal is to be working in the woods and have his own sugaring operation someday. Certainly not everyone who is exposed to sugaring winds up getting as involved as Anthony, but if you provide the opportunities, you may be surprised at what happens. In fact, Anthony's younger brother Chris is now following in his footsteps and keeping the project going strong.

The Wild Center, Tupper Lake, New York

Another major spin-off of the Shipman Youth Center project was the development of a community-based maple program at The Wild Center (The Natural History Museum of the Adirondacks) in nearby Tupper Lake. Their executive director, Stephanie Ratcliffe, saw the sugarhouse parked in front of the Golden Arrow on Main Street and got excited about the possibilities of doing something similar at The Wild Center. They wanted to start their own community sugaring operation, modeled after the SYC project in Lake Placid, but they didn't want to invest in building a new sugarhouse. In 2012 they rented the SYC mobile sugarhouse for the inaugural season. Even though the sap flow was dismal in the Adirondacks that year, they still had a very successful season. The Wild Center didn't make much syrup but did generate a tremendous amount of community interest and support. In community-based sugaring operations, success is a matter of perspective.

The Wild Center was built on sandy, abandoned farmland that had grown up primarily to pine trees. Since they have only a single tappable sugar maple on the entire 32-acre property, to develop a sugaring operation, they *had* to get the community involved. The Wild Center hosted two workshops during February and March that focused on their community-based program and maple sugaring in general. The programs were extremely successful, drawing in over 75 people from their immediate community and Greater Adirondack region. The Wild Center purchased 100 of the traditional, galvanized sap buckets, painted the attractive Wild Center logo on each one, and lent them out to nearby residents and schools to tap the trees on their land. Wild Center staff collected the sap on a regular basis and brought it back to their mobile sugarhouse located near the entrance of the museum. People came to watch the sap being turned into syrup, and each participant was given a pint of syrup for their efforts.

Building on their successes in the first year, The Wild Center applied for a grant from the Northern Borders Regional Commission to further develop the project throughout the region. The grant enabled them to build their own sugarhouse and equip it with energy-efficient processing equipment, including a 200 gph reverse osmosis and wood-fired evaporator.

FIGURE 9.9. The Wild Center painted its logo on 100 sap buckets that it provided to residents and schools in Tupper Lake. PHOTO COURTESY OF RICK GODIN

They hired two interns to work on maple programming in the Tupper Lake community and beyond. With grant funds, The Wild Center put on additional workshops at their facility and sponsored four maple schools in neighboring counties. Additional funds were secured to help the Paul Smith's College Visitor Interpretive Center (VIC) develop a demonstration sugaring operation. The VIC also put in a 2 × 4 evaporator and 200 gph reverse osmosis to process sap from trees located alongside their extensive cross-country ski trail system. They will also be processing the sap that nearby residents bring from the trees on their own properties. With the addition of the VIC project, community-based sugaring operations are expanding rapidly in the Adirondacks.

FIGURE 9.11. In 2013 The Wild Center constructed its own mobile sugarhouse with grant funds provided by the Northern Borders Regional Commission. Dave St. Onge explains the sugaring process to visitors. PHOTO COURTESY OF RICK GODIN

FIGURE 9.10. The Wild Center has also been able to generate some revenue by selling their attractive bottles of maple syrup in the gift shop and offering "spun maple delight" (their name for maple cotton candy) at special events. PHOTO COURTESY OF RICK GODIN

FIGURE 9.12. A postcard scene of the Burton Log Cabin and Sugar Camp conveniently situated on the village green.

"Pancake Town USA," Burton, Ohio

In 1931 the Burton Chamber of Commerce built a log cabin sugarhouse on the village green as "a unique publicity scheme" to get people to visit Burton and see how maple syrup is made. The original sugarhouse was modeled after the log cabin where Abraham Lincoln was born in Kentucky. As with most sugarhouses, they quickly outgrew it and put on five additions over the next 30 years. In the 1950s they started hosting large pancake breakfasts that became so popular, the local Rotary Club trademarked Burton as "Pancake Town USA." There are now four pancake breakfasts held simultaneously every weekend during March at separate locations. Each venue is sponsored by a different group and offers a slightly different breakfast, yet they all feature locally produced pure maple syrup. The Chamber has kept detailed records over the years and estimates that since the program started, over 500,000 people have come to Burton for pancake breakfasts, consuming roughly 25,000 gallons of syrup poured over nearly two million pancakes.

Every February the maple season kicks off with a tree-tapping ceremony on the village green. The production of maple syrup is carried out by an all-volunteer crew headed up by Amy and Mike Blair. Under their leadership, over 20 volunteers help to gather sap from approximately 2,000 taps spread out over multiple locations within a few miles of the Burton Log Cabin and Sugar Camp. Outside of maple season, the sugarhouse also serves as the chamber of commerce offices and visitor information center for Burton and surrounding communities. They keep their retail shop open year-round and make a lot of maple candy, cream, and sugar to sell to visitors. As a nonprofit organization, they use the proceeds to fund scholarships for two local high school students to attend college. They also supply all the holiday decorations during the winter months and flowers in summertime. Since they are able to sell more maple products than they can produce themselves, they also purchase syrup from nearby Geauga County sugarmakers. This innovative concept started by the Burton Chamber back in the 1930s continues to provide tremendous benefits for the community today. Given the successes that Burton has had in being "Pancake Town USA," think about how you could work with your local chamber of commerce and elected officials to promote the maple industry in your town or village.

"The Sweetest Little Town Anywhere Around," Shepherd, Michigan

The maple sugaring program in Shepherd got started in 1877 when the state government offered a tax break to landowners who planted maple trees in Michigan. The purpose of this program was to celebrate the nation's centennial and beautify villages. The residents of Shepherd participated in vast numbers. Fast-forward roughly 75 years and imagine streets now lined with large, magnificent sugar maples. During the spring of 1958 the trees were pruned, and people noticed that there was sap dripping all over the streets and sidewalks. Some folks got the idea of making maple syrup from the sap as a fund-raiser for recreational activities and facilities in the area. Soon after, a local attorney drafted the paperwork necessary to create the Shepherd Sugarbush Corporation as a nonprofit organization. Together with the local chamber of commerce, they held the first maple festival in 1959, creating the path for Shepherd to become the "sweetest little town anywhere around."

During the sugaring season, volunteers and students from throughout the community help gather sap and process it down into syrup. They drive a tractor with a large holding tank on a trailer and plenty of 5-gallon gathering pails in tow. There is an excellent and inspiring video of their sap collection and processing on their website www.shepherdmaplesyrupfest.org. The festival started as a one-day event serving a few hundred meals and has grown into a huge event serving

FIGURE 9.13. The large sugar maples that line the streets of Shepherd, Michigan, are tapped every year by community volunteers. PHOTO COURTESY OF BRUCE VIGNEAULT

FIGURE 9.14. How many towns do you know where a tractor and sap wagon cruises the city streets as volunteers help gather sap? PHOTO COURTESY OF JOHN MORGAN.

FIGURE 9.15. The Shepherd Maple Syrup Festival draws in thousands of people every year to celebrate the maple harvest with a wide variety of activities and events. Arnie Hammel drives their sap-hauling tractor down Main Street during the annual parade. PHOTO COURTESY OF BRUCE VIGNEAULT

11,000 pancake breakfasts over a three-day weekend. Not bad for a town of 1,500 people! Although the festival had its origins in maple sugaring, it has expanded to include many other family activities to draw people to Shepherd. Other aspects of the festival include craft shows, music and other entertainment, children's activities, museums, tractor pulls, classic car and tractor displays, parades, chain saw carving, and train and pony rides.

The proceeds fund a 10-week summer program for area youth and maintain a community park. The Shepherd Sugarbush Corporation recently purchased a plot of land where many of the events will be housed in the future. They also started a nursery of "sweet trees" that will be used for future plantings in the village as the older trees eventually die out. Many volunteers spend countless hours preparing for and managing the festival—it is truly a community effort, and everyone shares in the work and the rewards. However, none of this would have been possible without so many people planting sugar maple trees over a century ago and taking care of them over the years. If the Shepherd example isn't enough incentive to go out and plant some sugar maples, I don't know what is.

Maple Syrup Festivals

There are a number of highly successful festivals centered around maple syrup in the United States and Canada. Many of these have been around for decades and are only made possible with the continued support of a large number and wide variety of community volunteers. Although maple is the reason for the festival, they have all grown to include much more. Some festivals take place during sugaring season; others occur just after the harvest season when the weather is (usually) nicer and sugarmakers aren't as busy making syrup. I

FIGURE 9.16. Sugar maple seedlings are lined out in a nursery field until they get to a large enough size to serve as suitable planting stock for the village.
PHOTO COURTESY OF BRUCE VIGNEAULT

don't have the space available to describe all of them in great detail so I've provided links to some of the major festivals throughout the US and Canada. There are many others that are not listed here and the potential for more maple festivals throughout the country. If you would like to develop a maple festival in your town, I highly recommend checking out the websites for all these festivals and going to visit as many of them as you can. Be forewarned: While maple festivals are great events and do wonders to promote maple, organizing one is more like running a county fair than a sugarhouse! If you are up to the challenge and think you can rally others behind your cause, you may be able to get a festival started in your town.

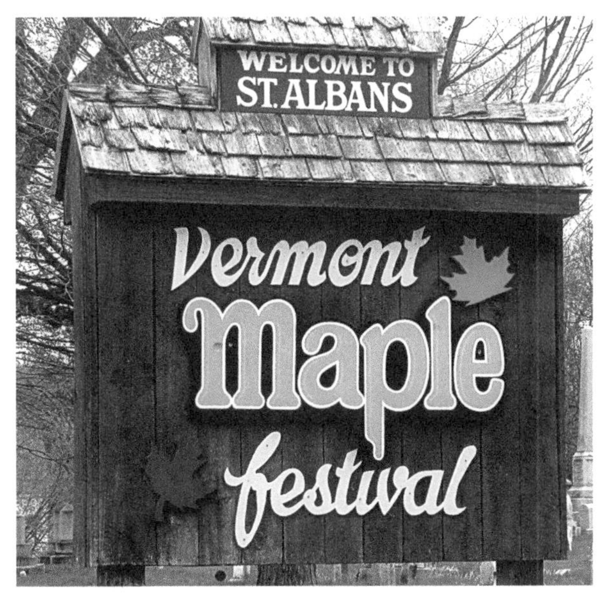

FIGURE 9.17. St. Albans is well known for its Maple Festival at the end of April every year, which coincides with the open house events at all the equipment dealers in northern Vermont.

United States

- Central New York Maple Festival, Marathon, NY: www.maplefest.org

- Hebron Maple Festival, Hebron, CT: www.hebronmaplefest.com
- The National Maple Syrup Festival, Medora, IN: www.nationalmaplesyrupfestival.com
- Vermont Maple Festival, St. Albans, VT: www.vtmaplefestival.org
- Highland Maple Festival, Highland County, VA: www.highlandcounty.org
- St. Johnsbury World Maple Festival, St. Johnsbury, VT: www.worldmaplefestival.org
- Vermontville Maple Festival, Vermontville, MI: www.vermontvillemaplesyrupfestival.org
- Pennsylvania Maple Festival, Meyersdale, PA: www.pamaplefestival.com
- Endless Mountains Maple Festival, Troy, PA: www.maplefestivalpa.com
- Geauga County Maple Festival, Chardon, OH: www.maplefestival.com

Canada

- Elmira Maple Syrup Festival, Ontario: www.elmiramaplesyrup.com
- Sunderland Maple Festival, Ontario: www.maplesyrupfestival.ca
- Festival Beauceron de L'Erable, Quebec: www.festivalbeaucerondelerable.com/en
- Plessisville Maple Festival, Quebec: http://festivaldelerable.com
- Bigleaf Maple Festival, Vancouver, British Columbia: www.bcforestdiscoverycentre.com
- St. Pierre Jolys Sugaring Off Festival (Manitoba maple), St. Pierre Jolys, Manitoba: www.museestpierrejolys.ca
- Maple in the County, Prince Edward County, Ontario: http://www.mapleinthecounty.ca
- Festival of Maples, Perth, Ontario: http://perthchamber.com/chamber-events/festival-of-the-maples/

CHAPTER 10

Where the Money Is: Marketing Your Pure Maple Products

If you just want to make some syrup for yourself, friends, and family, and you don't plan on selling any of your products or earning any money, then feel free to skip this chapter. On the other hand, if you would like to generate additional income from sugaring and would benefit from running a profitable operation, then this is probably the most important chapter in the book. The following pages include a compilation of the best information and advice I have ever seen or heard regarding marketing maple products. Many of the ideas presented came from other sugarmakers willing to share their experiences in how they have successfully marketed their products. Learning from others' successes will increase the profitability of your own operation while also helping to ensure the sustainability of the entire maple industry. The more marketing and promotion each of us can do in our own operations, the better off the entire industry will be.

Why Marketing Matters

The future development and overall sustainability of the maple industry is mostly contingent upon people's desire to produce and consume pure maple syrup and the economic forces that drive those decisions. With less than 1 percent of the available trees tapped for syrup production in the United States, we could easily produce more syrup, but we need to be able to sell it. Given that consumption of pure maple syrup is less than 3 ounces per capita in the US, 6 ounces per capita in Canada, and barely even measurable throughout the rest of the world, there is a tremendous opportunity to increase consumption of pure maple products. Although we have a great product, it's important to remember that maple syrup doesn't sell itself—we need to go out and market it. As long as demand stays

FIGURE 10.1. Although most of us would rather spend our time producing maple syrup, it is important to also devote considerable resources to selling our product. PHOTO BY NANCIE BATTAGLIA

high and prices remain at or near their current levels, syrup production can continue to expand at a rapid rate. However, if we focus too much on increasing production without a corresponding focus on marketing and promotion efforts, supply will outpace demand and syrup prices could drastically fall. For instance, would you be starting up a new sugaring operation or expanding your existing operation if bulk syrup prices were $2 per pound, there was a huge surplus of syrup already in reserve, and you had a hard time selling all your syrup at profitable levels? For those of us who just want to produce maple syrup on a small, hobby scale, bulk syrup prices really don't matter that much. But for anyone who plans on selling maple syrup on the open market, the overall market for maple syrup and its effect on pricing is of the utmost importance.

The maple industry has experienced rapid growth following a supply shortage in 2008 that wound up raising bulk syrup prices to record-high levels of $4 a pound. Favorable weather in 2009 produced a bumper crop, and the industry has been able to supply the markets ever since. Prices have come back down to between $2.60 and $3 a pound, yet whether prices will remain this high over the long term remains questionable. We are now producing more syrup than is sold in the market, and the "strategic reserve" of maple syrup in Quebec has grown to almost 100 million pounds after the 2013 sugaring season. My concern, which is shared by industry leaders, is that many sugarmakers could go out of business and production may fall if the surplus continues to grow and syrup prices fall much farther. At current bulk prices of $2.60 to $3, most large sugaring operations are profitable, and people have a good incentive to get started in syrup production or expand their current operations. However, if bulk prices fall to $2 or $2.20, many sugaring operations will lose their profit margins and there will no longer be a financial incentive to produce maple syrup. Thus, the extent to which the maple industry grows is limited primarily by economic forces of supply and demand that dictate pricing and profitability. Anything you can do to sell more pure maple products to a wider segment of society will not only help your business, but also help all producers of pure maple syrup.

It is worth noting that your marketing efforts will only help the maple industry if you focus on venues, products, and customers that are not currently being served by another maple vendor. If a customer simply buys a quart of syrup from you versus another sugarmaker or a grocery store, there is no impact on overall consumption levels. The only way your marketing and promotion efforts will help the industry is if you reach out to new markets and potential customers who may not already be purchasing pure maple products, or if you get existing customers of pure maple to buy even more. Given that the market share of pure maple is incredibly low, I believe the opportunities to reach new customers are significantly greater than competing for the existing consumers of pure maple. Our focus as an industry should be on getting a bigger slice of the overall pie for sweeteners, not competing over the extremely small piece that we currently have. The main objective of this chapter is to supply you with new ideas (and reinforce some old ones) to sell more maple products to a wider audience, thereby increasing the overall share for pure maple in the marketplace.

How to Sell Your Syrup: Bulk, Wholesale, Retail

There are three basic options for how to sell your maple syrup: in barrels through the bulk markets, in consumer-sized containers through retail and wholesale outlets, and processed further into value-added maple products such as cream, candy, and sugar. Chapter 11 deals entirely with value-added products while this chapter discusses sales strategies for pure maple syrup.

Bulk Syrup

Many of the sugarmakers I know love to produce maple syrup but don't care so much about selling it. If this sounds like you, then selling bulk syrup may be your best option. You can keep some syrup for your own use and sell the majority in a barrel as bulk syrup. Selling bulk syrup makes the most sense for large producers

FIGURE 10.2. A stack of empty barrels at Bascom's Maple Farm in New Hampshire. Bascom's is one of the largest buyers and sellers of maple syrup in the United States; many drums of syrup pass through their warehouses.

who don't have the time or desire to do a lot of marketing themselves. It also makes sense in areas where there is a lot of competition in the local marketplace, and in remote areas (such as northern Maine) where there are few people living nearby to purchase the syrup.

Putting your maple syrup directly into a large barrel and then selling it to another maple producer or bottling company is the easiest way to sell your syrup. If you don't have the time or interest in dealing with customers and marketing your products, simply selling syrup in bulk drums may work well for you (assuming bulk prices stay high). The disadvantage, of course, is that you typically get the least amount of money for your syrup when sold in drums. In recent years bulk prices have ranged anywhere from $2 to $3 per pound (which equates to $22 to $33 per gallon), depending on the grade, quality, and market conditions. Bulk syrup prices are currently being propped up by the Federation of Quebec Maple Syrup Producers, but if their quota and pricing system unravels and bulk prices fall, the profitability of your operation will be in jeopardy.

It is important to realize that the vast majority of maple syrup consumed throughout the world is purchased at grocery stores that are serviced by the large packers, and nearly all this syrup was originally sold by thousands of sugarmakers as bulk syrup. The three largest maple packaging operations in the United States are Maple Grove Farms and Butternut Mountain Farms in Vermont and Bascom's in New Hampshire. There are many large bottling facilities in Quebec and smaller ones scattered throughout the eastern US and Canada. These bottling companies purchase syrup from thousands of sugarmakers in 15-, 30-, or 40-gallon drums, and then blend different syrups together in large tanks before bottling. Many of the large sugarmakers (more than 2,000 taps) sell most of their syrup this way, usually in late April or early May as soon as the season is over. Sugarmakers may also sell bulk syrup to another sugaring operation that is not able to produce enough syrup to meet their market demand. These types of arrangements provide greater synergy in local syrup markets, allowing some producers to concentrate on

The Federation of Quebec Maple Syrup Producers: The OPEC of the Maple Industry

Quebec dominates the maple industry, producing over 75 percent of all the maple syrup in the world. As one faction of the Union of Agricultural Producers, the Federation of Quebec Maple Syrup Producers (FPAQ) has done much to advance the industry since it was created in 1966. There are now nearly 7,000 members that elect a 12-member panel to serve as the board of directors, overseeing all aspects of bulk maple syrup production and marketing for the province. By joining forces and putting adequate funds toward promotion and market development, the producers of Quebec have worked within this government-sponsored framework to advance the entire industry. In particular, by implementing a quota system, purchasing excess syrup for their strategic reserve, and setting the minimum prices for maple syrup sold in drums, the federation has stabilized bulk syrup prices throughout the entire industry.

All bulk maple syrup produced and sold in Quebec must go through authorized buyers for the federation. This provision has led to significant internal strife within the industry and the development of large black markets for maple syrup. Since the adoption of the Joint Plan in 2004, the federation receives a certain amount of money (it is now 12 cents per pound) for all the syrup and sets the minimum prices for bulk syrup that must be paid by authorized buyers in Quebec. This price is based on what the producers would like to receive for their output, not necessarily what the market can bear. Currently the prices are higher than they would be if left to the free-market forces of supply and demand. However, producers only receive 75 percent of the established price when they first sell their drums and are paid the remainder in installments as the syrup is sold on the open market. As the surplus grows and producers do not get paid their full price, an increasing number are becoming dissatisfied with the system. In order for the FPAQ to help maintain the stable prices that are encouraging expansion of the industry outside of Quebec, they must be able to sell the majority of syrup and keep their members satisfied. Only with a content membership and a profitable business model can the federation continue to serve as the OPEC for the maple industry.

producing the syrup and others to focus more on marketing pure maple products. Before you sell your bulk syrup to distant markets, it is first worth determining if there are other producers nearby who may be interested in purchasing it.

The main downsides of selling bulk syrup are that you receive the lowest returns for your efforts and that the price you receive is largely out of your control. When stored in a barrel, maple syrup is an agricultural commodity that can be bought and sold at roughly the same price. Individual producers have very little control over this price but rather must accept whatever the market dictates. Bulk prices are partially set by the federation and are also influenced by the general laws of supply and demand. Since Quebec produces roughly 80 percent of all the syrup and the vast majority of this syrup is sold in bulk markets, they have by far the strongest influence on bulk syrup prices. The federation establishes a minimum price that must be paid to producers in Quebec well before any trees are tapped. The price may wind up higher than this, but it cannot fall below the predetermined levels. The other major influence on bulk prices is the exchange rate between the United States and Canada. When the US dollar is strong, it's cheaper for Americans to buy Canadian syrup, and bulk prices fall in the United States. However, as the United States and Canadian currencies have equalized in value, there is no longer an economic incentive to purchase Canadian syrup. How that will play out in the future is anyone's guess. If you can predict how the United States and Canadian exchange rate is going to fluctuate, your time would be better off spent on Wall Street than in a sugarhouse.

How to Sell Bulk Syrup

Bulk syrup is almost always bought and sold on a weight basis, which can be confusing to people who normally buy and sell syrup on a volume basis (pints and gallons). Whenever bulk syrup is bought or sold, the barrel is first weighed on a large scale to account for (1) the density of syrup in the barrel and (2) how full the barrel is. A given volume of sugar weighs more than the same volume of water, so higher-density syrup (that contains more sugar) will weigh more than lighter-density. Therefore, higher-density syrups will command higher prices when computed according to weight. It is also necessary to know what the barrel weighs when empty (the tare weight); you simply subtract this number from the total weight to determine the weight of syrup in the barrel.

Whenever you sell bulk syrup, the buyer will open the barrel and check the syrup for color, flavor, and clarity. Lighter syrups usually command higher prices than darker grades, though the gap has been shrinking in recent years as more and more light syrup is produced. These days there is typically only a 5-cent-per-pound difference between grades; this number has been much higher in the past. Almost every buyer will also check the flavor of the syrup rather than just relying on the color. Off-flavors are becoming increasingly common in lighter syrups, especially those produced at the end of the season. Just because your syrup may have a light or medium amber color does not mean it will command those prices. Off-flavored syrups are graded as "commercial" and receive much lower prices. If you are unable to filter your syrup or do a poor job of filtering, expect to receive up to 50 cents a pound less for it. The large bottling companies will refilter all the syrup before they use it, but they don't want to purchase sugar sand and don't want to deal with unfiltered syrup unless they absolutely have to.

The type of barrel you put the syrup in depends on the grade of syrup and ultimate market. For table-grade syrup that will be put into a consumer-sized container, it is extremely important to use high-quality barrels that will not taint the syrup or impart an off-flavor. Stainless-steel barrels are the best option, and the price has fallen considerably in recent years as more companies compete for your business. Sometimes you can find used stainless barrels, but be sure you know the full history of what the barrels have been used for. As long as they were used for food—*not* chemicals—and you clean them thoroughly to remove any odors from the previous use, used stainless barrels may be acceptable. There are also some less expensive epoxy-lined steel barrels and onetime-use plastic liner barrels that can work well, though the plastic barrels are developing a poor reputation among many syrup buyers. Due to problems with rusting and possible lead contamination, the industry is moving away from galvanized barrels. In fact, as of 2012 the state of Vermont prohibited the packaging of syrup in galvanized containers. By the time you are reading this book, your state may have also adopted more stringent regulations. After all, maple syrup is a gourmet food and should always be stored in food-grade containers. If you have any galvanized barrels, and the company you are selling bulk syrup to is still accepting them, you can use those for all your commercial syrup produced at the end of the season, but it's best to not put any table-grade syrup in galvanized barrels.

The timing of when you sell your bulk syrup may have an influence on the price you receive for it. Some producers prefer to hold on to their barrels until November and December in the hope of selling the syrup for a higher price to other sugarmakers who may have run out and need syrup to fill holiday orders. When syrup is relatively scarce, this strategy may make sense. On the other hand, many sugarmakers sell the vast majority of syrup as soon as the season is over. The large packers are usually buying syrup at specific locations in April and May (often through individual sugarmakers), and you can get paid immediately for all your efforts over the past couple of months. The strategy you choose may depend on the nature of the markets and your overall need for income in a given year.

Finally, it is important to realize that selling bulk syrup only makes sense if your overall cost-of-production is less than the price you receive for your barrels. Since most large-scale sugaring operations can produce syrup for less than $25 per gallon, selling bulk syrup can be a profitable venture. On the other hand, for many smaller-scale sugarmakers, the total cost of production can be well over $30 per gallon. When this is the case, selling barrels is a losing proposition. Thus, in order to turn a profit from a relatively small-scale sugaring operation, it is essential to develop retail and wholesale accounts that fetch a higher price. The rest of this chapter focuses on how to set prices on consumer-sized containers and market your syrup to the rest of society.

Wholesale and Retail

Most sugarmakers sell at least some of their syrup in consumer-sized containers, whether through wholesale or direct retail markets. If you are willing to put the extra time and effort into bottling the syrup and marketing it directly to consumers or wholesale accounts, it is possible to gain much higher returns than simply selling bulk syrup. However, it is important to realize that any money you make in selling wholesale and retail syrup should be compared against your possible revenues of selling bulk syrup. This will give you a better sense of your profit margins for the time and money invested in putting syrup in retail-sized containers and marketing it to the public.

To calculate how much extra revenue you will gain from selling consumer-sized containers, I developed a spreadsheet that can be downloaded from the Chelsea Green Publishing website, www.chelseagreen.com/mapleguide. To use this, there are several variables you need to enter values for:

- Bulk price for the grade of syrup you are bottling—enter the price that you could have sold the barrel of syrup for *or* the price that you paid for it.
- Cost of containers and stickers—enter the price for each size container that you use and include the price of any stickers (they are usually less than a penny each).
- Wholesale and retail prices for syrup.
- Number of bottles filled; keep track of how many bottles of each size you fill when bottling a given batch of syrup.

Once you have entered in the necessary information, the spreadsheet calculates your profit margin for each container filled as well as the overall returns that can be realized once all the containers have been sold. Note that there are two worksheets on the Excel file—one is based on selling all the syrup retail whereas the other is a bit more complicated and allows for selling the syrup in a mixture of retail and wholesale outlets.

To show how this works in real life, I have included an example of the last time we bottled a 40-gallon barrel of dark amber syrup at The Uihlein Forest. For this example, I used the version that includes selling only in retail markets—we rarely sell any syrup to wholesale accounts. As you can see, our profit margins per individual container sold are highest on the larger-sized containers, yet we would make the most money if we sold all smaller-sized containers. For instance, we make $22.70 when selling a gallon jug of syrup and only $5.16 when selling a pint. However, once we have sold eight pints (the equivalent of a gallon), we will have made $41.31, nearly double the profit margins from selling one gallon-sized container.

The summary figures at the bottom of table 10.1 show how much additional revenue we generate by selling syrup in retail-sized containers. We could have sold the 40-gallon barrel for $1,157.31 on the open market, assuming it was completely full and the price of syrup was $2.60 per pound. Once all the containers have been sold, our total revenue would be $2,434. Subtracting out the container costs and stickers, our overall profit margin was $1,073.45, or $26.84 for each gallon of syrup sold in retail-sized containers. Clearly our revenues are much greater when selling in consumer-sized containers, but we must also consider the other costs incurred to determine if it really makes sense. After all, we don't generate any revenues when we package syrup in smaller containers; we only make money when we sell that syrup to others.

Being a part of Cornell University, we do not engage in any direct marketing of syrup, so our transaction costs are very low. We sell the vast majority of our syrup at the sugarhouse and through mail order to Cornell alumni. Given that we are located just outside the Olympic Village of Lake Placid, we get a lot of tourists who come to our sugarhouse to learn more about maple sugaring. We spend a great deal of time answering people's questions and showing them how we make syrup, and at

TABLE 10.1: Example of Profit Margins for Bottling Syrup in Consumer-Sized Containers

Barrel size	40 gal.
Bulk syrup price per lb.	$2.70
Bulk syrup price per gal.	$30.05
Revenue if sold in barrel	$1,201.82

Size of container	Price per container sold in retail outlets	Price to purchase empty containers	Quantity bottled		Total revenue	Cost of jugs	Profit per jug	Profit per jug on a gallon basis retail
			Number of bottles	Number of gallons				
gallon	$54.00	$2.37	18	18	$972.00	$42.66	$21.58	$21.58
½ gallon	$30.00	$1.84	15	7.5	$450.00	$27.60	$13.14	$26.27
quart	$18.00	$1.43	36	9	$648.00	$51.48	$9.06	$36.23
pint	$10.00	$1.22	20	2.5	$200.00	$24.40	$5.02	$40.19
½ pint	$6.00	$1.17	8	0.5	$48.00	$9.36	$2.95	$47.23
Totals			97	37.5	$2,318.00	$155.50		

Profit Calculations	
$960.68	Money earned by selling syrup in consumer-sized containers
$24.02	Additional revenue per gallon sold in consumer-sized containers
94%	Recovery rate of syrup in consumer-sized containers

the end of our tours, nearly everyone purchases some maple syrup. The exceptions are those who are traveling by plane; ever since the TSA ruled that you can't take maple syrup in your carry-on luggage, those traveling by air rarely buy anything. Since educating the public about the maple industry is part of our mission, I don't factor in staffing costs as part of our analysis. However, if I accounted for the employee wages spent giving tours and handling sales, our profit margins would drastically fall. For instance, when we spend close to an hour talking with visitors and they only purchase a quart or half-gallon container, the additional revenue may not even cover the staff wages. Thus, our most profitable venture is the self-serve cabinet located just outside the sugarhouse. We generate roughly a third of our retail sales from people who have come to the sugarhouse when nobody is around; they just take some maple syrup and leave the money in the sap bucket.

The reason I provided this example is to demonstrate why it is important to fully consider all the time you spend selling syrup and any other transaction costs (vendor fees at farmers' markets and craft shows, mileage on vehicles, and so on) before deciding if selling syrup in retail or wholesale markets makes sense for you. This requires a great deal more record keeping than most people want to bother with. However, if you are truly interested in whether selling syrup in consumer-sized containers makes economic sense, proper record keeping is a must.

In the example I provided above, we would make an extra $1,073 by selling the barrel of syrup in retail-sized containers. This may seem great, but just knowing that figure doesn't allow me to determine if it makes economic sense to do so. If it only took 30 hours of staff time to bottle the syrup and sell all of it, then I would be happy with this outcome. On the other hand, if doing all this tied up 60 hours of staff time, then I would have to reconsider whether it was worthwhile to do so. I encourage you to do this analysis for yourself the next time you bottle syrup—in particular, try keeping track

How Much Shrinkage Do You Get When Bottling?

One of the benefits of keeping good bottling records is that you can determine how much shrinkage occurs when transferring syrup from a barrel to retail-sized containers. For instance, when you take a 40-gallon barrel and fill up retail-sized containers, rarely will the total amount of syrup bottled equal 40 gallons. Your actual syrup recovery rate will vary based on how much syrup you put in the bottles, how much syrup you are able to extract from the barrel, and how much syrup was lost in your filtering medium. Many types of retail-sized containers are "full" when the liquid gets to the bottom of the neck. However, only filling to this level makes it seems as if you are shortchanging the customer, so many sugarmakers put in more syrup than legally required to make it appear full. It is also very difficult to get all the syrup out of a barrel, so you wind up losing some when transferring it to your canning unit. The large bottling companies will steam-clean barrels of syrup after they are emptied and then collect and boil down all the syrup-laden water into a new batch of syrup. Another source of shrinkage occurs during the filtering process, as some amount of syrup is always contained within the filter papers, cloths, or other medium. When you are done filtering, you can run hot water through your filter and then save that sugary water for reprocessing into syrup. Everything you can do to minimize the amount of sugar going down the drain will add to your overall bottom line.

of how much time it takes you to sell that much syrup as well as any other costs you may incur in doing so. Once you have done this analysis, you may determine that this is a great use of your time and pursue additional retail and wholesale markets. On the other hand, you may decide to raise your prices or start selling your syrup in bulk. The rest of this chapter explores various marketing strategies aimed at getting the most value for your time invested in producing and selling maple syrup.

Strategies for Selling Retail and Wholesale Syrup

One of my favorite quotes related to marketing was given by Dan Shepherd, a farmer I met in Missouri a few years ago. He said, "My father always told me, you can make a little bit of money doing what everyone else is doing, or you can make a lot of money doing what no one else is doing." As you may imagine, the Shepherds have focused on unique ventures that set them apart from everyone else in their region. In the midst of a corn- and soy-dominated landscape, the Shepherds grow pecans, raise buffalo, and focus on other consumer-friendly agriculture that draws people to their farm. Much of their success is due to the fact that they are growing unique products that people want and filling a niche in their community for agriculture-based experiences. If you happen to live in an area where no one else is making syrup, your sugaring operation will be unique and you'll be able to draw people to your sugarhouse to show them the wonders of maple production. For this reason, I think you have a better chance of developing a profitable maple operation in Illinois or Virginia than you do in Vermont and other northeastern regions. If you are the only game in town, everyone will come to you for their pure maple needs. However, just because you happen to live around a lot of other sugarmakers doesn't mean you can't differentiate yourself and offer your customers something unique and special. There are tremendous marketing opportunities with maple; the key is finding the right niche for your products and spending the time to fully develop your markets.

No matter what strategies you use to sell your syrup, make sure that whatever syrup you put in a consumer-sized container is high quality, adheres to all grading standards, and tastes great. Although this should go without saying, I want to emphasize the importance of keeping substandard syrups off store shelves. Unfortunately a lot of maple syrup winds up in the marketplace that does not conform to the standards of excellence people expect in a gourmet food such as pure maple syrup. Once people get an off-flavored or substandard maple syrup, this may turn them off from pure maple for a very long time. If you think there is anything wrong with your syrup, find another use for it besides packaging it in a consumer-sized container, and try to figure out what went wrong. If you sell it as bulk syrup to a large processor, it can be blended with other syrups to mask any off-flavor. Much of the substandard syrup also winds up as an ingredient in processed foods or as the 2 percent real maple syrup in an otherwise artificial pancake syrup.

What you'll find in this chapter is by no means a comprehensive list of the ways to market your products, but it is a good start. I'm sure there are plenty of excellent sales strategies that producers are using and I either haven't heard of them yet or didn't have space to include them in this book. I would encourage you to try some of these ideas for your own business and to also think creatively of other ways to sell your products. One important point that I want to emphasize is the need to always be on the lookout for new markets for maple syrup and other maple products. There are many places that do not carry pure maple products (but should), so look for opportunities that haven't yet been filled by other sugarmakers. It doesn't do the industry any good when one sugarmaker competes with another sugarmaker for the same shelf space. In order to grow as an industry, we need to expand our offerings and compete with other sweeteners, not other maple producers. Don't think of your neighboring sugarmakers as your competition; the real competition is the artificial pancake syrup and all the other highly processed, sugary foods that our society consumes in great abundance. If we can improve our market share of the sweetener industry by only a fraction of a percent, the pure maple industry would have a hard time just meeting the market demand.

Pure maple syrup is one of the easiest agricultural products to market. It has an interesting history and all the nostalgia of rural living and small-scale farming. People find the process fascinating; families love to come out to the sugarhouse to watch, smell, and taste the syrup being made. Besides honey, it is one of the only all-natural sweeteners available in the market, and there are numerous health benefits that we are now just discovering. With everything that maple has going for it, you would think there wouldn't be any trouble selling it. However, pure maple is the most expensive sweetener (by far), so marketing is essential to make people realize why the higher price is well justified. The following pages contain a sampling of the many marketing ideas that you can utilize to grow your business and make pure maple a bigger slice of the sweetener pie.

Marketing Birch and Walnut Syrup to Niche Markets

The Shepherds' logic of "doing things that no one else is doing" is also a great reason to produce birch and/or walnut syrup. Although it does take more time and fuel to process birch sap into syrup, the main reason prices for birch syrup are so high is that very few people are producing it. It's basic economics—when there is high demand and limited supply, prices go up. Even though the vast majority of people prefer maple syrup to birch syrup, prices for birch are usually three to five times those for maple. There is an ample supply of maple syrup and people can buy it in almost all grocery stores; on the other hand, birch syrup is extremely limited and only available in certain specialty stores or online. The same thing is true for walnut syrup. There is hardly anyone producing walnut syrup, so if you do the right kind of marketing, there is no doubt you can find lucrative outlets for this delicious and unique syrup. As with all new products, you will likely have to spend a lot more time educating potential customers about the unique qualities of birch and walnut syrup. Once you take time to do so, you'll have a hard time just trying to make enough of these syrups to meet demand.

Provide Syrup to a Local Pancake Breakfast

I am always surprised at the number of pancake breakfasts I see advertised that don't mention anything about pure maple syrup being served. These events are the low-hanging fruit for marketing syrup in your community. Of course nobody would advertise that they are serving artificial pancake syrup, but when it is pure maple syrup from a local producer on the menu, that could (and should) be a great selling point on the posters and signs advertising the event. Pancake breakfasts are usually put on to support a good cause. I recommend donating the syrup in exchange for being able to sell maple products at the breakfast. The attendees will know that you supported the cause by donating the syrup, and if they liked it, chances are many of them will want to purchase a bottle or two to take home with them.

Host a Pancake Breakfast at Your Sugarhouse

I recommend doing this at the end of the season to celebrate the end of another successful harvest. By the end of April or beginning of May, the weather is usually a lot nicer and people could sit outdoors on picnic tables. Not many sugarhouses are equipped to handle a large pancake breakfast, but most people have ample open space around their sugarhouse to set up a temporary eating venue. If you provide the space and the syrup, there may be other civic organizations to supply all the other necessary materials and staff to pull it off. You could let the organization keep all the profits, but you'll benefit by gaining tons of positive exposure and many customers for your maple products. One sugarmaker told me that he donated $700 worth of syrup for a pancake breakfast that the local firemen put on at his sugarhouse. While that may seem like a big donation, he sold $19,000 worth of syrup and other maple products that day. That is a return on investment anyone would gladly take.

FIGURE 10.3. The waiting line for the pancake breakfast at the Paul Smith's College sugarhouse that takes place at the end of April every season. PHOTO BY NANCIE BATTAGLIA

Organize a Maple Syrup Sale at Your Local School

Offer bottles of maple syrup at a wholesale price so that students can make money selling them and you get exposure throughout your community. Offering your maple products will be a much healthier and local alternative to selling candy bars or chocolate. It will help raise money for the class and get a lot of publicity for your operation. This strategy makes the most sense for sugarmakers who have children or grandchildren in a local school.

Set Up a Self-Serve Stand at Your Sugarhouse

If your sugarhouse is near a well-traveled road and easily accessible, you may want to put up a self-serve stand. People will often show up to buy maple products when you aren't there, and the self-serve stand allows people to pick up syrup at any time. You could pay someone to hang out all day and wait for customers, but unless you have a real store with a lot of traffic, this strategy won't be cost-effective. Even if some people take syrup without paying sometimes, the minor amount of theft that occurs is unlikely to amount to nearly as much as it would cost to staff a retail store all day.

We have a self-serve stand at our sugarhouse in Lake Placid. It is nothing fancy, just a simple wooden case, signage, and a sap bucket for people to put their money. We derive nearly a third of all syrup sales from this simple feature. Many local residents appreciate being able to pick up our syrup at any time, and visitors who come are pleasantly surprised to find syrup for sale under the "honor system." We have even had people come to our sugarhouse to get syrup just because they had heard of the self-serve window and wanted to see it for themselves. In our troubled world where most people lock everything to deter thieves, there are still many good people out there who appreciate being able to pay on the honor system. If you happen to live in an area where you think a self-serve stand would work, I highly recommend giving this type of sales strategy a try.

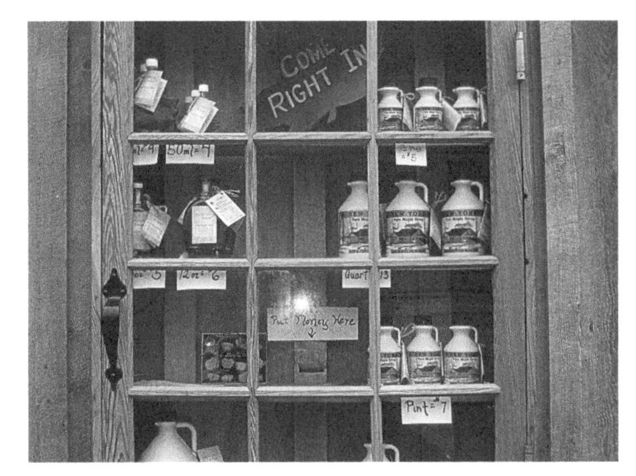

FIGURE 10.4. The self-serve window at Sugarbush Hollow in Wayland, New York. PHOTO COURTESY OF PETER SMALLIDGE

Feature Your Phone Number and Website Clearly on Your Bottle

This is especially useful for sugarmakers focused on developing their mail-order markets. By including language such as "Almost empty" followed by your phone number and/or website, you are making it much easier for customers to purchase syrup from you again.

Participate in Your Statewide Open House Event

Many states, such as New York, Vermont, Maine, and Pennsylvania, coordinate open house events for the general public during the sugaring season. In New York this event is called Maple Weekend; Maine refers to it as Maple Sunday, Vermont calls it Maple Open House Weekend, and Ohio holds the Maple Madness Driving Tour. Extensive advertising campaigns that are funded by all participants help to reach a much broader segment of society than any one sugaring operation could afford to do. All electronic and print media directs people to the participating producers that agree to be open for a certain weekend (or weekends) each March. If your state doesn't already have a coordinated event, you could work with other producers and make it happen. If you happen to live in a state that does already have a Maple Weekend or similar event, it would be well worth your while to participate. In New York, even with just over 100 producers participating, sales of maple products usually reach about $750,000 during the two weekend events in any given year. Given that there are hundreds of other producers who could (and should) also participate, the potential for growth is incredible.

If you decide to participate in your statewide open-house event, it is important to make the visitor

FIGURE 10.5. Maple Weekend activities in New York have been so successful that it now spans two weekends every year. PHOTO COURTESY OF THE NEW YORK STATE MAPLE PRODUCER'S ASSOCIATION

FIGURE 10.6. Musicians perform at the annual Fiddler's Fair that takes place at the Sugarbush Hollow sugarhouse and trails just outside Naples, New York. PHOTO COURTESY OF PAM MASTERSON

FIGURE 10.7. For many years, Cornell's Arnot Forest has held a Mother's Day Wildflower Walk & Pancake Breakfast to get people out to the sugarbush when the spring wildflowers, such as this white trillium, are in bloom. PHOTO COURTESY OF BRIAN CHABOT

experience as enjoyable, entertaining, and informative as possible. The open houses take place at a great time of the year when many families have cabin fever and are looking for a wholesome activity that gets everyone outside. The more family-friendly events and activities you can offer, the more people will come back year after year.

Host Festivals or Events at Your Sugarhouse Outside Maple Season

Sugaring season isn't the only time of the year that we should be attracting people to our sugarhouses. Many sugarhouses and sugarbushes are beautiful places that could be utilized for much more than just producing syrup in the spring. One example of this is the Fiddler's Fair held at Sugarbush Hollow in the Finger Lakes region of New York. By partnering with local organizations and offering the sugarhouse and sugarbush for the event, Chuck Winship has been able to consistently draw over 1,500 people to his sugarhouse each fall. Dozens of musicians from far and wide travel to Sugarbush Hollow and play at various stations along roads and trails in the sugarbush. People walk the trails and learn about maple production while taking in great music. They sell a wide variety of maple products all day long, and everyone learns about his sugaring operation.

Get a TV Chef to Cook with Maple Syrup or Sugar

Several years ago, a chef on the local news station in western New York demonstrated how to make a maple-apple pie using granulated maple sugar instead of white sugar in a traditional apple pie recipe. The chef was very enthusiastic about using the pure maple sugar and drew a lot of attention to the maple producer who'd supplied him with the product. This segment appeared during the holidays and led to a drastic spike in sales of pure maple sugar and other maple products during that time period. With the growing popularity of cooking shows, we need to get more TV chefs to demonstrate how to use maple on more than just pancakes. Sending samples

FIGURE 10.8. The Merle family all sporting the same attractive apparel for their Maple Weekend festivities. PHOTO COURTESY OF DOTTIE MERLE

of your maple products to different media outlets that feature cooking demonstrations could be one of the best investments you make.

Sell Apparel with Your Logo on It

Do you have an interesting and attractive logo? A catchy motto for your business? Having your logo embroidered on T-shirts, fleeces, hats, and other apparel can help spread the word about your maple products. Your friends, family, employees, and volunteers will be glad to wear the company apparel with pride, and your most loyal customers may also want to show support for their favorite sugarhouse. Make sure that whatever apparel you choose is high quality, otherwise this investment could backfire. If the company clothing looks tacky, customers may think of your maple products that way as well. Since I'm sure you want to project an image of quality for your maple products, be willing to invest enough money in quality apparel.

Offer Corporate Gifts

Many businesses give out gifts to their clients, partners, and associates during the holiday season. If you have any connections with local businesses or major corporations, offer them syrup and other maple products at your wholesale rate to have them give away for their annual gifts. Even if you don't make much money on the deal, it will put your syrup in the hands of many people who may have never had it before. I wouldn't offer the discounted prices right away, but if they express interest in using your maple products yet worry these might be too expensive, it will be worth your while to negotiate a price structure that is mutually beneficial.

Coordinate with Wedding Planners to Offer Your Syrup to Their Clients

Maple syrup is often a popular wedding favor given out to guests, especially for weddings taking place in the Northeast during the fall. Many producers highlight "wedding favors" in their brochures or websites and do a lot of business in this arena. One way to ensure additional business from wedding favors is to get to know the wedding planners in your region. Many couples hire someone to help them plan the wedding and take care of logistics. They pay this person good money to provide information and advice, and they will often do exactly what the wedding planner suggests. If you had one or more wedding planners recommending your maple syrup as wedding favors, chances are this aspect of your business would take off. It's a good idea to offer small bottles with attractive designs. Some people will go with the 50 or 100ml glass leaf, whereas others may opt for a heart-shaped bottle. Whenever I have been approached about doing wedding favors, I simply give the clients some catalogs and let them choose whatever container they would like. Of course this needs to be done well in advance to ensure you can order the containers and fill them before the big day, but I've found that most people planning weddings take care of this well in advance.

Put Company Stickers on Your Vehicles and Always Have Syrup Available for Sale

I know several producers who sell quite a bit of syrup just because people see the stickers on their truck. It catches their attention, so folks ask about it and usually wind up buying a bottle or two. You never know when you are going to run into somebody who wants some maple syrup, so make sure you always have a good stash in your vehicle. It also comes in handy if you get stuck on the side of the road and need some assistance from a Good Samaritan passing by. People doing you a favor will often feel awkward about accepting money, but they will be glad to take some pure maple syrup that you produced.

Pay People with Maple Syrup Instead of Cash

Whenever possible, try to barter with maple syrup instead of paying cash for goods and services. For instance, if someone does a job for you and would normally charge $60, see if they will take a gallon of syrup instead. People love to get maple syrup, and they will often tell their friends about the unique bartering that they did. It makes for a good story, and it will help spread the word about your business. Don't underestimate the value of syrup, either, as it usually "sells" for much more in the barter economy than it does on store shelves.

FIGURE 10.9. Larry Rudd's pickup truck clearly advertises his maple syrup business. Larry always has some maple syrup with him and has sold a lot of it over the years to curious people who noticed the decal on his truck. PHOTO COURTESY OF LARRY RUDD

Sell Maple Syrup Cookbooks

A number of cookbooks on the market today show people how to use maple syrup in a wide variety of recipes. Maple is not just for pancakes, and we need to do more to promote it as an ingredient in fine cuisine. You may make a little bit of money selling the cookbooks, but the real benefit will be seen when your customers buy a lot more syrup to make all the special dishes they read about in the cookbook.

Sell Maple-Themed Coloring Books and Activity Books for Kids

Or just give them away. The real advantage of doing this is to get kids interested in maple. The more they know about pure maple, the more they will want to consume it for the rest of their lives. Steve Childs has developed a *Maple Weekend Coloring Book* for kids from kindergarten through third grade and a *Maple Syrup Activity Book* for third through sixth graders. Both of these are available through the Cornell Maple Program, the New York State Maple Producers Association, and a variety of maple dealers.

Be a Vendor at Local Farmers' Markets, Craft Shows, Festivals, and the Like

This is certainly not for everyone. It takes a lot of patience and people skills to sit at a booth all day talking to people and making sales. However, if you have the personality to handle this type of work, it would be wise to get a spot at as many venues as you can find time for. In places like Vermont, it would be hard to find a farmers' market without at least one person selling maple products. In many regions of the country, however, there are plenty of farmers' markets without maple products. I would avoid trying to set up at any market that already has a maple vendor, though some of the markets in large cities may be big enough to

FIGURES 10.10 AND 10.11. Two publications available from the Cornell Maple Program are designed to teach children about maple sugaring and get them interested in the process. PHOTOS BY NANCIE BATTAGLIA

handle more than one sugarmaker. If you do find yourself at a market with another sugarmaker, try to find ways to complement each other. Rather than viewing each other as competitors, perhaps you can each offer slightly different maple products—there is a wide range of value-added products available to choose from, as described in the following chapter.

Create a Guest Book and/or Map at Your Sugarhouse

If you get a lot of visitors to your sugarhouse, I recommend developing a nice guest book. Having an attractive registry where visitors can leave their name, address, email, and comments will help build your customer database. You can use this list to send out announcements for special events or sales, such as during the holidays. The guest book can be placed near the register or next to an attractive map where people put a pin on their hometown. If your facility draws visitors from around the country (or around the world), it makes for an interesting display on how far people are willing to travel to get your syrup. People especially enjoy putting their pin on the map—it gives them a sense of connection to your sugarhouse and helps to create the authentic experiences that many people are looking for when traveling.

FIGURE 10.12. Horse-and-wagon rides through the sugarbush are a popular draw, bringing families out to your sugarhouse. PHOTO BY NANCIE BATTAGLIA

Offer Sleigh Rides at Your Sugarbush

If you have a nice trail system going through your woods and know someone who offers horse-drawn sleigh rides, this is a great way to draw families out to your sugarhouse. I recommend scheduling these during the beginning of sugaring season when you are more likely to still have a lot of snow. Toward the end of the season, when the weather is warmer and the snow is gone, offering horse-and-wagon rides is a good substitute. I know one sugarmaker who pays someone $1,000 to offer free horse-and-wagon rides all day to visitors during Maple Weekend. People greatly appreciate the free wagon ride and buy enough maple products to make the $1,000 investment for the team of horses easily pay off.

Join Your Local Chamber of Commerce

Joining your local chamber is an excellent way to connect with the other businesses in your region. Attending their meetings and events will provide great networking opportunities with restaurants, retail stores, and hotels in your region that could carry your line of maple products. With the increasing popularity of "going local" these days, it's important to make sure that all the other businesses in your region know what you have to offer. It is especially important to connect with the chamber during your open house events such as Maple Weekend; the more businesses you can get involved in promoting the maple theme, the more successful your event will be.

Hang Buckets with Your Logo on Trees Along the Main Roads Near Your Sugarhouse

People love to see the old metal sap buckets hanging on maple trees during sugaring season. Find out who owns the nice large maple trees planted along roadsides and in front yards throughout your town and nearby areas and offer to give them some syrup if they let you tap their trees. By placing a sticker with your logo on the bucket, everyone will know those are your buckets, and it will draw greater attention to your business. Expect to lose some to vandalism, but consider that just a cost of doing business for lots of good, free advertising. You'll probably also generate a lot more sales from the people whose trees you tap. The syrup you will offer them for the rights to tap their trees ($1 to $2 per tap in syrup value at a maximum) will pale in comparison with the amount they will want to purchase to get more of the syrup produced from their trees.

Get on Board with Social Marketing

I rarely go on Facebook, have never used Twitter, and only signed up for LinkedIn because a friend of mine asked me to. I spend enough time on the computer; I would rather conduct my social life in person. However, I know I'm missing out on a lot of information and possibilities as a result of this attitude. Social networking sites are becoming one of the most important ways to communicate and advertise in the modern world. If I was trying to promote a business, there is no doubt that I would be spending at least an hour a day on social networking sites. If you are like me and don't wish to spend more time on the computer than you have to, then find someone else in your family or a friend to manage your social networking accounts. Signing up is free—the only cost is your time—so if you enjoy doing this stuff yourself or know someone else who wants to manage your accounts, go for it.

Find a Reason to Send Out a Press Release

Local newspapers are always looking for good stories, so if you have something interesting going on that is newsworthy, definitely let your local media outlets know about it. Having an article appear in the paper about your business is the best free advertising you could ask for. It is *much* more effective than placing an advertisement, and you don't have to pay for it.

Give Out Maple Candies for Halloween

If you live in an area where you get trick-or-treaters, or if you have friends and family in a nearby city or village who do, consider giving out small pieces of maple candy in lieu of the usual assortment of regular candy. You could also find candy molds with a Halloween themes that provide a unique, healthier treat for kids to enjoy. Be sure to wrap the candies in small cellophane bags and put a personalized sticker on the bags to seal them and identify your business. The average American consumes 27 pounds of candy bars in a given year, yet only a small percentage of people ever gets to eat a pure maple candy. Halloween presents a great opportunity to try to do something about that.

Invest Time and/or Money in a Quality Website

It is important to remember that maple syrup is a high-end, high-quality, luxury food item. The website for your business should be indicative of the quality of your products. If you don't have the money to invest in a professional website or the time and skills to develop it yourself, try to find a (good) student who needs to develop a website for a class project. A well-designed and informative website will do wonders for your business; many customers expect it these days, and if you don't have one it could cost you a lot in potential sales. That being said, just because you have a nice website, don't expect it to drive a lot of your sales without additional

FIGURE 10.13. People love to see buckets hanging on maple trees. If it isn't obvious what sugaring operation the buckets belong to, you may want to stencil your company name or logo on the buckets. PHOTO BY NANCIE BATTAGLIA

Get Real Maple: An Innovative Marketing Campaign

Arnold Coombs is one of the top marketers of maple syrup in the world. He's the main salesman for Bascom's and represents a variety of brand name maple products. In October 2011 he started an online marketing campaign using social media to inform consumers about the differences between pure maple syrup and its artificial competitors. There is now a www.getrealmaple.com website, Facebook page, and Twitter account. By May 2013 he had over 40,000 followers on Facebook and growing strong. He did not launch this to promote his own brands of syrup, but rather pure maple syrup in general. All efforts are geared toward informing people about the differences between pure maple and the other "table syrups" that try to imitate the real thing. In particular, the websites do an excellent job of exposing Pinnacle Foods for their Log Cabin "All Natural Table Syrup" that is packaged in the traditional plastic syrup jug. One of the videos shows people being interviewed on the street about the new Log Cabin syrup and what they thought the jug contained. Of course many of the people incorrectly guessed maple syrup, whereas some of the savvier folks realized that it was a knockoff. Interestingly enough, Pinnacle Foods is currently being sued in Vermont for violating the Consumer Protection Act as part of a class-action lawsuit for misleading consumers into purchasing their syrup. If the lawsuit is successful, Pinnacle Foods may have to pull this product from Vermont grocery stores and pay damages. Stay tuned.

This kind of generic promotion is key to advancing the maple industry forward. There is a lot of confusion among consumers when it comes to pure maple and its artificial competitors. We have a great story to tell and we need to tell it! Using social media is an excellent, inexpensive way to get the word out. Significant research has been conducted on the health benefits of maple syrup, and the Get Real Maple campaign has been at the forefront of getting this new information out to people. Educating people about maple and all its benefits is the key to ensuring sustainable growth in the maple industry—the more people know about maple, the more they appreciate it and want it in their diets. We should all be grateful to Arnold and his team for their efforts and join them in the campaign to educate the world about pure maple syrup.

FIGURE 10.14.

FIGURE 10.15. This Log Cabin "All Natural" Table Syrup contains only 4 percent maple syrup and is mostly rice and cane sugar. In taste tests, held at Lake Placid, New York, the vast majority of people strongly objected to the flavor. The main concern is that consumers will purchase this syrup thinking it is pure maple. PHOTO BY NANCIE BATTAGLIA

promotional efforts. There are hundreds of websites where you can purchase maple syrup on the Internet, so simply spending a lot of money on yours will not always pay off in terms of sales. However, if you are doing a lot of outside marketing, you will need a good website to support all your other marketing efforts when customers go to your website to learn more about your company.

FIGURE 10.16. Jamie Rogers shows a group of second graders how to make maple syrup at Heaven Hill Farm in Lake Placid, New York. PHOTO BY NANCIE BATTAGLIA

Accept Credit Cards

Although it can be a pain to handle the extra paperwork and you have to give a percentage of the sale revenue to the credit card company, accepting credit cards is a surefire way to boost your retail sales. This is especially true for sugarmakers who sell at farmers' markets, craft shows, fairs, festivals, and so on. Most people do not carry a lot of cash with them, so the amount they are able to buy will be limited without a credit card option. You may want to establish a minimum amount for people to be able to use credit cards, as the extra hassle and fees may not be worth it for someone to just buy a pint of syrup. The technology for accepting and managing credit cards has advanced tremendously—to the point that most sugarmakers I know can now collect payments with their smartphone.

Give School Group Tours

If the local schools in your area can still go on field trips, it would be in your best interests to give as many tours as you can. Everyone loves going to see maple sugaring operations—the more kids you can educate about how maple syrup is made, the more syrup you will sell to those kids' parents. If at all possible, try to schedule the school field trips within a couple of weeks of your upcoming open house event. Many of the kids will want to bring their parents out and show them the cool place they got to see on their field trip. In addition to the good act of teaching kids about maple, your bottom line will also receive a boost from these extra efforts.

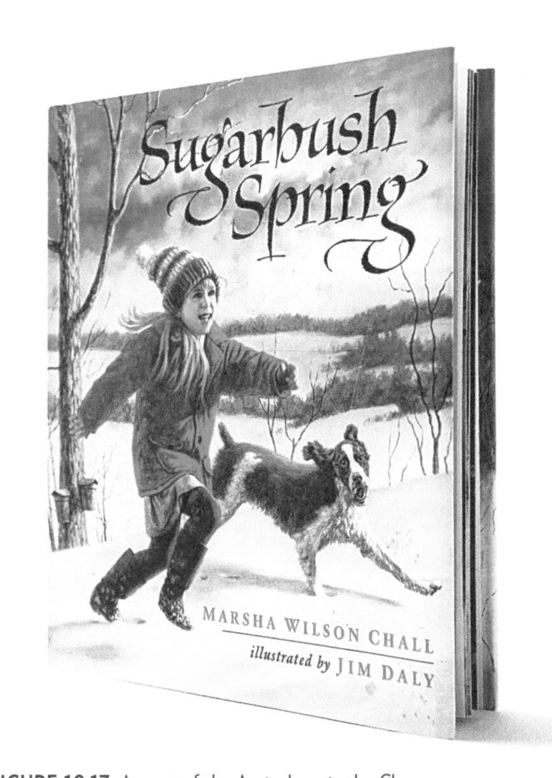

FIGURE 10.17. As part of the Agriculture in the Classroom program, we read *Sugarbush Spring* to third graders in all the local schools. The readings took place the week before Maple Weekend and led to our highest attendance ever for our open house event. PHOTO BY NANCIE BATTAGLIA

Contact Tour Bus Operators

If you live in an area that has tour buses come through, especially during the fall leaf-peeping season, call the operators and offer to be a stop on their tour. Sugarhouses with nearby trails are especially well suited since it gives people an opportunity to take a walk into the sugarbush

Shipping Maple Syrup with the USPS

Most shipping companies charge their customers by the volume and weight of the goods being delivered. Maple syrup is heavy (11 pounds per gallon), so it can cost a lot of money to ship it across the country. Most of the people who order syrup through the mail (at least from us) live in Florida, Arizona, or California. Apparently maple syrup isn't as readily available in these states, and those who moved away from the Northeast want to receive a jug of their hometown sweetener when living far away. Whereas it used to cost a lot of money to ship syrup, with the flat rate and regional rate boxes from the US Postal Service, it is now fairly inexpensive to do so. The best deal to ship to the West Coast is putting 2.5-gallon jugs in a medium flat-rate box; they fit in perfectly, and it only costs about $11 to ship. If you sign up for an online account and print the labels from your home office, the driver will pick up your packages and you save about 50 cents on postage.

to see how sap is collected. I have given many tours to bus groups in Lake Placid, and the operators I have dealt with have all been pleasantly surprised that the tour was free. Even without charging for the tours, these groups are usually the most lucrative events of the year. After less than an hour of showing them around our sugarhouse and taking them for a short walk to see the sugarbush in all its fall glory, I then spend another 20 or 30 minutes selling hundreds of dollars' worth of maple products.

Take Advantage of Your Local Markets and Identity

It is usually a good idea to play up your local and regional identity in your marketing efforts. For certain customers, the fact that they are supporting a maple producer in a given town, county, state, region, or country will mean a great deal. You may also be able to highlight the "terroir" of your syrup—the unique soils and climate of your region—but I caution you against taking this too far. I don't think it does the industry much good when producers advertising their "single-source" syrup as being so much better than maple syrups that are blended together from many regions. Whenever possible, we should be comparing our pure maple syrup with other sweeteners, not with other pure maple syrup.

Become a Member of Your Local Farm Bureau and Cooperative Extension

This will allow you to develop key contacts with other farmers and sugarmakers in your area. It will also literally put your business on the map, as these types of organizations often print hard-copy and online maps of the farmers in a given county or region. Connecting your business with other efforts to support the local food economy is a great idea!

Encourage Relatives to Partake in Maple Royalty or Ambassador Programs

Chances are that if you are reading this book, you wouldn't qualify as a contestant in maple royalty or ambassador contests. However, if you have any teenage relatives, encouraging them to participate in these programs is a great way to get them involved in your sugaring operation. Not only will their efforts help to promote the entire maple industry, but having an official maple ambassador associated with your sugarhouse will also bring additional publicity and marketing opportunities to your operation.

Develop and Distribute Attractive Business Cards

Although you don't need a business card to develop a successful sugaring operation, for the small cost of

FIGURE 10.18. Our original jug design focused on the Adirondacks, but since most of the Cornell syrup is sold on campus, we developed a new jug design to incorporate the Cornell name and iconic clock tower. PHOTO COURTESY OF BRIAN CHABOT

FIGURE 10.19. Developing a custom design on glass containers is a great way to sell high-quality syrup. You can easily recoup the additional cost of the glass containers by selling them at a premium price. There are many people who don't like the idea of plastic jugs and are glad to pay a little extra for a nice glass container. PHOTO COURTESY OF ARTISAN PRINTING OF VERMONT

FIGURE 10.20. The Black Rooster Maple Company in Keene, New York. PHOTO COURTESY OF JOSHUA WHITNEY/EASTBRANCH DESIGNS

developing one you will likely earn your money back many times over with increased sales and visibility. If you want people to come to your sugarhouse, put a map on the back. And if you offer more than just maple syrup, it's a good idea to include a listing of all the different products that you offer. Make sure your business card is professionally done, attractive, and always keep a good supply of them in a safe place in your wallet or purse.

Develop Customized Jugs for Your Maple Syrup

If you are selling a lot of maple syrup in retail and wholesale containers, it usually makes sense to develop your own customized containers. There is an initial setup fee, but if you are ordering and filling enough bottles, the additional revenue and exposure you will get from having your own brand of syrup will easily cover the additional expense of a custom printing. The two leading suppliers in the maple industry of custom jugs are Sugarhill Container and Bacon Jugs for plastic jugs, and Artisan Printing of Vermont for glass containers.

Develop an Attractive Logo

And use it. Your signage and logo is an indicator of the quality of your product. Take advantage of the mystique, lore, and heritage of maple syrup, the beauty of maple trees, the maple leaf, or whatever else you would like. One of the best logos I have seen is from Black Rooster Maple Company; it utilizes the natural image of roosters with the maple leaf embedded between them.

Use an Email Listserv

If you have a guest book, online ordering system, or other ways of getting customers' email addresses, this provides an excellent opportunity to send interesting and timely emails to them. If you send out an email, be sure to use the bcc feature to avoid revealing people's email addresses to everyone on the list. Don't overload them with emails, either; only include timely, important, and interesting information that people may want to know. Make it easy to unsubscribe, and choose a subject line that makes people want to click on and open your email.

Sell Christmas Trees, Especially Cut-Your-Own

Going to the country to pick out a Christmas tree is a favorite tradition for many families. If you have an open field by your sugarhouse, I highly recommend growing Christmas trees and offering cut-your-own sales. This will attract people to your sugarhouse during the holidays, which of course is the biggest retail season of the year. While many avoid going to the mall, taking the family to a sugarhouse and Christmas tree farm is a vastly different and much better experience. Your sugaring business will help draw in people for Christmas tree sales—and selling Christmas trees will also boost sales of maple products. Look for these types of synergies between different ventures whenever possible.

Promote the Health Benefits of Pure Maple

There is a tremendous amount of information now available on the health benefits of pure maple syrup. We need to do a better job of capitalizing on this information and getting it to consumers to reap the full benefits of the research. At the keynote address of 2013 New York State Maple Conference in Verona, New York, David Marvin of Butternut Mountain Farms discussed the marketing opportunities for pure maple and how maple syrup fits directly in line with many of noted author Michael Pollan's "food rules." First off, Pollan warns us about "food-like substances" and advises people not to eat anything that their great-grandmother wouldn't identify as food. This is no problem for maple syrup, as our ancestors have been consuming maple sap and its derivative products for centuries. Another rule is to not eat anything that contains high-fructose corn syrup (HFCS). Obviously pure maple represents the opposite end of the spectrum from all the HFCS-laden imitators. Pollan also advises not eating anything with more than five ingredients, and we all know that the only ingredient in maple syrup is maple sap.

Another one of Pollan's food rules is to not eat anything that displays health claims on the packaging. After all, healthy, wholesome food shouldn't need descriptors telling us how good it is for us. Although health claims are hardly ever found on maple products, this is one of Pollan's rules that we may want to break. Most people don't realize how much healthier pure maple is for you than most other sweeteners; we should do a better job advertising that. I got a chance to meet Michael Pollan a few years ago when he gave a lecture on the UVM campus. We discussed maple syrup briefly, and he thinks that people should be consuming a lot more of it as opposed to other sweeteners. If we could get more food celebrities to state this publicly, we may not be able to make enough syrup to supply the market demand.

Have 6 Million Pounds of Maple Syrup Stolen

They say in the media world that any publicity is good publicity. Well, the now-famous 2012 maple syrup heist from the Quebec strategic reserve has undoubtedly helped bring maple syrup into the national conversation! People were amazed to learn that there is a strategic maple syrup reserve in Quebec and were baffled at the amount of syrup stolen. It was described as being enough to cover 180 million pancakes, and of course there were many puns about the "sticky situation." Everyone seemed to be talking about the maple syrup heist—it was even featured in a segment on *The Daily Show with Jon Stewart*—enhancing maple's "cool factor" among youth. I have since talked with two screen-writers who are working on projects based on this story. Given all of the media attention, I would be very surprised if there wasn't a large spike in syrup sales as a result.

Partner with Nearby Ski Resorts

In March and April ski areas throughout the Northeast are winding down their ski seasons as sugarmakers ramp up syrup production. When the sap is flowing, there may be ways of partnering with a local ski resort to market maple products during sugaring season. Vermont has developed a Ski & Maple Map that depicts all the ski areas and a wide selection of sugarhouses that are open to the public. If people can spend $80 to $100 on a lift ticket and afford a weekend vacation, certainly they can also spend the extra money to buy real maple syrup instead of the cheap, imitation stuff. There has also been a surge of interest among ski centers in developing maple operations on their own properties. Titus Mountain in northern New York just put in a 6,000-tap sugaring operation and is poised for future expansion. I expect more ski areas to follow suit in the coming years.

Sponsor a Maple-Themed Fund-Raiser Dinner

Sugaring season is an excellent time of the year to host a maple-themed event. If you are involved with a nonprofit organization or know somebody who is, offer to donate the maple products for a fund-raiser dinner to benefit that organization. One great example of this is the Berkshire Grown March Maple Dinner (http://berkshiregrown.org/march-maple-dinner) held every year in Stockbridge, Massachusetts. All the dishes feature maple syrup in one way or another, promoting the maple industry while educating people that maple

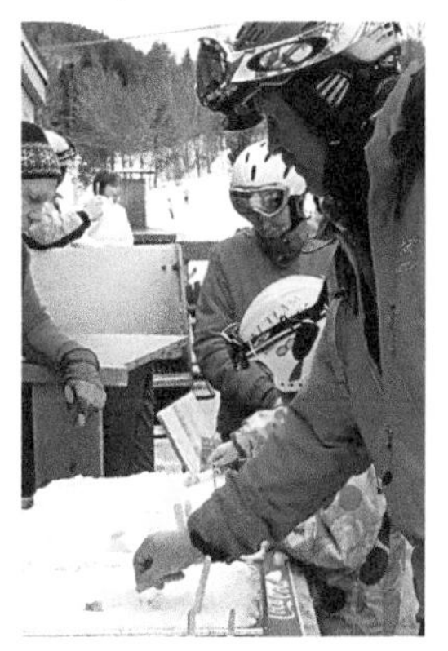

FIGURE 10.21. The Lake Placid Rotary and Shipman Youth Center teamed up to offer maple cotton candy and sugar-on-snow to skiers at Whiteface Mountain. PHOTO COURTESY OF JON FREMANTE

FIGURE 10.22. When the mother of this adopted tree at the Paul Smith's College VIC called to check on it during a cold spell, Sean Frantz wrapped a blanket around it and read the tree a story to comfort it. PHOTO COURTESY OF BRIAN MCALLISTER

can be used for more than just pancakes and waffles. By sponsoring the event with your maple products, you will also earn the goodwill of everyone who attends the event, boosting your future sales.

Adopt-a-Tree Programs

Providing someone with the opportunity to adopt a maple tree is an innovative marketing technique that has recently been introduced into the maple industry. In exchange for their purchase, people usually receive a certificate of adoption and variety of products that came from "their tree" over the course of the year. There are at least three companies that are successfully promoting this concept; a brief overview of each is provided below.

Rouge Maple

In 2012 Dino Di Pancrazio and Julie DeBlois got started in the maple syrup industry by developing an adopt-a-tree program. Dino was familiar with programs in Italy where people adopt an olive tree and receive about a liter of olive oil each year. After moving to Quebec, he decided to apply the adoption concept to the maple industry and started a new business venture called Rouge Maple. After all, Dino believes that "a good maple syrup is like a good olive oil, albeit on

FIGURE 10.23. The attractive packaging and clever marketing of Rouge Maple has given them an excellent boost in developing their sugaring enterprise. PHOTO COURTESY OF DINO DI PANCRAZIO

FIGURE 10.24. Tonewood has put together a very nice adoption package that includes their signature "maple cube" of pure maple sugar. PHOTO COURTESY OF ANDREW WELLMAN

the other end of the taste spectrum" and that every kitchen should have some. Rouge Maple utilizes attractive packaging and has catered to high-end customers willing to pay more for organic, luxury foods. Their adoption program provides an extra cachet for people who want to feel good about their purchases. Adopting anything has a very positive connotation—and who wouldn't want to take care of a maple tree and receive delicious maple syrup in return?

Their adoption program also provided a unique marketing angle for a press release announcing their business launch—it was picked up by dozens of media outlets and got them a lot of free publicity. Within the first year, their syrup was selected for gift baskets for the Country Music Awards, Oscars, and Grammy Awards. They also produce a line of gourmet maple products including salad dressing, barbecue sauces, and mustards that have done very well. Dino and Julie devoted significant effort toward developing a high-quality website, and they sell all their products through www.rougemaple.com. Given the success that Rouge Maple has had with their adopt-a-tree program and marketing maple syrup as a luxury item to high-end clients, I'd be surprised if more companies didn't pursue this strategy in the future.

Tonewood

Dori Ross grew up making maple syrup as a hobby with her family on a farm in rural Ontario. She now lives in the Mad River Valley of Vermont and started a tree adoption program through her new business venture, Tonewood Maple, in 2012. Not being a producer herself, she decided to partner with neighboring sugarmakers who all have long histories of sugaring the same property for many generations. Her mission and message to preserve small-scale maple sugaring in the Northeast while helping to save the trees and traditional Vermont lifestyle resonates with a lot of people. By adopting a tree and telling the story of the families whom the syrup comes from, Tonewood is providing customers not only with maple syrup, but also with the satisfaction of helping preserve the image and heritage of Vermont that they love. Tonewood donates a portion of all proceeds to UVM's Proctor Maple Research Center through the 1% for the Planet program. This is all part of the Tonewood message: "Adoption provides an opportunity to support

FIGURE 10.25. Tonewood donates a portion of their revenues to the Proctor Maple Research Center through the 1% for the Planet program. This lets people know that they are supporting not only private sugarmakers in Vermont, but also the researchers that help the industry thrive.

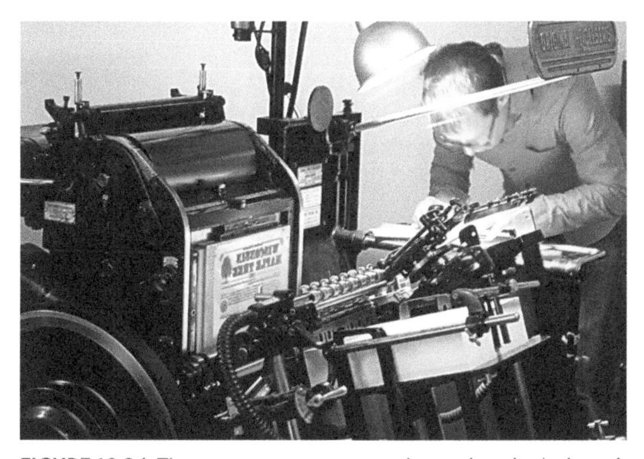

FIGURE 10.26. The antique printing press used to produce the Anderson's adoption certificates helps give their program an authentic, old-fashioned image that resonates well. PHOTO COURTESY OF STEVE ANDERSON

talented craftsmen and protect the environment, while indulging in a sweet treat." This message, coupled with attractive packaging and wide selection of other gourmet maple products, has helped launch a very successful business. The maple industry could use a lot more people with the vision and marketing skills of Dori Ross.

My Anderson's Maple

Steve Anderson is serious about marketing maple syrup. He is already well known for the Anderson's Maple sponsorship of NASCAR and has recently started a tree adoption program. The My Anderson's Maple program presents an opportunity for people "to get to know a real, live Sugar Maple tree in the Wisconsin Northwoods and become part of a sustainable family tradition." People who buy in to this program become "honorary sugarmakers" and are invited to tour the sugarbush, visit their tree, and receive some maple syrup, pancake mix, and maple popcorn. The Andersons also send email updates on the harvest and let folks know when the leaves on their tree are at peak color. They partnered with local students to put their trees online through the use of GPS units and Google Earth so that people could easily find them. Another nice marketing angle is their use of an antique printing press to do the printing of their adoption certificates, which they proudly display on their website.

Taking Maple Marketing to the Next Level

All the ideas presented up to this point can be easily implemented by nearly all sugarmakers. They don't require much initial investment and may add greatly to your bottom line. If you want to take your marketing to the next level and are willing to make larger investments in time and resources to sell your maple products, there are more strategies you could pursue.

Develop a Pancake House or Sugarhouse Restaurant

If you really want to get serious about marketing maple syrup at your sugarhouse, one of the best ways to go about doing this is developing a restaurant on site as part of your sugaring operation. Be forewarned: Most new restaurants fail, and making syrup is nothing like running a restaurant! It requires a completely different set of skills and abilities, a lot of regulations and reporting requirements, and you may even have difficulties getting it approved by your town zoning board. But if you have the right location and can get the right people to help run it, starting a pancake house/restaurant at your sugarhouse could be a great investment.

In Quebec there is a long tradition of celebrating the maple season with a trip to a *cabane à sucre* (sugarhouse restaurant) during March and April. The vast majority of these establishments are only open for 8 to 10 weeks while the sap is flowing, yet they do a tremendous amount of business during that time. As part of an economic analysis of the Quebec maple industry, ÉcoRessources Consultants estimated that the province's 320 *cabanes à sucre* generate approximately $140 million over the eight-week period in which they are open. People go there to enjoy maple-themed meals, watch the sugaring process, listen to live music, and participate in outdoor activities such as sledding and sleigh rides. Given their popularity, a well-known food columnist for the *New York Times* traveled to Quebec a

FIGURE 10.27. Many sugarhouse restaurants, such as the one pictured here, are highly successful, but they need careful planning and management to become moneymakers rather than money pits.

few years ago to write about these unique culinary adventures. One of the most famous was recently featured on a *CBS Sunday Morning* segment—the reporter and the sugarmaker both did an excellent job highlighting the wonderful atmosphere and experience.

Based on the success of the sugarhouse restaurants in Quebec, a graduate student at the University of Maine developed a GIS analysis of potential locations for sugarhouse restaurants in Maine.[1] Given the importance of the French heritage to the success of *cabanes à sucre* in Quebec, Veronique Theriault first determined the number of residents with French heritage for a given township. She then plotted on the map the maple producers who belong to the Maine Maple Producers Association and/or participate in Maine Maple Sunday, as these are the operations most likely to cater to tourism. Her analysis found 40 sugarhouses within 40km of a large enough French population to support a sugarhouse restaurant. Certainly not all of these locations would develop such a facility, but this study did illuminate the fact that there is a large potential to develop *cabanes à sucre* here in the United States.

If you are thinking of developing a restaurant at your sugarhouse, it is important to first develop a thorough business plan for how the facility will operate and draw in potential clients. I believe that having a local population with French heritage is important—but not necessary—for the development of a successful business. Nearly everyone loves maple syrup and would welcome the idea of eating at a *cabane à sucre*, whether they have French ancestry or not. If this is an idea you may want to pursue, I encourage you to first visit other sugarhouse restaurants in the United States and/or Canada. Winton Pitcoff recently published an article in *Farming: The Journal of Northeastern Agriculture* on the longstanding tradition of sugarhouse restaurants in western Massachusetts. Most of the establishments described in this article are only open seasonally, when the sap is flowing and occasionally when the leaves are changing. Opening just on weekends during these busiest times may be your best option, especially if your place isn't located along a main road and getting to it would be challenging. Destination visits are great for weekend excursions, but not during the week when people are usually busy at work.

One possibility is teaming up with a local entrepreneur/restauranteur who may want to develop the restaurant at your sugarhouse. You could provide the site and let somebody else do all the work. You will want to spend sufficient time identifying potential collaborators and make sure you iron out all the details before going forward, but if you can find the right partner, it will be a win–win for everyone. As the sugarmaker, you will have an excellent outlet for your syrup and will draw many more people to your sugarhouse than you would otherwise. The entrepreneur will have an excellent venue for a successful restaurant; it won't be just another place to eat on the strip but rather a unique destination that people look forward to going to.

Develop a Maple-Themed Bistro and Café in an Urban Setting

By far the best marketing I have ever seen for maple products is through the Citadelle Cooperative's Maple Delights Bistros—check out their website at www.mapledelights.com. They have four dedicated stores that focus entirely on maple products in Montreal, Quebec City, Vancouver, and the Montreal airport as well as their own counters within two other cafés in Quebec. I personally visited and was greatly impressed with the bistros in Montreal and Quebec City—being inside them, it's hard not to get excited about maple. Two of their bistros also contain maple museums to educate people about the wonderful history and process of maple syrup production. In addition to selling a full line of maple syrup and value-added maple products, Maple Delights offers a variety of beverages, pastries, and desserts that all utilize maple syrup or sugar as a main ingredient. In 2011 the Citadelle received a prestigious award from the North American Maple Syrup Council for this excellent marketing venture, a testament to the success of their operations.

In a similar vein, Mary Hilton established the Maple Run Emporium in 2010 on Main Street in the quaint little college town of Potsdam, New York. Although she doesn't have the customer draw of a major metropolitan area, her store takes value-added maple products to a whole new level. Maple Run Emporium features a wide variety of maple-related gourmet foods as well as maple-themed kitchenware, gifts, home furnishings, bath and beauty products, and other fine wares. Mary's mission "is to showcase the world of maple in a gourmet grocery and gift shop environment that inspires customers to explore and create with maple foods, tools, scents, and furnishings." She has done a wonderful job so far and

FIGURE 10.28. A maple themed bistro/café in an urban setting, such as the Citadelle's Maple Delights Bistro, is an excellent way to showcase the variety of pure maple as a gourmet sweetener. PHOTO COURTESY OF THE CITADELLE COOPERATIVE

FIGURE 10.29. Maple Run Emporium in Potsdam, New York, features a wide array of pure maple products as well as kitchenware and other gourmet items.
PHOTO COURTESY OF MARY HILTON

has grand plans for how to expand her business. If you have a unique maple-themed product that you think she may want to carry, I encourage you to visit her website at www.maplerunemporiums.com and learn more about her company and exciting vision for the future.

As seen with these examples, maple syrup presents an incredible opportunity to develop retail stores and cafés in urban settings. Pursuing this concept requires a completely different set of skills and is certainly not for everyone. In fact, the vast majority of sugarmakers should never even consider pursuing this type of marketing strategy. However, if you really want to get serious about marketing maple and live close enough to a major population center, opening up a maple-themed bistro or café in a Main Street setting could be a very successful venture. I see no reason why every metropolitan area on the East Coast shouldn't have a maple bistro, café, or whatever you want to call it. If you have any inclinations of doing this yourself, I would encourage you to visit one of the Maple Delights bistros and/or the Maple Run Emporium—it will be well worth the trip.

Build a Mobile Sugarhouse and Take It Places

One of the best ways to market maple is by bringing the sugarhouse to the people. Building an authentic-looking sugarhouse on a trailer frame and towing it behind a truck is an excellent way to bring the rural experience of maple sugarmaking to urban and suburban areas. This presents a tremendous opportunity to market maple syrup and value-added products to large and diverse audiences. When people see the sugarhouse, it catches their attention and sets you apart from all other vendors in a given area.

There are different types of mobile sugarhouses that serve different types of purposes and audiences. Some are built almost exclusively for syrup production and educational purposes. For instance, the Vernon-Verona-Sherrill High School FFA program developed a mobile sugarhouse that is often taken down to New York City to show inner-city students how pure maple syrup is made. They don't use this venue to sell maple products, but certainly they could. On the other hand, the Vermont Maple Sugar Makers' Association has a mobile sugarhouse that it takes to large venues such as the Big E and has been very successful at promoting maple through this joint marketing venture.

FIGURE 10.30. The Vernon-Verona-Sherrill FFA chapter has their own mobile maple sugarhouse, which they move throughout New York to teach other students about the art and science of maple syrup production. PHOTO COURTESY OF KEITH SCHEIBEL

FIGURE 10.31. This mobile sugarhouse was built to accommodate syrup production as well as retail sales and food vending. It contains an electric panel hookup, two separate sinks for washing hands and dishes, washable floors and surfaces, a refrigerator/freezer, and a food service cart inside. The cupola is easily lowered to bring the overall height down for transport. PHOTO COURTESY OF JOHN CRANLEY

If you decide to build a mobile sugarhouse, I recommend a structure that can also be used to prepare and serve food. Check with your local health department to get a list of the features you will need to include to make your mobile sugarhouse an approved kitchen facility. If you plan for it from the beginning, you can build your sugarhouse to easily accommodate all the necessary features. It's important to realize that most of the places you would take your sugarhouse to are *not* good venues for selling syrup in a container. At the fairs, festivals, and special events where a mobile sugarhouse work well, people are much more interested in purchasing items that they can consume on site than carrying around a jug of maple syrup all day. Just having an approved facility to make pancakes and waffles will greatly help your sales, but the real money to be made in the maple industry is in producing and selling value-added maple products. A mobile sugarhouse that doubles as a kitchen is one of the most economical and practical ways of marketing these items.

Final Thoughts

If you get serious about marketing maple, chances are you will develop a greater demand for maple products than you can currently meet yourself. This is a good problem to have. In the long term you may want to expand your operation by getting more sap out of your existing trees, adding more taps on your own land, leasing nearby sugarbushes, or buying raw sap from nearby landowners. There is also an easy short-term remedy: Whenever you have a gap in the amount of syrup you have produced yourself and what you can sell in the marketplace, you can usually purchase bulk syrup to fill your market demand. Any marketing professional will tell you that the last thing you want to do is create demand for a product and not be able to supply it. There is a chance that your potential customer will purchase maple syrup from somewhere else if you are out, but it is also possible that the person will just forgo purchasing maple products altogether. Since maple syrup is not a necessity, people will often go without it if it's not readily available. Buying and repackaging bulk syrup provides better synergies between those who are good at producing large amounts of syrup and those who are better positioned to market the syrup, thereby increasing the overall market for pure maple products and helping the industry grow.

I occasionally hear from sugarmakers that they won't purchase outside syrup to repackage, either because it is deceptive to the consumer, or because they wouldn't be able to purchase syrup of as high a quality as what they make themselves. For sugarmakers who are concerned about deceiving customers, bear in mind that most customers just want to get some maple syrup or other maple products—the vast majority of them don't really care if the syrup was originally produced by yourself or another sugarmaker. Your main concern should be making sure that any barrels of syrup you buy in meet the same quality and taste standards that your customers are used to, so be sure to fully evaluate and taste each barrel before purchasing. If you have steady customers every year that absolutely only want your syrup, then be sure to set enough of this aside and start bottling purchased syrup before you completely run out. Or if you are very concerned about deceiving customers, then simply tell them the truth: The demand for your products is greater than you can currently meet, so you are also selling the syrup of other sugarmakers whom you have carefully selected to ensure the highest quality standards.

While I can understand not wanting to deceive customers by bottling bulk syrup purchased from another sugarmaker, it is ill advised to believe that you could never buy bulk syrup that meets the same standards as yours. There is certainly a lot of substandard and off-flavored syrup out there, yet there is much more high-quality syrup available in barrels than there is off-flavored syrup. Once you start doing some research, I'm sure you'll find more than enough syrup that tastes just as good as (or maybe even better than) yours. While we all like to think that our syrup is special, the reality is that all pure maple syrup is special and there are thousands of sugarmakers who make excellent syrup. There are minor differences in the flavors of maple syrup based on the individual site characteristics and processing methods of sugarmakers—this is part of the terroir of maple. However, the similarities in syrup from different sugarmakers far outweigh the minor differences in flavor due to terroir and processing style. Consumers have a hard enough time telling the difference between different grades of syrup, let alone the same grade of syrup produced at two different sites.

CHAPTER 11

Value-Added Products: The Path to Profitability

In addition to syrup, there are a wide variety of products that you can make from the sap of maple, birch, and walnut trees. Although we all enjoy tapping trees and producing syrup, one of the best ways to actually earn a profit is by using the syrup to produce and sell value-added products. Nearly everyone has heard of the standard value-added products such as maple candies, maple sugar, and maple cream. Less well known are the dozens of other products that utilize pure maple syrup, sugar, and cream as a main ingredient. There are plenty of good resources available on the finer points of making these value-added products, so I have not devoted valuable space in this chapter to the processing end. Rather, the focus of this chapter is to introduce you to the wide variety of maple and birch products and give you a better idea of what is possible. The more we can use maple and birch syrups as sweeteners in other products, the better off our own bottom lines and the entire industry will be.

Given the high profit margins that are possible when making and selling value-added products, it surprises me that only a small percentage of sugarmakers actually include these in their product line. To be fair, making syrup is a lot different from producing confectionary products from the syrup, and it takes a certain skill level to properly produce, package, and market these products. Most of the sugarmakers I know would rather be out in the woods tapping trees and cutting wood than in the kitchen making maple candies. However, even if you don't have the time or interest in producing value-added products yourself, there may be other family members or friends who would like to do this as part of your overall sugaring operation.

If you want to do this yourself or have an interested friend or family member, you'll want to pick up a copy of the *Maple Confections Notebook* developed by Steve Childs, my colleague at Cornell.[1] Steve has been doing research on the best techniques for making value-added products over several years and has amassed a wealth of useful information in this manual. He also puts on Maple Confections workshops throughout the eastern United States every year; try to attend one if you can.

FIGURE 11.1. The *Maple Confections Notebook* developed by Steve Childs at Cornell is a must-have for anyone serious about making value-added maple products. PHOTO BY NANCIE BATTAGLIA

Maple Cream, Candy, and Sugar: The Traditional Value-Added Products

Cream, candy, and sugar are the traditional value-added products manufactured entirely from pure maple syrup. To make these, you simply boil the syrup to a higher temperature, thereby removing more of the water, and then agitate the thickened syrup after different cooling periods. Each product has its own specifications for the temperature range to boil the syrup to and the amount of cooling that should take place before agitation. You can purchase specialized machines to make these products on a commercial scale, yet you can also make them at home with ordinary baking equipment.

There are plenty of good resources out there on how to make maple cream, candy, and sugar, so I won't take up valuable space in this book describing those methods. However, the economics of making and selling these products is not well understood, so I have devoted substantial time to them. After all, maple cream, candy, sugar, and the like are only value-added products if you price them at a high enough level to earn extra money from producing and selling them. To help you understand how much additional money you can earn from various maple products, I developed some Excel spreadsheets that can be downloaded from the Chelsea Green Publishing website, www.chelseagreen.com/mapleguide. To use these for yourself, you need to keep track of how much of a particular item you make in a given batch and how much syrup you used to produce it. You also have to record your retail and wholesale distribution of these products as well as your pricing and packaging costs. Finally, if you keep track of the time you spend making and marketing these products (as well as any other marketing expenses), you can also determine your hourly wage for making and selling value-added maple products.

FIGURE 11.2. Maple cream is a delicious, spreadable maple product that can be used for a variety of baking and confectionary applications, or just enjoyed as a snack. My favorite ways to eat maple cream are as a dip for pretzels and spread on toast. PHOTO COURTESY OF ARTISAN PRINTING OF VERMONT.

FIGURE 11.3. A large batch of maple candies drying in preparation for being packaged. PHOTO COURTESY OF BRIAN CHABOT

FIGURE 11.4. Prior to the 20th century, the most commonly produced maple item was maple sugarsince it was easier to store and preserve. Maple sugar is now a luxury item that is marketed as the local, natural, and healthier alternative to heavily processed cane and beet sugars. PHOTO BY NANCIE BATTAGLIA

Maple Candies

If you are set up at a fair, festival, or anyplace where people are consuming food on site, these are among the best products you can make. They are a pure, natural treat that many people love, and since you can't usually find them at the grocery or convenience store, people often search for them at fairs and festivals. Most candies are often made in the form of a maple leaf, though you can purchase candy molds in all different types of shapes and styles, depending on your target market and the time of the year. If you can make them on site and let people watch you boil the syrup and fill the molds, you won't even have time to stock the shelves before they are sold out.

Maple candies can also be one of the more profitable value added products. Table 11.1 analyzes a recent batch of candies made by a sugarmaker in New York. Taking 1.5 gallons of syrup that was only worth $45 as a bulk commodity, she was able to transform this syrup in to almost 700 pieces of candy that sold for nearly $300. When you factor in the cost of packaging and depreciation on the candy machine, this sugarmaker was able to earn more than $200 in additional profits from this venture. Since it took about four hours to do all the work, the sugarmaker made over $50 per hour for turning maple syrup into candy—not bad for an afternoon job! Just because she was able to earn this kind of money does not mean that you will do the same, of course; it all depends on how efficient you are at converting syrup into candy and what prices you sell the candies at.

You can determine how efficient you are in making candy by simply dividing the weight of all the candy made by the weight of the syrup used to produce the candy. Given the fact that we have to boil off a significant amount of water in order turn syrup into candy, the theoretical maximum conversion rate is 73 percent. However, your actual conversion rate will often be much less due to overfilling of the molds, spilling candy between the molds, and having syrup stick to your pans and equipment. I know many sugarmakers who sell these scraps as "maple crumbles" for people to put in their coffee and tea or use them as samples; your actual

Testing for Invert Sugar Content in Syrup

If you are going to produce value-added products, it is important to use syrup with invert sugar levels appropriate for what you are making. Invert sugars are glucose and fructose that are created through microbial activity on sap before it is boiled. They are also referred to as "reducing sugars" and have a large impact on how syrup behaves when processed further into candy, cream, or sugar. A small amount of invert sugar content is desirable and helps provide the characteristic maple flavor, but too much can cause syrups to not crystallize properly. Typically light-colored syrups produced at the beginning of the season have lower invert sugar levels than dark syrups produced later on, but syrup color is not always a perfect indicator of invert levels. I recommend testing any syrup that you plan on using for value-added products according to the following procedure:

Make a 1:10 dilution of maple syrup and water using a gram scale. Place a small paper cup on the scale and then "zero" it out. Add 10 grams of syrup to the cup and then 90 grams of water until the scale reads 100 grams. It doesn't matter that you get exactly 10 grams of syrup to start with; what is important is that the final solution contains 10 percent maple syrup and is well stirred. You can then test the amount of invert sugar content in the solution with an ordinary glucose meter that is used to measure blood sugar levels for diabetics. You don't need an expensive meter, just a basic one that provides a numerical output of mg/dL for whole blood readings. Insert a test trip into the 10 percent syrup solution and place it on the meter. Whatever the reading is, multiply it by 0.02 in order to determine the percentage of invert sugar. For instance, a reading of 50 would equate to 1 percent invert sugar content since 50 × 0.02 = 1. Once you know the invert sugar content, you can then decide whether the syrup is suitable for a particular purpose.

TABLE 11.1: An Example of the Economics of Making Maple Candy Provided by an Experienced Sugarmaker in New York.

Number hours spent making and packaging candy	4
Gallons of maple syrup used	2.1
$/lb for maple syrup	$2.70
Cost of candy machine	$1,600.00
Lbs of candy made over the life of machine	2,000
% of candy given as samples	3%
Machine cost/batch	$11.47
Measured conversion rate	72%
Syrup to candy conversion rate (sold)	61%
Overfill rate	15%

CANDY SIZE	NUMBER OF CANDIES MADE	TOTAL OZ OF CANDY	
⅓ oz	695	229.35 oz	
		14.3 lbs	what is sold
		16.875 lbs	what was measured

Products	Number sold		Prices		Packaging cost	
(ENTER DIFFERENT ITEMS HERE)	RETAIL	WHOLESALE	RETAIL	WHOLESALE	(PER UNIT)	TOTAL
6 oz (16 pieces) of ⅓ oz	43		$7.00		$0.15	6.45

total revenue	$301.00
total profits	$213.21
profit/hour	$53.30
profit/gallon of syrup converted to candy	$101.53

* does not include the cost of electricity or vendor fees

FIGURE 11.5. You need to be careful when filling candy molds in order to minimize waste—as with all tasks, practice makes perfect. PHOTO COURTESY OF PETER SMALLIDGE

FIGURE 11.6. Maple candies can be crystal-coated to extend their shelf life and make them more marketable. PHOTO COURTESY OF BRIAN CHABOT

waste can be much less if you maximize the utility of the scraps. In this example, the actual weight of all the candy produced from the 2 gallons of syrup was nearly 17 pounds, whereas the calculated weight of 695 pieces of 1/3-ounce candies is just over 14 pounds. Thus, the actual conversion rate was 72 percent when looking at the amount of candy produced but only 61 percent when considering the amount of candy sold. In this example, the 1/3-ounce candies being made were actually closer to 0.4 ounce in weight, resulting in an overfill rate of approximately 15 percent. If the molds were filled to have only 1/3-ounce of candy in them, the sugarmaker could have produced another 120 candies (and bring in an additional $60 in revenue). I would strongly encourage you to determine your overfill rate by weighing how much candy you produce in total versus the theoretical weight of all the candies produced and sold.

Granulated Maple Sugar

Although granulated sugar is now considered one of the value-added products, it is important to remember that nearly all maple sap used to be converted into maple sugar. Maple sugar is easier to store than syrup and will not spoil, so Native Americans and early settlers produced much more sugar than syrup. It wasn't until cane sugar became less expensive and more readily available around the turn of the 20th century that maple syrup overtook maple sugar in total production. Even though cane and beet sugar cost *much* less than maple sugar, the market opportunities for maple sugar have never been better. I highly recommend reading all about maple sugar in the *Confections Notebook* and then making it yourself.

If you just put some jars of maple sugar on your display table, chances are you won't sell much of it. People generally know what maple syrup is and how to use it, yet many are surprisingly confused about maple sugar and its potential uses. To sell maple sugar, you have to tell people what it is and have them sample it. But don't just have them eat sugar all by itself. I think the best way to sell maple sugar is by offering samples of maple cheesecake dip made with pure maple sugar (see the sidebar). This is simple to make and shows people one of the many ways to use maple sugar when cooking and baking. It can be replaced 1:1 with other granular sugars in all recipes, and if you convey the message that maple sugar is the local, healthier, better-tasting, and more sustainable sugar for the Northeast, I'm sure you will sell a ton of it.

Maple Cheesecake Dip

Many people enjoy a maple cheesecake dip produced with pure maple sugar. The recipe is easy to follow: Just mix 8 ounces of cream cheese, 8 ounces of Cool Whip, and 8 ounces of maple sugar together in a blender. It's so easy even I can do it! At a taste testing of nearly 500 people at Empire Farm Days, 95 percent said that they liked it "extremely" or "very much."

Maple Cream

This is by far my favorite maple product. It is smooth and creamy, with a wonderful, enhanced maple flavor. Despite its winning qualities, it has never gained the foothold in the marketplace it deserves. One of the issues is the confusion caused by its name—you've probably heard stories or seen for yourself people taking a sample of maple cream and rubbing it on their hands as a moisturizer. To avoid this some producers call it maple butter, but this leads many to believe that it is a dairy product. (As a quick aside, maple sugar whipped into butter is unbelievably good!) A few sugarmakers have decided to call it maple spread, and I know one company that is adding fruit flavors such as raspberry and strawberry. Whatever you call it, I recommend packaging the cream in a glass jar. That way people can see what they are buying, and as long as you are making a quality, appealing product, you'll sell a lot more of it that way at higher profit margins.

Another issue with maple cream that some see as a detriment is the separation of syrup from cream that

FIGURE 11.7. Several companies are now selling a gear-driven cream machine that makes the cream-making process much easier. One strategy you might consider is purchasing a cream machine in partnership with other nearby sugarmakers. Rarely will you all need to use it on the same day, and as long as you all take care of it properly, everyone will benefit. PHOTO COURTESY OF BRIAN CHABOT

occurs over time. The Food Venture Center at Cornell worked with sugarmakers in New York to develop a shelf-stable cream that won't separate and is fairly easy to make, yet very few sugarmakers follow this recipe (you can read more about it in the *Confections Notebook*). Personally I don't think the separation should be viewed as a bad thing—you can actually market it as a positive attribute. Peanut butter brands that contain natural oils at the top cost a lot more money than the more heavily processed Peter Pan and Jif varieties that don't require any stirring. The brands with oil at the top are perceived as more natural and wholesome (because they are), and thus many consumers are willing to pay the extra price. Since maple cream is made entirely from pure maple syrup, it is perfectly natural for some of the syrup to separate out. I tell people they can stir it back in, pour it on top of their pancakes, or just drink it straight! We are charging people a premium price for maple cream, so it makes sense to market it as a natural, healthy sweetener made entirely from pure maple syrup. After all, I doubt you can turn Aunt Jemima into a maple-flavored cream—and even if you could, who in their right mind would ever want to eat that?

Since not many people know what cream is, I always recommend sampling it at special events. The easiest way to do that is to have a 1-pound jar open with a tray of pretzel sticks next to it. Make sure the cream is warm and creamy, otherwise you'll have a lot of broken pretzel sticks in your jar! The sweet and salty of the pretzels and cream is a winning combination; I've only found a few people in this world who don't like it. Without sampling, your cream will likely sit on the shelf and eventually get moldy, so sacrifice a jar and let people try some—you'll be glad you did!

TABLE 11.2: Economic Analysis of Making Maple Cream

Number of jars filled		Cost of jars	Price of jars		Number of jars sold		Total revenue
			RETAIL	WHOLESALE	RETAIL	WHOLESALE	
	Number hours spent making and packaging cream	3					
	Gallons of maple syrup used	2.59					
	Lbs of maple syrup used	28.75					
	Oz of maple syrup used	459.984					
	$/lb for maple syrup	$2.70					
	Price of cream machine	$1,600.00					
	Lbs of cream made over the lifetime of machine	1,000					
	Percent of cream given out as samples	5%					
1.5	½ lbs of cream produced	$1.00	$8.00	$6.00	16		$128.00
18	1 lb jars of cream produced	$1.15	$14.00	$10.00	9		$126.00
	Value of syrup used to make cream	$77.62					
	Cost of jars	$22.20					
	Cost of machine/batch	$30.00					
	Total revenue	$241.30					
	Total profits	$111.48					
	Profit/hour	$37.16					
	Profit/gallon of syrup converted to cream	$43.04					
	Shrinkage from syrup to cream	65%					

* does not include the labor cost of marketing or electricity and vendor fees

It is fairly easy to determine the economics of making maple cream. Simply keep track of the amount of syrup you used to make the cream and your total output and sales. Table 11.2 provides a typical analysis of making cream on a gear-driven cream machine. Although the theoretical shrinkage rate from syrup to cream is 77 percent, in reality it will be difficult to achieve this. Cream often sticks to the sides of your equipment, and we often put more cream in the jars than we theoretically have to. However, it is much better to give your customers a little bit extra cream rather than having them feel as though they were shortchanged. In this example, the additional profit earned by converting a gallon of syrup to cream was $43. It's worth noting that this profit margin accounts for the cost of buying the machine and the jars as well as the opportunity cost of selling the syrup in bulk markets.

Sugar-on-Snow

Also known as "jack-wax," sugar-on-snow is one of those quintessential sugaring traditions. To make this you simply boil syrup to a higher temperature (18 to 40°F above the boiling point of water) and then pour it onto packed snow or crushed ice so that it rapidly cools and crystallizes. The lower the temperature you bring the syrup to, the more taffy-like it will be; higher temperatures will result in a glass-like consistency.

Maple Soda

Making maple soda is a fairly simple process that could greatly increase your sales at fairs, festivals, and any venue where you are selling products for immediate consumption. We drink a lot of soda in this country,

Safety with Value-Added Products

Boiling syrup to very high temperatures and then moving the extremely hot syrup from one vessel to another can be dangerous. Spilling incredibly hot syrup on your bare skin will lead to extremely painful and debilitating burns. This recently happened to a Cornell employee when she was making "maple suckers" at the Arnot Forest. The pain was described as much worse than childbirth, and she missed 12 weeks of work as a result. We now have a policy at Cornell in which anyone working with hot syrup must wear appropriate gloves, apron, safety glasses, long sleeves, long pants, and quality work shoes. Following these safety precautions is especially important when moving hot syrup, though it makes sense to wear the appropriate apparel at all times.

FIGURE 11.8. A common way to serve sugar-on-snow at a *cabane à sucre* in Quebec.

and using maple syrup as the sweetener may be one of its best uses. Considering how often the average person eats pancakes versus how often they drink soda and other beverages, it's a wonder why we spend most of our efforts marketing pure maple syrup as a pancake topping. Marketing maple syrup in soft drinks is an excellent way to increase consumption of pure maple (which is currently only 3 ounces per capita). Someone would only have to drink two or three cups of maple soda over an entire year to get the same amount of maple syrup in their diet as the average American consumes annually. Considering that Americans drink almost 100 gallons of soda in a given year, only about 2 percent of us would have to consume maple as our soda of choice 7 percent of the time in order for the consumption of maple syrup in the United States to double. To put this another way, for every 50 people that you know, do you think at least 1 of them would like to drink maple soda? If it was actually available for purchase, do you think they would choose a maple soda once out of every 15 times that they have a soft drink, even if they had to pay a quarter or two more for it than a typical soda? I certainly know my answer; the only question I have is, "Why we aren't doing more to market maple sodas?"

Being able to offer an all-natural, great-tasting soft drink without any high-fructose corn syrup or artificial sweeteners and preservatives presents a huge market opportunity for maple syrup. There are plenty of people who like to indulge in sweet, carbonated beverages and would prefer a local, healthier option. Maple soda is one of those guilty pleasures that you don't have to feel that guilty about, and it tastes great as well! However, since I am more than a little biased on this, consider some research that Steve Childs did with maple soda at Empire Farm Days in 2007.[2] Steve made a maple soda with dark amber syrup that contained 14 percent sugar by weight (similar to a Mountain Dew) and served it to over 500 people. The results were very favorable: 70 percent indicated that they liked it "extremely" or "very much," whereas 24 percent liked it "moderately" or "slightly." The Massachusetts Maple Producers Association also experimented with maple soda at the

FIGURE 11.9. Steve Childs sampling maple soda at Empire Farm Days, where 70 percent of the people surveyed claimed to like the soda "very much" or "extremely." PHOTO COURTESY OF BRIAN CHABOT

Big E in 2012. They sold roughly 800 cups at $2 each and received a great deal of positive feedback—most of the customers compared the taste to cream soda. If I worked in marketing for the food and beverage industry and was able to get those results for a new beverage made with all-natural ingredients, there is no doubt that I would be pursuing this product aggressively.

The challenge of developing maple soda on a larger scale will undoubtedly be the cost and availability of high-quality, good-flavored maple syrup in sufficient quantities for a major manufacturer to want to develop this product. They can get HFCS, cane sugar, and other natural or artificial sweeteners at a fraction of the cost of maple syrup and just continue to sell their existing sodas at high profit margins. The only company I have discovered that is making shelf-stable sodas with maple syrup as the sweetening agent is the Vermont Sweetwater Bottling Company. They have been in business since the 1990s and have developed a successful line of maple sodas and seltzers, yet their company is a very minor player in the overall soda market. It's worth noting that they use 1 ounce of maple syrup in their 12-ounce soda bottles, equating to 5.5 percent sugar by weight (much lower than the maple soda that Steve made for Empire Farm Days). I have tried the Vermont company's soda and it tastes great—there is an excellent maple flavor, yet the production cost is low, as is the calorie content. Since many people who would choose a pure maple soda are health-conscious, limiting the calories makes a lot of sense. Even if maple sodas never become a major part of the soft drink industry, you can still take advantage of this opportunity by developing this type of beverage yourself.

If you sell maple products at fairs, festivals, farmers' markets, or any venue that involves on-site consumption of food and beverages, then you should seriously consider making maple soda. There are three ways to go about making the soda, depending on the scale at which you plan to operate:

1. Purchase large containers of seltzer and then mix it with maple syrup to achieve your desired

Economic Analysis of Making and Selling Maple Soda

On the Maple Products Calculator spreadsheet available from the Cheslea Green Publishing website, www.chelseagreen.com/mapleguide, one of the worksheets allows you to determine your costs and profit margins for selling fresh maple soda. Table 11.3 contains a typical example, though I encourage to use the calculator to fill in your own values and see if making and selling maple soda makes sense for you.

As you can see, making maple soda can be a very profitable venture. In this example, you would still be earning $26 per hour for making and selling fresh maple soda. You could pay someone $10 an hour to help run this machine at your sales booth and still make $16 an hour. That may not sound very impressive, but keep in mind that the additional profit for each gallon of syrup sold as soda is $135. Of course the more sodas you sell in a given time period, the more your hourly wages go up. I encourage you to offer samples in 2-ounce cups; the cost of doing so will be minimal and you will likely develop some loyal maple soda customers as a result. It's also important to keep everything cold—warm soda of any flavor doesn't taste great, even pure maple.

TABLE 11.3: Economic Analysis of Producing Maple Soda for On-Site Consumption

Raw Material Prices	
$2.70	Bulk price of maple syrup
$0.99	Cost of 2-liter seltzer bottle
$0.01	Cost of ice per oz
$0.09	Cost of cup and straw
Batch Size	
4	Liters of seltzer used
28	Oz of maple syrup
Batch Values	
11%	Sugar in soda (by volume)
$8.54	Total cost of batch
163	Total oz produced
Prices and Proportions	
$2	Price of cup of maple soda
8	Oz of soda in cup
$0.42	Cost of soda in cup
4	Oz of ice in cup
$0.04	Cost of ice in cup
Profit Margins	
$1.45	Profit margin per cup of maple soda
$135.43	Profit per gallon of syrup converted to soda
18	Number of sodas made/hr
$10.00	Labor rate/hr (what you pay someone to make and serve the soda)
$16.13	Profit made per hour

* analysis does not include the cost of the mixing container, vendor fee, or samples.

consistency. This requires the least amount of initial investment, though your cost to produce each cup of soda will be higher since you have to purchase the seltzer. To avoid making a big mess, be sure to add the syrup slowly to the seltzer and not the other way around!

2. Buy a home carbonation unit and make individual batches of maple soda. These are readily available, do not cost much money, and are relatively easy to use. In 2012 alone, Americans purchased 1.2 million units to make their own sodas and seltzers. The main advantages are that it can save you the cost of purchasing seltzer

and will help boost sales by being able to demonstrate how you make the soda. However, it can be fairly labor-intensive and you may not have time to run this machine and help customers, so you'll want to have someone on site who is specifically devoted to making the sodas. For especially large events, the home-scale units won't be able to keep up with demand, so you would need multiple units going at once.

3. Buy a commercial soda maker and adapt it to handle maple syrup. The NYSMPA recently experimented with this at the state fairgrounds in Syracuse, with mixed results. Maple soda is a popular item, but they had difficulties adapting a regular soda maker to handle maple syrup. If you can get this to work, all you will have to do is pull the trigger on the nozzle and watch maple soda come out of the fountain. These machines usually come with four to eight different heads for the various types of sodas that are often served. With this type of setup you can have maple sodas with differing flavors and sugar concentrations to match the desires of a wider range of consumers. Once someone figures out how to adapt these machines to make maple soda in an efficient manner, I'm sure the market for it will take off.

If you think maple soda is a good idea but don't have the time or desire to market it yourself, I encourage you to seek out restaurants that may want to pursue it. There are many chefs and bartenders who would be glad to offer maple soda as a specialty beverage in addition to their traditional offerings. One of the issues restaurants have to deal with is the fact they typically don't own their soda machines; rather, they lease these from Pepsi or Coke and can only use soda mixes purchased from the company. There are ways around this obstacle, so suggest that they try it out during March as a seasonal beverage when the sap is flowing and people are thinking more about maple products. I have no doubt that many restaurants will turn you down and say they aren't interested. However, I am equally certain that if you are persistent and choose the right establishments, you will find a great number of chefs to try it out. Once they get positive feedback from customers, you will have developed another great market for your syrup. Home-scale carbonators are also becoming increasingly popular as manufacturers have made them less bulky and more user-friendly. People who use these are willing to spend extra time and money on more natural, homemade, and authentic food and beverages—exactly the kind of customers we are looking for. As sugarmakers, we need to invest more resources in marketing maple syrup as the healthiest and best-tasting sweetener to use in a home carbonation machine.

FIGURE 11.10. The maple soda being produced by Vermont Sweetwater Bottling Company. Each 12-ounce bottle is sweetened with nothing but 1 ounce of extra-dark pure maple syrup. PHOTO BY NANCIE BATTAGLIA

Maple Cotton Candy

Like the vast majority of adults, I don't find regular cotton candy appealing. Eating pure sugar off a cone with lots of pink and blue food coloring just doesn't look, sound, or taste very good. However, by blending

TABLE 11.4: Economic Analysis of Making Maple Cotton Candy

$7.50	Wholesale price per lb for maple sugar
$0.47	Value of maple sugar (per oz)
25	Lb size of cane sugar bag
$15.00	Price of cane sugar bag
$0.04	$/oz cost of purchasing cane sugar
4	Total oz of sugar per bag
20%	% of mix that is made of maple sugar
$0.50	Total cost of sugar per bag (considering the value of wholesale maple sugar)
$0.03	Price of plastic bag
$0.01	Twist tie
30	Number of bags made/hr
$4.00	Retail price per bag
$1,400.00	Purchase price of cotton candy machine
10,000	Number of bags made over the life of the machine
0.14	Machine cost per bag sold
5%	% of product made given out as samples
$3.30	Profit per bag of maple cotton sold
$99.01	Profit made per hour
$478.54	Profit per gallon of syrup converted from bulk syrup to cotton candy

* Does not include the cost of converting the syrup to granulated sugar.

FIGURE 11.11. Mike Bennett makes maple cotton candy outside a mobile sugarhouse at the Lake Placid I Love BBQ Festival.

pure maple sugar with cane sugar, and leaving out the artificial dyes, we can make a maple cotton candy that tastes great and appeals to a much broader spectrum of society. Whenever you are making maple cotton at fairs and festivals, be sure to offer people samples. Parents are more willing to let their children have the "all-natural" version of maple cotton candy, and once they try a sample, they usually buy a bag for themselves as well! Although maple cotton is best served fresh at special events where people eat it right away, some sugarmakers are having great success with putting it in hard plastic containers for extended shelf life. Rather than limiting the market to just on-site consumption, proper packaging can keep the maple cotton fresh for people who want to relive the fair experience year-round. The fact that you can now buy maple cotton at Whole Foods is a testament to the incredible market appeal for this "all-natural" treat.

What really gets people excited about making maple cotton candy is that it is the most profitable way to sell maple syrup. By converting the syrup into granulated sugar, mixing it with cane sugar, and spinning it into an airy, fluffy treat, you can earn a large return on your investment. Table 11.4 provides an economic analysis of a typical maple cotton candy operation. After accounting for the cost of the machine, packaging materials, and the maple and cane sugar, the additional profit per gallon of syrup converted to maple cotton candy is nearly $500, assuming you are selling a 4-ounce bag for $4 and put in 20% pure maple sugar in the mix. Note that this is based on purchasing maple sugar at $7.50 per pound. If you were to calculate your profits based on converting your bulk syrup to granulated sugar, the profit margin could be greater. Even if you're only making 30 bags per hour (you can actually do more than that), you're still making about $100 per hour. If you are concerned about spending $1,400 on a new unit, consider this expense as simply a few gallons of syrup or a couple days of your time.

Maple as an Ingredient

Given that pure maple is a natural sugar, it can be used in more foods than I could list in this book. Some of the

FIGURE 11.12. Maple balsamic salad dressing is extremely easy to make. Simply mix ½ cup olive oil, ½ cup maple syrup, and ¼ cup balsamic vinegar, then add in whatever herbs and spices you like. This is the only salad dressing we use in our household, and I've eaten a lot more greens ever since my wife introduced me to this delicacy. PHOTO BY NANCIE BATTAGLIA

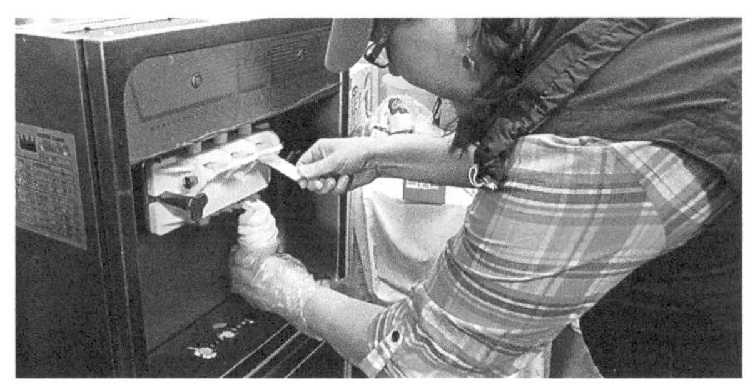

FIGURE 11.13. If you sell maple products at fairs, festivals, and other special events, I highly recommend getting a soft-serve ice-cream machine. Simply add pure maple syrup to the vanilla mix for a delicious maple ice cream (also called a maple cremee in Vermont). If you don't have a soft-serve machine available, offering pure maple syrup on top of regular vanilla ice cream is also a winning combo.

FIGURE 11.14. Paul Smith's College in the Adirondacks specializes in culinary arts and forestry. They often have the culinary students create various dishes featuring pure maple syrup produced at the college. PHOTO BY NANCIE BATTAGLIA

FIGURE 11.15. Several companies are now selling maple coffees and teas that use pure maple as a flavoring agent. PHOTO COURTESY OF THE NEW YORK STATE MAPLE PRODUCERS ASSOCIATION

FIGURE 11.16. If you like peanut butter, you will love maple peanut brittle. This is one of the more popular items produced at Mapleland Farms in Salem, New York. PHOTO COURTESY OF MARY JEANNE PACKER

FIGURE 11.17. All types of nuts can be coated with pure maple syrup, the most common being peanuts, walnuts, cashews, and pecans. Here peanuts are being mixed with maple syrup before being placed in the oven to bake. PHOTO COURTESY OF BRIAN CHABOT

FIGURE 11.18. The best desserts require the best ingredients, and nothing surpasses maple in terms of flavor, versatility, and all-natural goodness. PHOTO BY NANCIE BATTAGLIA

FIGURE 11.19. I've never had a barbecue sauce that didn't taste better with pure maple in it. The Maple Barbecue Sauce offered by Whiteface Mountain Gourmet Maple Products is a top seller throughout the Adirondacks and beyond. PHOTO BY NANCIE BATTAGLIA

FIGURE 11.20. There are very few dishes that don't taste better with a little maple pepper. Since this is relatively hard to find, we have several customers who come to our sugarhouse specifically to get maple pepper. PHOTO BY NANCIE BATTAGLIA

many items that feature pure maple syrup or sugar as one of the main ingredients are maple popcorn, maple fudge, maple granola, maple ice cream, maple milk shakes, maple lollipops, maple jelly, maple coffee, maple tea, maple peanut brittle, maple-coated nuts, maple slushies, maple-glazed meats, maple mustards, maple salad dressings . . . the possible uses for maple are limited only by your imagination. There is a saying that the only thing pure maple syrup isn't good on is the floor! While this is mostly true (don't try it on your pizza), pure maple does serve as a great complement and sweetener for a diverse array of food and beverages.

To illustrate how maple can be used as a premier ingredient in almost any dessert, consider the case of Vancouver chef James Coleridge, who recently developed a gelato using pecans and maple syrup. He entered his creation at the 2010 Gelato Festival in Florence, Italy, and won both the public choice award and the judge's award.[3] Coleridge took his "old-world style of training in Italy and did a new-world flavor" that was emblematic of his Canadian heritage. Those of us who know about the superior flavor of maple syrup versus other sweeteners should not be surprised that he was able to win this competition by incorporating pure maple syrup. As we continue to market maple syrup throughout the world, it is important to develop new uses for maple syrup in traditional cuisines. What other desserts or dishes can you think of that would be better with pure maple?

Brewing, Fermenting, and Distilling with Tree Sap and Syrup

The sap and syrup from maple and birch trees can be used for brewing, fermenting, and distilling into a wide variety of alcoholic beverages. Since maple and birch sap basically consist of water and sugar, they are excellent raw materials for brewing, fermenting, and distilling. Although sap and syrup are relatively expensive raw materials, the premium price that the finished products command in the marketplace makes it well worth the higher cost of production. While homebrewers have been using maple and birch for decades to make various wines, beers, and liquors, an increasing number of businesses are now commercializing these products. The following paragraphs highlight some of the many companies that are successfully turning maple and birch sap and syrup into profitable alcoholic beverages. It is far from an all-inclusive list, but does provide an overview of what is possible.

Maple Sap Ale

Making beer from the last run of sap is a forgotten tradition that is now making a comeback. Many years ago, rather than making really dark and potentially off-flavored syrup from the yellowish, bacteria-laden sap at the end of the season, some sugarmakers used this sap to brew a strong beer. They would boil it down partially and then add yeast, hops, and perhaps some raisins or sugar, then stick it in a barrel for a few months until it was ready. The tradition of sap beer was popular enough that Vermont artist John Cassel recorded a song all about sap beer in the 1970s. This practice was mostly forgotten for decades, yet is being revived today and may get a boost from popular media stories that have recently highlighted sap beer being brewed in Vermont.[4]

Kevin Lawson specializes in small batches of unique, fine beers at his Sugarhouse Brewery in Warren, Vermont. He has brewed beer with fresh sap, but his preferred method is to use the partially boiled-down sap that comes off the evaporator at the end of the season. Many sugarmakers will chase the last of the syrup out of the pan by feeding in permeate or water when there is no more sap to boil. Once the contents of the evaporator become diluted to the point that it doesn't make sense to keep boiling, many sugarmakers will just dump the remaining contents. When Kevin heard about this, he asked a couple of sugarmakers to skip this process altogether. When there is no more quality sap to boil, they simply drain the pans and bring the contents to Kevin. With everything mixed together, the sugar content usually ranges between 10 and 14. Kevin now turns this into the Maple Tripple Ale, a fine beer that

FIGURE 11.21. The Lake Placid Pub & Brewery has had great success in their experiments with maple sap ales. PHOTO BY NANCIE BATTAGLIA

FIGURE 11.22. This may look like an ordinary fermentation tank at the Lake Placid Pub & Brewery, but Kevin Litchfield is filling it with 360 gallons of fresh maple sap to be brewed into a maple sap ale.

recently won the Silver Cup at the 2012 World Beer Cup for Specialty Beers. Kevin makes it a point to only get the partially boiled sap from a couple of sugarmakers who stop making syrup before the sap starts to turn buddy or off-flavored. If you try this at home, be sure that whatever liquid you draw off the evaporator tastes and smells good. If you have already reached a point in the season when you are producing an off-flavored commercial-grade syrup, then your beer probably won't be winning any awards!

The Lake Placid Pub & Brewery also experimented with maple sap beer in 2012 with great success. Kevin Litchfield, the head brewer, simply replaced the 360-gallon kettle of water with sap and then added in 5 gallons of maple syrup at the end for a maple sap pale ale. Not an overly sweet beer, it had just a hint of maple flavor. The beer was extremely popular and sold out in a few short weeks, much sooner than other seasonal varieties. Given the success of the maple sap ale, Kevin used maple sap to replace water for their signature Ubu ale in 2013. The sap added a little extra sweetness and potency to the beer, making their most popular beer even better.

Maple Beers

Whereas maple sap ales can only be made for a brief period every year, beers that include maple syrup can be brewed year round. The microbrewery business in the United States has been exploding in recent years with a focus on small batches of unique beers. By using just a small amount of maple syrup, you can make a slightly sweet beer that tastes great and has excellent marketing appeal. This trend seems to be catching on, as there are a number of breweries that now include a maple beer in their assortment of craft brews. Sam Adams, one of the largest craft breweries in the US, goes through a lot of maple syrup when making their Maple Pecan Porter. Although I doubt anyone along the lines of Budweiser will ever be brewing with pure maple, there are over 2,000 microbreweries in the country, so if you have one near your sugaring operation, see if they want to use your syrup for a seasonal special. Maple beers seem to be especially popular during the autumn when maple

FIGURE 11.23. Mark VanGlad of Tundra Brewery produces a MaPale Ale with the maple syrup, grains, and hops grown on his farm in the Catskills. Because he produces all the ingredients himself, he is able to sell the majority of his beer at producer-only greenmarkets in New York City. PHOTO COURTESY OF DONNA WESSEL

leaves are at their peak colors; they should become even more popular in March when the sap is flowing throughout the Northeast.

Birch Beer

Although many of us have probably heard of birch beer, few people actually know what it is. Historically, the sap from black birch (*Betula lenta*) trees was used to produce a non-alcoholic, carbonated beverage similar to root beer. There were also some regions that created alcoholic versions of the product, but this was the exception rather than the norm. The vast majority of birch beer on the market today uses artificial flavoring and sweeteners to create a carbonated soda with a unique flavor reminiscent of wintergreen. Some manufacturers claim that they use birch oil that has been distilled from the sap of birch trees, though this seems highly unlikely. I suspect most of the flavoring agents are chemically synthesized or come from the bark and other parts of the tree rather than the sap.

The big opportunity that I see with birch beer is using birch sap to replace water in regular beer recipes. Although there is confusion over what birch beer actually is, the name is common enough that it invokes people's curiosity. Using clear, fresh birch sap from the beginning of the season in lieu of water allows brewers to create a unique "birch beer" that has excellent marketing appeal. The birch sap probably won't change the flavor of whatever beer is made with it, but since birch sap is a spring tonic full of minerals and nutrients, it does make the beer a bit healthier than your standard beer and provides an excellent marketing angle. The Lake Placid Pub & Brewery used 360 gallons of our birch sap to create a birch beer in 2013; it was very successful, and they are already looking forward to doing it again next year.

Birch Wine

There are several companies throughout the world using birch sap as one of the main ingredients in wines. One of these is Sapworld, which is owned and operated by Craig Lewis in Newfoundland. In addition to other birch-derived products, he has developed and copyrighted Springwine, aka Lady of the Woods. Its marketing appeal is based on the fact that he uses birch sap from the pristine forests of Newfoundland that is full of minerals and nutrients. Sugar is added to the birch sap until the sugar content is high enough for fermenting into wine. According to their marketing materials, this wine . . . "shows clear and bright in the glass; provides a delicate fragrance of soft apple and pear; sits well in the mouth with pleasant acids balancing the evident fruit sugars; persistent flavors of ripe pear, Mandarin orange, and Delicious apple blend harmoniously and without bitterness." This certainly sounds good, and with the natural origin of birch trees, it makes for a winning combination.

Boreal Bounty is another company from the Manitoba region that specializes in birch wine and other boreal forest products. Started in 2005 by Doug Eryou in conjunction with the D. D. Leobard Winery of Winnipeg, they developed a wine called Tansi derived from birch sap. In addition to their standard birch sap wine, they also have a wide range of products that use the extracts of other boreal trees and plants. Their list of wines includes birch sap mixed with cherry, lingonberry, sea buckthorn, cranberry, and Saskatoon. They also utilize the sap of boxelders—the only species of maple growing in Manitoba.

Maple Wine

While touring different sugarbushes in Quebec a few years ago I met Alberto Milan, the owner of a Canadian wine company. This was during the height of the recession when many businesses were suffering, yet Alberto's sales were soaring. He provided us with a brilliant rationale for why he decided to focus on using maple for wine rather than as a pancake topping. According to Alberto, "When the economy is good, people have lots of money and they like to celebrate and drink alcohol . . . and when the economy is bad, people are upset and like to drown

FIGURE 11.24. An assortment of birch wines with boreal fruit flavors. PHOTO COURTESY OF FRANK FIEBER

FIGURE 11.25. *Venerable* is a great name for a maple wine—it has connotations of strength and makes great use of the French word for maple: *erable*. PHOTO COURTESY OF ALDO NEYRA

FIGURE 11.26. Maple ice wines are extremely sweet and do make an excellent after-dinner drink. People expect ice wines to be sweet, so marketing a maple version makes sense, as everyone associates maple with being sweet. PHOTO COURTESY OF ALDO NEYRA

FIGURE 11.27. Two of the most popular mead varieties produced with maple syrup by the Saphouse Meadery in New Hampshire. PHOTO COURTESY OF CRIS DOW

their sorrows by drinking alcohol." So no matter what happens with the economy, the sale of alcoholic beverages continues to grow. Sales have been particularly strong in China and are now expanding into the US market through a distributor under the name of Maple Connoisseur. His company produces a table wine, sparkling wine, and ice wine all made with pure maple syrup. The process involves diluting the syrup with water and then starting the fermentation process until most of the sugar is consumed by yeast. Venerable is the table wine and is not nearly as sweet as you would imagine it to be, though the ice wine definitely packs a punch.

Maple Mead

I was first introduced to maple mead by Jeff Moore, a seventh-generation sugarmaker from Windswept Maples Farm in Loudon, New Hampshire. He went to Paul Smith's College in the Adirondacks and spent a lot of his free time in the spring helping at our sugaring operation in Lake Placid. I've gotten to know Jeff and the Moore family well over the years; my wife and I visited their farm during our honeymoon in summer 2011. As a wedding gift, they gave us a couple of bottles of mead that was produced from their maple syrup by the Sap House Meadery in Center Ossippee, New Hampshire. We were both greatly impressed with the quality of the mead, so I reached out to the owners to learn more about their company.

Sap House Meadery was started in 2010 by two cousins, Ash Fischbein and Matt Trahan. Both worked in the restaurant business for many years, but had grown tired of cooking and wanted to do something different. Since they were both avid homebrewers, they decided to start making wine with local ingredients from their region. There aren't any grapes in New Hampshire, but there is certainly a lot of maple syrup! After visiting 35 sugarhouses all over the state, they were especially impressed with the operation at Windswept Maples and started to get their syrup exclusively from them. Most of their meads are not strictly made with maple syrup, but rather use a blend of maple syrup with honey—this is known as an acerglyn. Many people expect maple mead to be excessively sweet, and depending on how it is made, it certainly can be. However, by properly controlling the sugar concentration and fermentation, it is possible to make maple-syrup-based meads that are relatively dry or semi-sweet. After years of experimentation, Ash told me he found the perfect balance that seems to work for them. They have already won awards for their signature Sugar Maple mead at the International Wine Festival in the Finger Lakes of New York. They also produce a variety of other meads utilizing local ingredients, including their Hopped Blueberry Maple, Blackberry Maple, and Peach Maple.

Meaderies have been gaining in popularity in recent years. Ash told me that when they started, there were only 35 meaderies in the country and 2 in New England. In the past two years, 6 more meaderies opened in New England and there are now over 100 in the United States. Mead is considered "other wine" by the federal government, and there are some hassles in dealing with the bureaucracy of creating and selling alcoholic beverages. However, if you are willing to go through all the red tape, there are excellent opportunities in turning some of your maple syrup into another valuable drink.

Maple Liquors and Spirits

Maple syrup can be distilled into hard alcohols or can be added to finished products as a flavoring agent. Craft distilleries are becoming increasingly popular as many states enact legislation that supports these operations, rather than forbidding them as they have in the past. Since maple

FIGURE 11.28. Cabin Fever whiskey uses pure maple syrup to provide a unique marketing angle and superb flavor. PHOTO BY NANCIE BATTAGLIA

syrup is so expensive, it makes much more sense to use it as a flavoring agent rather than simply as a source of sugar for distillation purposes. However, with the increasing amount of off-flavored, commercial-grade syrups being produced at the end of the season these days, I suspect that this could change. Thanks to advances in vacuum tubing and spout designs, many sugarmakers are still gathering plenty of sap at the end of the season when the syrup being produced is no longer fit for human consumption. This type of syrup is perfect for distilling into vodkas and other spirits, since none of the off-flavor will come through in the finished product. You can then add some high-quality syrup for flavoring to have a 100 percent pure maple spirit.

Crown Royal recently came out with a maple "finished" whiskey. The advertising and label on the bottle are very deceptive, leading people to believe that it contains pure maple. In actuality, this is simply a Crown Royal whiskey with "natural" maple flavors added at the end of the process. The whiskey has gotten a lot of attention and could have done a lot to promote pure maple, but unfortunately it turned out to be just another score for artificial flavoring. Interestingly enough, the parent company of Crown Royal also recently bought out Cabin Fever, a small company in New Hampshire that is making maple-flavored whiskey with pure, Grade B maple syrup. What started as a small hobby in the Robillard family garage is now a very successful business that is being marketed throughout the world by Diageo.

Final Thoughts: Buying and Selling Value-Added Products

Making and marketing value-added products is a lot different from producing maple syrup. It is certainly not for everyone! However, even if you don't have the time, interest, or desire to make your own value-added maple products, don't let that stop you from selling them. There are many sugarmakers and companies that offer (at wholesale prices) a variety of maple products to sugarmakers who then sell them in their own retail markets. The common ones include maple sugar, cream, and candy, but there are many more. You can get these with just a generic label on the package, with your own private label, or in completely unmarked boxes. You can then add your own label or unique packaging and mark up the price for retail sales. Selling these products will also help draw in customers who wish to purchase the value-added products and pick up some of your syrup in the process.

FIGURE 11.29. This advertisement from Merle Maple Farms shows some of the many products that you could have produced for you under a private label. There are several sugaring operations that have expanded their offerings in recent years to supply the growing markets for value-added maple products. IMAGE COURTESY OF KATE ZIEHM

CHAPTER 12

Tree Sap: Nature's Energy Drink

The first year I tried making maple syrup was pretty much a disaster. My father, brother Jeremy, and I tapped several trees on our property in Lake George in an attempt to produce maple syrup. We didn't do nearly enough research beforehand and consequently burned many of my mother's favorite pots. To say that our syrup took on a smoky flavor would certainly be an understatement! The lone bright spot in our adventure was discovering how delicious the sap was—fresh from the tree or partially boiled down into an even sweeter, golden liquid. After several failed attempts at making syrup, we gave up on that aspect altogether and just drank the sap. Whatever sap we didn't drink fresh got boiled down for 20 to 30 minutes on the stove until we had created "Adirondack Sweetwater." While I certainly love pure maple syrup, drinking the sap is what really got me hooked on sugaring.

Maple sap is a natural spring tonic and has been used by people for centuries throughout the world. To understand why, picture the following scenario, which was quite common for the Native Americans and the first settlers to North America . . .

Imagine that you truly live off the land, deriving all your sustenance from what you can grow, forage, hunt, and trade for. The winters are colder, longer, and you have to brave the elements for several months just to survive until the next spring. You may have some maple sugar remaining from the previous spring's efforts and occasionally come across a wild beehive loaded with honey, but you rarely have any sweet foods. Imagine that you don't have a warm, comfortable home to stay in, a car to get around with, or any of the conveniences of modern life. You've been eating a lot of the same, relatively bland food, and supplies are starting to diminish.

By the beginning of March, the nights are still frosty but the days are turning warm and sunny. Sap is now flowing in the maple trees, and you head out to gather the fresh, sweet, nutritious liquid. The trees produce more sap than you can possibly drink, and you boil off the excess water to make sugar for year-round use. The entire community comes together to celebrate the passing of winter and arrival of spring. Almost everyone shares in the work and the rewards. No wonder maple trees and the spring sap flow were so important to our ancestors.

The above scenario would be hard for most people to endure today. These days we have easy access to

FIGURE 12.1. Drinking sap right from the spout is a favorite activity for children (and many adults)!

sugary foods and beverages at any time. A sweet drink in the springtime is no longer a luxury to be waited for after a long winter, but rather a fixture of our everyday diets. In the United States and many other countries, there are more people concerned with dieting and consuming less sugar than there are people just trying to get enough food to sustain themselves. Obesity and its related health problems are now the biggest problem we have related to food, not starvation or scarcity. We have become spoiled by our excesses and largely out of touch with the seasonal nature of food.

Our dysfunctional relationship with food can easily be seen when examining how people first react to the idea of drinking maple sap. Many think of tree sap as the gooey stuff that comes out of pine trees, dropping onto their cars and causing a sticky mess when they park underneath them. To be fair, the thought of eating pine sap also disgusts me, and since we make turpentine out of pine sap—hardly an edible product—this feeling is well justified. However, when people do get a chance to try some maple, birch, or walnut sap, nearly everyone's opinion is drastically altered. They are surprised that it is basically just like water with a hint of sweetness. Indeed, maple sap is usually 98 percent water, 2 percent sugar, and loaded with minerals, nutrients, enzymes, antioxidants, phenolic compounds, and more. Walnut sap has about the same sugar content as maple, whereas birch sap usually contains usually 1 percent sugar or less.

When I first developed the outline for this book, I was just going to include a couple of pages about maple sap in the chapter dealing with value-added products. After all, I think fresh sap is the most valuable product you can get out of a tree. But the more I thought about and researched this subject, the more I realized that an entire chapter should be devoted to tree sap. What really changed my mind was reading through the Proceedings of the 1st, 2nd, and 3rd International Symposiums on Tree Sap Utilization. Researchers from around the world gathered in Japan in 1995, 2000, and 2005 to discuss their latest research and development projects. Although there was some mention of maples, almost all the papers were focused on birch sap—not as a raw material for producing syrup, but as a healthy beverage all by itself.

As the old saying goes, "If life gives you lemons, make lemonade." Although maples are found throughout the world, they are most abundant in eastern North America. Just as we have extensive maple forests, the northern portions of Asia and Europe contain various species of birch, and the people of Eurasia have done an excellent job utilizing the trees they have. It is truly fascinating to read about all the research and development with birch sap throughout the temperate world. Most of the papers were written and presented by scientists in China, Japan, Korea, Russia, and European countries. In other regions of the world, birch sap is used almost exclusively for drinking whereas our utilization of maple sap has been largely relegated to boiling down to syrup. This may be due to the fact that it requires so much more birch sap to make the same amount of syrup, and the flavor of birch syrup is considered by most people to be inferior to maple. Nevertheless, we have a lot to learn about using tree sap as a healthy beverage, not just as the raw material for making syrup.

Although birch sap has a much larger following throughout the world, there are cultures that do consume maple sap on a significant basis. The people of Korea have a long history of drinking tree sap and celebrate it each year with a festival called Namakje. The tradition is especially prevalent in the Kurye Province, which contains the highest mountain in the country, Mount Chiri. Jong Soo Woo from the Mount Chiri Climbers Association described the history of the festival and Koreans' connection to drinking sap at the 1st International Symposium on Tree Sap Utilization in 1995. Many Koreans believe that tree sap is a "mysterious and holy life-water from nature" and they drink it while praying for good health, family prosperity, and enough summer rainfall to ensure adequate harvests in the fall. The tradition goes back more than 1,000 years, to at least to the ninth century. It was abandoned for roughly 40 years in the early 1900s after the fall of the Chosun dynasty but was resurrected when the Republic of Korea was established in 1946. Today over 30,000 people attend the festival each year to drink sap and partake in the many events.

Acer mono is the maple tree most prevalent in Korea, and its Korean name is *gorosoe*, which means "the tree

FIGURE 12.2. A sap-drinking contest at a gorosoe festival in South Korea. The contestant who drinks a liter of sap the fastest wins the prize. PHOTO COURTESY OF KIYOUNG LEE

FIGURE 12.3. People drinking gorosoe sap in a Korean "sugarbush." There has been much more interest in drinking maple sap since an article appeared in the *New York Times* highlighting the Korean sap-drinking industry. PHOTO BY PARK JIN-HEE

that is good for the bones." There is an interesting legend of how this name came to be. In the ninth century, a Buddhist monk named Do-sun spent 35 years helping to build a temple near Baewoon Mountain. One winter he had spent several months sitting cross-legged in meditation. When he eventually achieved the spiritual enlightenment he was searching for and tried to get up, his knees would not bend, so he tried grabbing a tree to help stand up. The tree trunk broke and sap flowed out readily. Do-sun drank all the sap and was soon able to stand up straight; ever since then this tree has been known as gorosoe—"the tree that is good for the bones." With modern science, we now understand that the high mineral levels in the sap, most notably calcium, potassium, and magnesium, are why it is so good for us and our bones. More research still needs to be done, but the medicinal properties of maple sap are becoming more apparent every year.

Today there is a vibrant market for both maple and birch sap in Korea. In fact, the sap from *Acer mono* was highlighted in a *New York Times* article by Sang-Hun Choe in 2009 titled "In South Korea, Drinks Are on the Maple Tree." The article illustrates the weekend retreats people take in the mountains, drinking as much as 5 gallons of sap in a day. People sit on heated floors, in an environment similar to a sauna, so that they can drink as much sap as possible. As one participant described it, "You keep drinking while, let's say, playing cards. Salty snacks like dried fish help because they make you thirsty. The idea is to sweat out all the bad stuff and replace it with sap."[1] Choe also describes how the large amounts of calcium contained in the sap make this treatment especially beneficial for those with osteoporosis.

The market for tree sap in Korea is in full swing during the tapping season of March and April. During the week of Gyonchip (one of the seasonal divisions occurring in early March), maple sap prices are roughly $10 per gallon, about double the price of maple sap at other times of the year.[2] This difference is also seen with the sap of birch trees, as the $8-per-gallon price during the week of Gokwu (mid-April) is two to four times that charged the rest of the season. According to a 2007 government report, that year 1.3 million gallons of sap were harvested from nearly 800,000 taps. Although their production per tap is extremely low, they value the sap at a high price and sell the vast majority of it directly to consumers. Most of the sales are arranged through the Internet, and the sap is delivered directly to customers' homes. Picture the old milk truck in America, but containing maple or birch sap instead. I've even had Korean immigrants who live in New York City come to our sugarhouse in Lake Placid just to purchase maple sap. They were surprised and impressed that I knew what gorosoe was, and they purchased 15 to 20 gallons of raw sap on two separate occasions. I charged them $5 a gallon for the raw sap, which they

FIGURE 12.4. Tapping *Acer mono* in South Korea. Note the small diameter of this tree and the large number of tapholes it already has. In North America, a small tree of this size wouldn't qualify for even one taphole, let alone two. Since maples are not as prevalent in Korea and these small trees only produce a limited amount of sap, the Koreans wouldn't even consider boiling off 40 gallons of this all-natural, healthy beverage just to get 1 gallon of maple syrup. PHOTO COURTESY OF KIYOUNG LEE

FIGURE 12.5. Bottled sap for sale at a gorosoe festival in South Korea in 2013. At the current exchange rate, these prices equate to about $10 per gallon of sap. Since it normally takes 40 gallons of sap to make 1 gallon of syrup, this equates to approximately $400 per gallon of syrup. PHOTO COURTESY OF KIYOUNG LEE

considered a bargain! Considering that it equated to $200 per gallon of syrup, we made out quite well.

The tree sap industry is certainly not restricted to Korea. Russians drink a lot of birch sap, and several Russian scientists have devoted tremendous time and energy to exploring the effects of birch sap on human and animal health. The majority of this research took place from the 1960s to '80s, though it has been revived in recent years. In 2005 a group of Russian researchers published a synthesis of the literature dealing with the most important species of birch throughout Russia.[3] There are 15 different species of birch in Russia, but only 4 are primarily used for sap production. The authors touted the health benefits and cultural traditions involved with drinking birch sap. They cited over 150 references in their article, a testament to the importance of birch trees and birch sap to the Russian culture. In fact, Russians have such a strong affection for their birch trees that people often compliment women by comparing their beauty to that of birch trees.

In a paper presented at the first symposium in 1995, another researcher[4] highlighted the traditional medicinal uses of birch sap in five countries (Table 12.1). Although many of these medicinal benefits have not yet been confirmed by modern science, it is highly likely that at least some of these purported benefits are legitimate. The fact that many cultures spanning diverse regions of the globe utilize birch sap for the same ailments lends additional credibility to the claims. It's just a matter of time before modern science proves how and why birch sap is so good for us.

Reading about the medicinal properties of birch sap convinced me even further that there must be more to maple sap that we just don't know about at this point. While there are unique compounds in birch sap that are responsible for some its health benefits, the sap of other trees—including maples and walnuts—is also likely to have its own beneficial compounds. Since there is an abundant supply of birches and Norway maples (*Acer platanoides*) in Europe, some scientists in Latvia recently conducted experiments to determine the chemical composition of their birch and maple trees. Both species exhibited favorable results, though nearly 50 percent more phenolic compounds and roughly twice as many antioxidants were found in the maple sap as in the birch sap.[5] There is clearly much more to tree sap that we have not yet discovered; further research is warranted to determine all of the multiple health benefits.

There has been some scientific research concerning maple sap in recent years in Korea. Scientists there have

TABLE 12.1: Traditional Curative Properties of Birch Sap in 5 Countries*

	Japan	Korea	China	Finland	Russia
HYPERTENSION	X	X		X	
URINARY PROBLEMS	X	X	X	X	
GOUT	X		X	X	X
DECREASED WORK CAPACITY		X	X	X	X
GASTRITIS		X	X	X	X
KIDNEY PROBLEMS		X	X	X	
SCURVY			X	X	X

* Adapted from M. Terazawa, 1995.

been performing various experiments using mice and rats to determine the impact of consuming maple sap on a number of health issues. One of their first experiments involved feeding different amounts of maple sap to mice that exhibited osteoporosis-like symptoms due to a low-calcium diet.[6] For seven weeks, different groups of mice were given a variety of beverages as part of their diet, including springwater, 25, 50, or 100 percent maple sap, and a special beverage with high levels of calcium. At the end of the study, researchers examined their bones for size and density. The high-calcium solution yielded the greatest benefits, whereas the 50 and 100 percent maple sap diet also produced very favorable results. Mice that only had 25 percent sap or springwater suffered the greatest negative effects on bone density and size. To understand why this could be the case, the authors examined the mineral contents of the pure springwater and maple sap. The maple sap had 37 times as much calcium, 16 times as much potassium, and 3.7 times as much magnesium as the springwater. In an obvious conclusion, the authors "propose that the sap of *A. mono* may improve low-calcium diet induced osteoporosis-like symptoms by augmenting mineral levels in the body."

The same scientists conducted additional studies to assess the medicinal benefits of *Acer mono* and *A. okamotoanum,* another species of maple found only in Korea. Maple sap is known among Koreans as a hangover cure, so they set out to see if maple sap decreased the serum alcohol and acetaldehyde levels after acute ethanol treatment in rats.[7] In simple terms, they got a bunch of rats drunk, gave some of them maple sap and others water, and then determined how the rats recovered. Those that consumed maple sap fared much better, and researchers attributed this to the abundance of minerals found in the sap. Another study examined whether maple sap would have beneficial effects on hypertension in rats. The rats that had 50 or 100 percent maple sap experienced lower blood pressure levels (among other benefits), and the authors attributed this effect to the additional potassium in the sap.[8] Finally, they noticed that the sap of *A. okamotoanum* smelled very much like ginseng, so they collected some roots of this maple and did chemical analyses. They discovered that it contains two of the same pyrazines (aromatic organic compounds) found in ginseng root, which helps explain the similarities in odor. According to Professor Eui-Bae Jeung, they have only discovered these pyrazines in *A. okamotoanum* and not in any other species of maple. Given the Koreans' affection for ginseng, it is not surprising that they also prefer the sap from this tree.

Although these studies are useful, they were all conducted in a laboratory setting with animals. What is truly needed are clinical trials with human subjects.

The Federation of Quebec Maple Syrup Producers has been sponsoring research looking at the health benefits of maple sap and syrup. Although most of the studies have focused on syrup, they found that the sap contains nearly 50 vitamins and minerals, amino and organic acids, polyphenols, and phytohormones. In particular, the abscisic acid present in the sap may have tremendous health benefits; more research is currently being conducted in this field. Once there is applied research examining the effects of consuming maple (and birch) sap on human health, it is likely that many more people will want to drink these all-natural beverages.

FIGURE 12.6. Many people enjoy drinking sap right out of the bucket. If you do this, just be careful that your buckets meet the highest standards for safety and cleanliness. PHOTO BY NANCIE BATTAGLIA

Drinking Sap: Fresh or Pasteurized

Some people enjoy drinking sap fresh from the tree, while others prefer to boil it for a brief period to kill any bacteria or yeast. Since it is certainly possible for harmful bacteria to be found in sap, the cautious solution is to pasteurize it before drinking. However, this will kill all the bacteria, both good and bad, thereby precluding possible consumption of probiotics that are important for human health.

According to the current definition put forth by the World Health Organization, probiotics are "live microorganisms which when administered in adequate amounts confer a health benefit on the host." Most probiotics are a strain of *Lactobacillus*, a microorganism found in yogurt and other dairy products. However, for people who are lactose-intolerant, are allergic to dairy, or choose to follow a vegan diet, getting probiotics into the system can be challenging. Researchers in Canada have been exploring the feasibility of sterilizing sap and then reintroducing probiotics to create a healthy beverage capable of delivering probiotics to people who cannot consume dairy products.[9] A private company has received a patent on this method, so you may see it in a grocery store near you at some point in the near future.

It is worth noting that maple sap is basically sterile inside the tree; it is not until it is exposed to the atmosphere or comes in contact with collection equipment

FIGURE 12.7. If you are planning on drinking sap without first pasteurizing it, a clean 1-gallon jug with a small hole cut out to let it hang on the spout works great. Simply take it off the spout, stick it in the fridge, and replace it with a new jug. The best jugs to use for this purpose are ones that previously contained water, although other food-grade containers can be used after cleaning and rinsing. PHOTO COURTESY OF JEREMY FARRELL

that it picks up various strains of bacteria. Luc Lagace, a maple researcher with Centre ACER in Quebec, has spent considerable time and resources along with his colleagues identifying the bacterial communities commonly found in maple sap. They recently used advanced technology to identify a wide array of bacterial communities found at the taphole, with *Pseudomonas* and *Rahnella* the most commonly occurring genera.[10] Although it is possible that probiotics could become introduced into the sap, it is also possible that harmful bacteria could make their way in. For this reason, I always recommend filtering and pasteurizing the sap before drinking—just to be safe. In the same way that municipalities implement a "boil water" policy whenever there is a water main break, I also always recommend sterilizing the sap before drinking. I would feel terrible if someone wound up with contaminated sap (based solely on their collection practices) and then became ill themselves or made other people sick by serving contaminated sap to them. While there is a good chance you could drink raw sap your entire life and never get sick from it, when you are serving sap to other people (or recommending them to try it), it is always best to err on the side of caution.

All that being said, my favorite way to drink sap is right out of the bucket or while dripping out of the spout. There is something special about being outside on a warm, late-winter day, feeling the sunshine on your skin and drinking the sap straight from the tree. When you've been working hard tapping trees and gathering sap, there is no better way to quench your thirst than drinking fresh sap. Once you have done this, it is easy to see why it was such a sacred tradition for our ancestors in so many countries. The mind is a powerful thing, and I think the experience of getting outside to tap trees and gather sap is one of the main reasons why it is so good for us. The experience may do as much good for your body as the actual vitamins and minerals that the sap contains. This is especially true for people who generally eat a healthy diet anyway. After all, many of us aren't lacking in vitamins or minerals, but rather those authentic experiences that help tie us to the natural world. Just as eating a tomato that you grew yourself is a lot different from buying one at the grocery store, drinking the sap from your own trees can be a compelling experience that helps to deepen your connection to the land. Food scientists have created sap-like beverages full of vitamins and minerals, but buying one of these "energy drinks" is not nearly the same as drinking fresh sap from trees you have tapped.

While I am certainly a strong proponent of drinking sap, I only drink it in the beginning and middle of the season under the most sanitary conditions. I stop drinking sap once it gets warm and the sap takes on a yellow tinge. At this point in the season, the spouts and collection equipment have been contaminated by bacteria and yeast, and the sap begins to take on a sour flavor. This sap would certainly have to be boiled before you drink it. I usually don't drink sap out of our tubing system, but rather always out of buckets or bags. Even if the tubing weren't contaminated with bacteria and yeast (which I know it is), by the time the sap gets to the collection tank it may have a bit of a plastic flavor to it. Pasteurizing the sap would kill any bacteria and yeast, but I rarely pasteurize sap before drinking it myself.

If and how you decide to drink sap is a personal choice. If you are willing to take the risk of consuming something that could have harmful bacteria in it, then you may consider drinking fresh sap. However, if you are concerned about possibly consuming harmful bacteria yourself or serving bacteria-laden sap to others, then boiling the sap beforehand is a must. Finally, unless you want to drink a sap-based tea or coffee, I would cool the sap down before drinking after you have pasteurized it. Just as most people prefer cold water to warm water, cold sap is an especially refreshing drink. I personally don't enjoy drinking warm sap (nor do I like warm water), but there is nothing more refreshing and rejuvenating than cold sap on a warm spring day.

Carbonated Sap

If you are a fan of carbonated beverages, you can make a fresh maple seltzer simply by using maple sap instead of water in a home carbonation machine. There are plenty of different models on the market that are relatively inexpensive and easy to use. Simply substitute maple sap

FIGURE 12.8. The maple seltzer currently being produced by the Vermont Maple Sweetwater Bottling Company. I don't normally like the taste of seltzer, but this maple sap seltzer is phenomenal. PHOTO BY NANCIE BATTAGLIA

(or birch or walnut sap) for water and use the carbonation machine as directed. With your own carbonator, you can control the level of fizziness in your seltzer and make it fresh whenever you want. I highly recommend this to anyone who likes carbonated beverages. As an added bonus, you'll also be able to make your own maple soda (as described in detail in chapter 11).

There is a company in Vermont that has been instrumental in commercializing the concepts discussed above. Brothers Bob and Rich Munch applied for and received a patent in 1995 to create the products that they market through Vermont Sweetwater Bottling Company. Their patent covers the methods of pasteurization, filtration, concentration, and bottling of carbonated maple sap. A 2009 article in *The Atlantic* featured their successful business, which now sells roughly 10,000 cases each year.[11] In addition to carbonated maple sap, they also make a pure maple soda and several other flavored soft drinks. Although not all of their beverages utilize maple sap and syrup, these two products provided the impetus to develop their thriving company.

Cooking with Sap

The sap from maple, birch, and walnut trees can be used in place of water in nearly all recipes. We use maple sap to make beef stew, and the carrots, onions, potatoes, beef, and broth all taste just a little bit sweeter. We also steam vegetables with sap and then use the boiling sap to make rice. The nutrients that leach from the veggies wind up in the sap and are absorbed by the rice, making it even more delicious and nutritious. Sap can also be used in place of water when making coffee and tea, eliminating the need to add processed sugar. Oatmeal is another food that greatly benefits from using sap in lieu of water. The uses for tree sap are only bound by your culinary instincts and curiosity.

Betsy Folwell at *Adirondack Life* magazine recently wrote an article about maple sap and its culinary uses.[12] She provided several recipes using sap, including Drunken Beans, Fau Phox, and Sugarbush Bagels. This article prompted a couple of restaurants in Lake Placid to utilize maple sap in their specials featured in March over the course of Maple Weekend in 2012. The fresh sap certainly added a unique flavor to the normal winter specials and was greatly appreciated by all who tried it. I have since come across several restaurants that are cooking with maple sap and getting good press coverage as a result.

Koreans have experimented with making a soy sauce using sap from their maple trees instead of regular water in the recipe.[13] They found that maple-sap-based soy sauce had roughly twice the amount of minerals as ordinary soy sauce, and it was especially high in calcium, potassium, and magnesium. Glucose and galactose were the only sugars found in ordinary soy sauce, whereas the maple sap soy sauce also had fructose and sucrose. Of the 14 amino acids found in regular soy sauce, 3 of them appeared in lower abundance in the maple sap soy sauce whereas 11 were found in greater abundance. Overall the authors concluded that using maple sap instead of water when making soy sauce results in a healthier and more nutritious product.

Commercializing Tree Sap

To date there has been much more commercial activity with bottling and selling birch sap than there has been with maple sap. A quick Internet search for "birch sap" reveals a wide variety of purveyors throughout Europe, Russia, and Asia. There are many websites where you can buy pure birch sap or other beverages that utilize birch sap as the main ingredient. Most of the products use citric acid to preserve the sap, and many come with added sugars. The main obstacle that most companies encounter is preserving tree sap for year-round consumption while still maintaining the flavor and health benefits of fresh sap. The other challenge is trying to supply a market for 52 weeks when the sap is only running for less than a month. This requires a lot of warehousing and strategic planning to ensure a steady supply at an affordable price. It also requires a great deal of marketing and outreach to teach the public about maple and/or birch sap. Most people are originally skeptical of drinking tree sap, so you need to offer samples and do whatever outreach is needed to get people to understand why maple and birch sap are incredible, all-natural beverages. Once people taste the sap and discover that it is just like pure water with a hint of natural sweetness, they will gladly buy it. However, if you just put it on a shelf somewhere and hope somebody will try it, chances are you won't sell much.

FIGURE 12.9. Pure maple sap and three fruit-flavored varieties offered by Troll Bridge Creek Inc. in Ontario. COURTESY OF KEITH HARRIS

FIGURE 12.10. Keith and Lorraine Harris of Troll Bridge Creek Inc. offering samples of their KiKi Maple Sweet Water at a health food store in Ontario. PHOTO COURTESY OF TASHA JEFFRIES

Maple Sap

The last few years have seen a surge in activity with bottling maple sap. In 2009 Keith Harris retired from his corporate job and started Troll Bridge Creek Inc. with his wife, Lorraine, in Ontario, Canada. Keith decided to use his science and business background to start an entirely new venture in bringing maple sap to the marketplace as an all-natural, healthy beverage. Within a year Troll Bridge Creek Inc. developed the KiKi Maple Sweet Water brand and bottled thousands of gallons of maple sap in 12-ounce glass bottles. They have since greatly expanded production and branched out to include blueberry-, strawberry-, and cranberry-flavored versions that have pure fruit juice added to the maple sap. For a couple of years they also had lemon-ginger and lemon-mint flavors, but these were not as popular so are no longer offered. The response from health food stores has been remarkable; over 150 outlets across Canada now carry their products. Keith is also in conversations with a number of Asian businesses to export their maple sap to Korea, China, and Japan. Over the past year he has been working with

FIGURE 12.11. Sweet Run is one of the new "maple waters" being launched in 2013. PHOTO COURTESY OF GABRIEL VRISHAKETU PELLETIER

FIGURE 12.12. We're all used to seeing these types of tankers hauling milk away from a dairy farm, but as the maple water business continues to develop, we'll see a lot more such trucks hauling fresh, clear sap away from large sugarhouses like this one. PHOTO COURTESY OF GABRIEL VRISHAKETU PELLETIER

researchers at Conestoga College in Ontario to develop additional processing techniques to preserve maple sap for year-round consumption. The idea is to then license this technology to others so that sugarmakers throughout the United States and Canada could also bottle and sell pure maple sap as a healthy beverage. Stay tuned for further developments in coming years.

With the financial support of Agriculture and Agri-Food Canada, the Quebec Department of Agriculture, Fisheries and Food, and the Conseil pour le développement de l'agriculture du Québec, the Federation of Quebec Maple Syrup Producers spent the last seven years developing a method to preserve the sap such that it can be stored for 18 months at room temperature and still maintain its original quality. They are now licensing the technology to three companies that are certifying the maple water under the NAPSI process (NAPSI is an acronym for "Natural Authentic Pure Sterile Integral"). Seva, Oviva, and Maple3 are the different brands that you may already be seeing on grocery store shelves in Canada. The sap is being bottled aseptically using the same type of Tetra Pak containers that can keep milk from spoiling without refrigeration for several months. In February 2013 the federation put out a press release announcing the launch of maple water. It appears that the sap sourcing and sales will be limited to Canada for the time being, but we may be seeing it in the United States in coming years.

Although the federation claims that the maple water will be "a worldwide first," this fails to acknowledge the efforts of Troll Bridge Creek in Ontario, all the gorosoe harvesters in Korea, or another Canadian entrepreneur who had also been working on a similar process. In 2011 Gabriel Vrishaketu Pelletier started a company to preserve and bottle maple water for year-round sales. I recently tasted some of the sap that Gabriel had bottled in 2012 and was impressed with how good it tasted, even though it had been nearly a year since it was packaged. We all know how quickly sap can spoil and turn sour if left untreated, yet the process he used was able to maintain the freshness as if the sap had just flowed fresh out of the tree. Gabriel is also using aseptic processing technology with Tetra Pak containers, but since he developed this idea separately, he doesn't have to license the federation's technologies and certification process.

In 2013, Gabriel moved beyond the R&D phase into commercial production and sales under the brand name Sweet Run. One of Gabriel's marketing strategies is to provide bottled sap (that can be stored at room temperature) to other sugarmakers to sell along with their syrup and other maple products. You can learn more at www.sweet-run.com.

Not all of the maple sap commercialization is happening in Canada. Here in the United States, I have come across several restaurants and health food stores selling maple sap in various forms over the past several years. Most notably, in 2012, Feronia Forests, LLC, a sustainable forestry company and certified B-corporation with timberland holdings in Massachusetts and New York State, started researching various processes to extend the shelf life of maple sap. After positive developments in their first year, Feronia bottled enough maple water in a shelf-stable manner to run a regional test market in the summer of 2013. They are planning a commercial launch of maple water in the spring of 2014 under the label *Vertical Water*. Feronia's *Vertical Water* will soon provide added growth to the subsegment of all-natural functional waters category, much in the way coconut waters have done over the past few years.

Birch Sap

Rather than being processed into syrup, the majority of birch sap collected in the world is used as a beverage. Most of it is converted into what's known as a "Forest Drink" (or a similar translation) that involves adding sugar to the sap and preserving with citric acid and/or heat treatment for year-round consumption. Although there is a good market for this type of beverage, a company in Finland is taking a different approach. Susanna and Arto Maaranen have developed a unique method of preserving birch sap without having to heat, freeze, or refrigerate the sap. Their company, Nordic Koivu, is able to keep the birch sap in a natural state and therefore maintain all the health benefits of fresh birch sap for everyday use throughout the year. They have not yet patented their technology, because doing so would reveal the trade secrets that they have spent years of research and development to discover.

They first got the idea of bottling birch sap in 1995 and came out with their first non-heat-treated product in 2001. The Maaranens did a tremendous amount of research in starting this company and published a book in 2003 based on their acquired knowledge and experiences. They named it *Koivunmahla: malja luonnolle ja terveydelle*, which means "birch sap: a toast to nature and health." Their business has steadily grown over the years—they now ship birch sap to 22 countries throughout the world. Their products are mostly found in health food and specialty stores, and are advertised as gifts for business purposes. The largest markets are in Europe and Asia, and the Maaranens have just begun working with companies to export their products into North America.

FIGURE 12.13. The glass bottle from Nordic Koivu is so decorative, simple, and elegant that it has been featured in art displays in Finland and France. This company clearly understands the market potential and demand for a natural health drink and is capitalizing on its unique ability to preserve birch sap in its natural state. PHOTO COURTESY OF NORDIC KOIVU, LTD.

Can Drinking Birch Sap Prevent Birch Pollen Allergies?

Just as many people believe that consuming local honey can ease the symptoms of local pollen allergies, some people also believe that consuming birch sap could help people who suffer from birch pollen allergies. In fact, many people with birch pollen allergies have contacted Nordic Koivu to get some pure birch sap. They consumed the birch sap before and during the time of the year when birch pollen is plentiful. The results were promising, so Nordic Koivu decided to do more internal testing on this subject. They gathered a small, unofficial test group consisting of people who are allergic to birch pollen and had them drink birch sap for three years when birch pollen counts were high. According to their website, "Each member of the test group felt that regular consumption of birch sap had a dramatic effect in alleviating their allergy symptoms." Because internal testing by a company selling a product is not scientifically valid, Nordic Koivu is now partnering with university professors to conduct more scientific studies. By the time you are reading this, there may be peer-reviewed literature evaluating this theory—stay tuned.

Nordic Koivu notes that there is strong and rising interest among the cosmetic, food, and beverage industries for using birch sap as an ingredient. Many companies are working on new product development utilizing natural ingredients—and of course tree sap fits the bill. The Maaranens are now producing birch sap in large containers for several industrial clients according to their individual, specific requirements. Because they are able to collect and preserve high-quality birch sap in its natural form, without any heat treatment or preservatives, there are a multitude of uses for their product.

The temperatures are usually much warmer when birch sap is flowing as compared with maple. Therefore the sap is more likely to spoil and needs to be collected and processed carefully and quickly. Nordic Koivu has developed a custom-made sap collection system, utilizing stainless steel and a special type of plastic that allows the company to maintain the highest-quality sap for as long as possible. They are also working on a project to have subcontractors collect birch sap, which they would then deliver to the plant for processing and bottling. They originally experimented with having another company collect sap for them; once that proved successful, they have expanded to include another four or five sap collectors. This allows them to focus on processing, bottling, and marketing the sap without having to worry about gathering it. By having trained people gather the sap with customized and specific materials, they can also ensure a high-quality product.

Final Thoughts

Whereas commercializing tree sap for year-round consumption is a difficult venture, you may have better luck marketing sap as a seasonal product with a limited shelf life. Our relationship with seasonal beverages in America is highly varied. Some beverages are seasonal even though they don't have to be; others should be seasonal but are now produced year-round. For example, Americans are used to having eggnog available only during the holidays, though it can easily be made at any time of the year. Many breweries feature seasonal beers, but the ingredients can be sourced at any time from local or distant places. Milk used to be a seasonal product in the Northeast since farmers dried their cows off during the winter when pasture was no longer available. The flavor of the milk was also seasonal, as it changed throughout the year depending on what the cows were eating. Now milk is available in the same homogenized form with the same basic flavor year-round. Apple cider used to be produced only in the fall, but we can now store apples in controlled environments and press them into cider at any time.

As our food system has evolved over the past century, many of us have lost touch with the seasonal nature of food. However, since tree sap only flows during a limited time of the year and is difficult and expensive to preserve, it may work best as a seasonal product. With the rise of CSAs, year-round farmers' markets, and other venues for local food distribution, getting fresh, minimally processed maple sap to the market is much easier than it used to be. There are many people who would love to drink sap as a seasonal "spring tonic." If you can find a way to supply fresh, properly processed and packaged sap to them in an economical manner, then you can certainly develop a successful business.

We first started selling sap to our local health food store (Green Goddess) in Lake Placid in 2011. The store provided us with quart- and pint-sized ball jars that we filled with filtered sap that had been heated to at least 160°F for several minutes. We were able to use our canning unit and the filtering system that we normally bottle syrup with, so there weren't any additional equipment expenses. We got paid $2 for filling the jugs; Green Goddess charged $3 with an additional $1 deposit on the glass jar. Since there are 4 quarts in a gallon and it normally takes 40 gallons of sap to make 1 gallon of syrup, we were paid an equivalent price of $320 per gallon (of syrup) for our sap. There was certainly extra time involved in filtering, heating, and bottling the sap, but the price we received more than paid for the extra time involved. Over 90 percent of the quart bottles that we delivered sold within three or four days, and we took the few bottles that didn't sell within that time period back to the sugarhouse to be processed into syrup with the rest of our sap. This type of sales and distribution system has worked well for us; if you can figure out a way to make it work in your community, your customers will be happy and your bottom line may also significantly improve.

It is important to realize that the market for beverages is *much* greater than the market for sugary syrups. Whereas we need water to survive, most of us should cut back on our sugar consumption. Furthermore, people rarely consume pancakes and waffles, and if they do have them for breakfast, the vast majority use the fake stuff as their topping of choice. Even if they did choose pure maple, the amount of syrup consumed would still pale in comparison with the overall beverage consumption. The sap of maple and birch trees provides a natural tonic consisting mostly of pure water with small amounts of sugar, minerals, and other

FIGURE 12.14. Quart and pint containers of maple sap have been a big hit at the local health food store in Lake Placid with just a simple tag affixed to a glass jar. PHOTO BY NANCIE BATTAGLIA

FIGURE 12.15. The display of coconut waters at a popular grocery store in New York City. Soon enough I expect these shelves to also be filled with maple and birch water. PHOTO COURTESY OF DANIEL GRATTAN

beneficial compounds. Removing almost all the water to produce sugary syrups may not be the best use for these all-natural, delicious, and nutritious beverages.

To envision the possible market demand for tree sap, consider the growth of coconut water in recent years. Coconut water is the liquid inside a young coconut. People have been drinking it for centuries in the tropics—and it is indeed a healthy, refreshing beverage. However, it wasn't until some savvy marketers started bottling it and selling it to health-conscious consumers in North America that it started to really take off. Coconut water can now be found in stores throughout the world and has developed into a huge industry with overall sales much greater than pure maple. The health benefits of coconut water are nearly identical to those of maple and birch sap; it is described as a super-hydrating beverage containing significant amounts of electrolytes with only a limited amount of natural sugars. The claim to fame for coconut water is that one serving contains more potassium than a banana. To be fair, coconut water does have more potassium than maple or birch sap, but it also has a lot more sugar and doesn't taste nearly as good. It also comes from tropical plantations and must be transported long distances for us to enjoy it in the United States.

On the contrary, harvesting sap from maple and birch trees can be carried out with minimal environmental impact and helps to retain ecologically important forests throughout the world. It is truly the local, seasonal spring tonic for temperate regions. By drinking the sap from maple and birch trees, we can get nearly the same health benefits as coconut water at a fraction of the cost, and without the associated environmental impact of clearing tropical forests to plant coconut trees and shipping the final product halfway around the world. People should be drinking coconut water in tropical countries, but wherever maple and birch trees thrive, the sap that flows through them in the spring should be our first choice. With the right marketing and business development planning, I believe the market for maple sap as a beverage will eventually outperform the sales of pure maple syrup. Likewise, rather than boiling our birch sap down into syrup, it is likely that most of the birch sap will continue to be used primarily as the pure, all-natural beverage that it is.

CHAPTER 13

Managing Your Sugarbush for Optimum Sap Production

Healthy trees and vibrant sugarbushes are the backbone of any sugaring operation. Through the process of photosynthesis, maple, birch, and walnut trees are able to utilize sunlight, CO_2, and water to produce oxygen and sugars. By tapping these trees in the spring and taking some of their sap, we are able to produce delicious and nutritious syrups. When trees are stressed, not growing much, and producing fewer sugars, our yields of sweet sap are substantially reduced. Even though this seems obvious, I've found that many maple producers tend to take their sugarbush for granted, especially when everything seems to be just fine. When there aren't any major pests or disease outbreaks to worry about, or ice storms, floods, or other environmental disasters, sugarbush management may be one of the last things on your mind. However, under closer scrutiny you may discover that your trees are not performing at their maximum potential or even growing fast enough to allow you to comply with conservative tapping guidelines. Trees need to put on enough new wood each year to ensure that your tapping is sustainable. Without adequate growth, you will eventually wind up drilling into stained columns of wood from previous tapholes, thereby limiting your sap flow in future years.

Practicing good forestry requires a lot more skill and knowledge than you can learn from one chapter in a book. Since I can't teach you everything there is to know about proper forest management (nor do I know everything!), the main objective of this chapter is to persuade you of the need for deliberate and careful stewardship of your forest to maintain healthy trees and high sap yields for many years to come. I provide a general overview of the basic principles behind sugarbush management, but am only able to scratch the surface on key aspects of silviculture. As you read through the following pages, keep in mind the specific dynamics of your woodlands and how these principles apply to your management objectives. Although sugarbush management can involve activities besides timber harvesting, cutting trees is the main tool we use to manipulate the forest. Timber harvesting isn't inherently good or bad; rather it is how it is applied in a given situation that determines its usefulness and long-term effect on the forest. Being able to determine which trees to cut, understanding why they should be removed, and knowing how to harvest them in a safe and efficient manner will go a long way toward ensuring a sustainable future for your sugarbush.

Assessing the Resource

Before taking any deliberate management in your sugarbush, it is important to fully assess your forest resources. If you have at least 20 acres, I strongly suggest hiring a forester to conduct a forest inventory for your property and develop a management plan based on your ownership objectives. Although some state forestry agencies offer this as a free service to landowners, you will likely have to pay a private consulting forester if you want a detailed inventory and plan. It is usually well worth the cost to develop a management plan, especially if you have a significant amount of woodland. Having an approved management plan may help to reduce your income taxes and property taxes (depending on your state) and will allow you to apply for state and federal cost-share programs when these are available.

FIGURE 13.1. Conducting a forest inventory provides an opportunity for you to get to know your trees very well. Diameter tapes are often used to get precise measurements on the size of your trees. PHOTO BY NANCIE BATTAGLIA

FIGURE 13.2. Point sampling with a prism or angle gauge can provide you with reliable inventory data in a relatively short period of time, but it requires more skill and training to carry out properly. PHOTO COURTESY OF PETER SMALLIDGE

In an ideal situation, you (as the landowner) would have the time, skills, interest, and knowledge to conduct the forest inventory yourself. There are resources out there to teach you how to do it, and you could also hire a forester to help you collect and process the data. One of the best websites to help teach and guide you through the process is sponsored by the USDA Cooperative State Research, Education, and Extension Service.[1] If you take the time to learn how to do the inventory yourself, you will get to know your land in a way you probably never would have otherwise. After you have walked every part of your property and measured countless numbers of trees, you will be more invested in the process and will have a greater incentive to properly manage your resources.

Conducting a forest inventory will provide you with an excellent overview of the status of your woods and all the species growing there. However, if you are only interested in whether you have enough potential taps for a viable sugarbush, you don't necessarily need to go through the time and expense of conducting a full forest inventory. The process for collecting, processing, and interpreting the data on potential taps is relatively easy. You should first clearly delineate all of your property and stand boundaries on a detailed map. A stand is a distinct portion of your property where the forest type, structure, and age are essentially the same. For instance, on a 40-acre woodlot, you may have 28 acres of sugarbush that has been established for many decades, 8 acres of young woods that developed from an abandoned field, and 4 acres of evergreens that were planted 30 years ago. Each of these units would be considered its own stand, and you would want to do separate inventories and management strategies for each.

Once you have your property and stand boundaries established, you'll need to establish a sampling pattern in which you put in one sample plot for every acre within each stand. For larger properties or stands greater than 25 acres, your sampling intensity can be reduced, but realize that the more plots you put in, the more confident you can be with the results. There are advanced statistical techniques to determine the sampling intensity for a given confidence interval, but going through those calculations would be unnecessary for this purpose. Finally, it is important to be as random as possible when determining your sample points. Most of us gravitate toward the areas that have a lot of large sugar maples, but doing this will skew your results by overestimating the number of taps on your land.

Once you have your sampling plan figured out, there are two ways of assessing the number of potential taps. One uses a fixed area plot; the other, a variable radius plot. The variable radius plot requires the use of an angle gauge or prism and is more complicated to learn, so check with your local Cooperative Extension office to see if there are any training sessions available. A

TABLE 13.1: Calculating Number of Taps per Acre

Date	8-Apr-13
Stand number	1
Plot size	1⁄20 acre

	Plot 1	Plot 2	Plot 3	Plot 4	Plot 5	Plot 6	Plot 7	Plot 8	Plot 9	Plot 10
Number of 1-tap trees recorded	4	3	2	4	3	1	5	4	4	3
Number of 2-tap trees recorded		1	2		1	2				1
Number of taps/acre per plot	80	100	120	80	100	100	100	80	80	100
AVERAGE NUMBER OF TAPS PER ACRE FOR THE STAND			94							

A sample data sheet used to figure out the number of taps in a potential sugarbush. In this example, the producer took 10 sample points and wound up with an average of 94 taps per acre. You can make up your own data sheets to resemble this one.

conceptually simpler but more time-consuming method is to establish 1⁄20-acre fixed radius plots. Working with a partner, have someone stand at the plot center with a tape measure (or any piece of heavy string, rope, tubing, or the like) that is 26.4 feet long and will not stretch when pulled tight. The other person can measure out the 26.4 feet and tally all the tappable trees that fall within the radius of the circle. Alternatively, if you have to work by yourself, just drive a stake in the ground and have the string or rope affixed directly to the top of the stake. Since a circle with a radius of 26.4 feet is exactly 1⁄20 of an acre, you simply multiply the number of tappable trees you count by 20 to get an estimate of the number of potential taps per acre. If you have trees above 18" in diameter that you would put two taps on, count those trees twice. By repeating this procedure many times throughout your sugarbush, you can average all your results together to get a general sense of the number of potential taps on your land.

Types of Sugarbushes: Even-Aged and Uneven-Aged

Before you do any management in your sugarbush, one of the fundamental concepts that you need to understand is the difference between even-aged and uneven-aged stands. The management strategies vary greatly for these types of stands, so it is important to determine the age structure of your sugarbush. Even-aged woodlots originated after agricultural land was abandoned or after widespread, intensive cutting of existing forests. Most of the woodland we have today originated following agricultural abandonment or clear-cutting 50 to 100 years ago; if your property has not had any significant logging activities in recent decades, chances are it would be considered even-aged.

Following agricultural abandonment or clear-cutting of existing forest, initially tens of thousands of seedlings became established per acre. These trees competed for sunlight and resources for many years in the race to the canopy. The genetically superior and faster-growing trees were able to shade out the slower-growing, genetically inferior trees. Over several decades, the dense thickets gradually thinned themselves through natural competition and mortality to include a few hundred trees per acre or less. Some of the smaller trees were able to survive in the lower canopy, especially shade-tolerant species such as sugar maple. Thus, in a typical even-aged forest, you may find sugar maples with wide ranges in diameter and height. It is common to find 6"-diameter maples that are the same age as the 16"-diameter trees that you are tapping.

This is a difficult concept for many people to comprehend. When we see two trees of different sizes, we usually

FIGURE 13.3. Although they are different sizes, all the maple trees in this photo are the same age; some just happened to grow faster than others. Many of these trees will die out over time, and the disparity in sizes will persist. PHOTO COURTESY OF PETER SMALLIDGE

FIGURE 13.4. This sugarbush would be considered uneven-aged: The maple trees occupying various portions of the canopy and understory are different ages as well as different sizes. PHOTO COURTESY OF WWW.MORGUEFILE.COM/ARCHIVE/DISPLAY/764437

assume that the larger tree is simply older than the smaller one. In fact, landowners are often convinced to harvest their large trees when they are told that "thinning out the old ones will help the younger ones grow up." However, as Dr. Ralph Nyland, my silviculture professor at SUNY-ESF, used to state, "Once a runt, always a runt." In even-aged stands, the smaller trees may be genetically inferior to the larger ones and will probably never achieve the growth rates that the larger ones do. Even if you remove competing trees in the overstory, their growth may only improve slightly. In even-aged stands, cutting the larger, superior trees to open up more growing space for the small, inferior stems is not an appropriate or sustainable forestry practice.

Although most of our forests are considered even-aged, there are many uneven-aged stands that originated following cutting, windthrow, or natural mortality. Some of the most common examples of uneven-aged stands are sugarbushes that have been tapped continuously by several generations of sugarmakers. Periodic disturbances, such as timber harvesting, icestorms, windthrow, and the like create gaps in the canopy that allow new stands of saplings to develop. Since sugar maples are shade-tolerant, they can become established under the canopy of existing trees and then shoot up to the canopy when an opening appears. As sugarmakers, we must realize that the trees we are tapping today won't live forever, so we need to think about the future forest if we hope to ensure the long-term sustainability of our sugarbush. As long as young saplings are growing vertically with a clear central leader, these maples are prime candidates for canopy replacement in

FIGURE 13.5. If you look up in your canopy and don't see much daylight, then your trees have little room to grow and would benefit from thinning.

uneven-aged stands. However, if they become stunted for too long in the understory, the crown may flatten out and they will never be good replacement trees. Management of uneven-aged forests can be very complex, so if you think your sugarbush is uneven-aged, I highly recommend getting the assistance of a skilled forester to maintain adequate stocking over time.

Thinning Your Sugarbush

To understand the importance of managing your forest, consider a familiar example that is easy to relate to from the garden. Imagine you plant a row of carrot seeds and never get around to weeding them or thinning them out. When you go to harvest the carrots at the end of the summer, all you will have is a bunch of spindly, small carrots that aren't worth dealing with. However, if you take the time to pull the weeds and remove some

FIGURE 13.6. Whenever you cut any trees in your sugarbush, it is useful to examine the stump to look at the annual growth patterns. If you notice that the annual rings are significantly closer as you get toward the bark or if you have a hard time telling where the annual rings are, then it is likely that your sugarbush would benefit from thinning. PHOTO COURTESY OF PETER SMALLIDGE

of the carrots during the summer, the outcome will be much different come harvesttime. You will have a bountiful harvest of large, delicious vegetables. Now think of your maple trees as the carrots growing in a garden—just on a much larger scale. If you never do any cutting, you will have too many trees for optimum growth and your sap-producing maples will be stunted. You may have a lot more trees to tap, but those trees will not be performing at their full capacity or giving as much sweet sap as they could. For a given acre of land, would you rather have 100 healthy maple trees growing fast and developing large, wide crowns or 150 or more mediocre trees that are crammed in together?

So how do you know if you need to do some thinning? With a row of carrots, it's pretty easy to distinguish the weeds, and the seed packet tells you what the proper spacing should be. With a forest, determining the optimum spacing is more difficult—as is distinguishing the "weeds" from crop trees. If you look up into the canopy of your sugarbush during the summer and don't see much daylight, chances are that your trees are too crowded and would benefit from thinning. Likewise, if the forest floor is dark in the summer and you have very little growth in the understory, then you'll probably want to do some thinning. A professional forester can be very helpful in assessing the need for specific management activities such as thinning and marking trees to be removed. Before you get started, there are some general guidelines that you should follow whenever you start cutting trees in your sugarbush:

- **Safety first: Don't cut any trees unless you have the appropriate training, tools, and personal protective equipment (PPE).** Tree cutting is an extremely dangerous activity, and you must keep safety as the number one priority at all times. Accidents happen when you least expect them, so you need to be prepared and alert at all times with the proper protection in case something goes wrong. Even if you are only planning on cutting a couple of trees, be sure to put on your PPE.
- **Purchase a quality chain saw and take proper care of it.** A good chain saw is expensive but will pay for itself many times over in increased productivity and added safety. New chain saws come with many features that make cutting wood safer and easier on your

FIGURE 13.7. Peter Smallidge, New York State extension forester, demonstrating the minimum personal protective equipment that you should have while operating a chain saw—a hard hat with ear and eye protection, chaps, and gloves.

FIGURE 13.8. If your chain saw looks anything like one of the collectibles displayed here, then it belongs on a wall like this and nowhere near a tree that needs to be cut. PHOTO COURTESY OF PETER SMALLIDGE

body. Be sure to take the time to properly maintain the saw—and never cut with a dull or loose chain.

- **Never do any cutting when you are tired.** Cutting trees is dangerous enough when you are alert, let alone when you are tired. If it is getting toward the end of the day and you don't have a lot of energy left, leave the sawing for another day. The only time I have gotten seriously injured when cutting trees was at the end of the day when I was tired and just trying to finish bucking up the last tree. Our motor skills are not nearly as sharp when we are tired, and that is when we tend to make mistakes.
- **Don't operate in the woods when it is wet.** If you operate any sort of heavy machinery in the woods when the soils are saturated, you could do serious damage to the root systems of your trees. Maples have the vast majority of fine roots in the upper 6" of soil. Driving tractors and skidding logs over wet soil could do serious damage to the root systems of the trees you are trying to help by thinning. The best time to operate is when the soil is dry or completely frozen. Having a nice blanket of snow on top is helpful, but only if the soils are frozen underneath. If you can plan your harvesting operations for the dead of winter, chances are your maple trees will be a lot happier. The next best option is cutting trees during the summer or early fall when the soil is relatively dry.
- **Cut the worst trees first.** It is fairly easy to identify the worst trees to cut out of your sugarbush. Target the undesirable species as well as trees that have bad form, cankers, disease, dieback in the crown, scars on the trunk, and the like. This focus will include cutting some poorly formed maples and other species. Once you have all the worst trees removed, then you can start making the hard decisions on what other trees to cut.
- **Never remove more than a third of the total volume in a single cutting cycle.** If you cut out too many trees at once, you may wind up doing more harm than good. Opening up the canopy too much could lead to problems with windthrow, sunscald, epicormic branching, and increased susceptibility to drought. The "thinning shock" that could ensue may lead to slower growth among your residual trees for several years until they recover from the treatment. Rather than cutting a lot of trees all at once, it is

FIGURE 13.9. Operating a tractor or skidder during mud season could cause serious rutting and water-quality issues, not to mention damage to the root systems of your main sap-producing trees. PHOTO COURTESY OF PETER SMALLIDGE

FIGURE 13.10. Trees with weak forks are good candidates for removal before one of the main limbs eventually breaks off.

FIGURE 13.11. A freshly thinned stand of pole-sized sugar maple. The remaining trees now have room to expand their crowns and put on additional growth.

FIGURE 13.12. Keeping high-value hardwoods such as black cherry as part of your sugarbush will help add to the diversity and beauty of your land while also providing a source of marketable logs for the future. PHOTO BY NANCIE BATTAGLIA

better to cut out fewer trees at more regular intervals, perhaps every five years to ten years.

- **Strive for some diversity.** As maple producers, we are primarily interested in helping the sugar maples on our properties grow and develop. However, it is important to include a component of other high-value hardwoods to add diversity to your woods. Monocultures tend to suffer more during pest and disease outbreaks, so try avoiding a pure sugar maple stand if possible. However, don't try to retain diversity by keeping conifers/evergreens in your sugarbush. Conifers provide habitat for squirrels, and you'll wind up with a lot more damage to your tubing if you don't cut them out.
- **Start your thinning early when the trees are relatively young.** I think the best time to start thinning a developing sugarbush is when the trees are 5 to 6" in diameter. You could start at a younger stage of stand development, but it will require cutting a *lot* more trees and you won't be able to do much with the material that you are cutting out. By the time the trees are 5 to 6", they will have developed a main trunk with at least 8 feet of clear wood for future tapping and it's easier to tell which trees to favor. They are still in the young, juvenile stage and respond very well to thinning at this size. By reducing the competition among your best sap-producing trees when they are still in the sapling or small pole-timber stage, they will grow much faster and get to tappable size much sooner. Spending time thinning in young stands is one of the best investments you can make in the woods.
- **Test sap sugar concentrations on maples that you are considering thinning.** Not all maples are created equal and some produce sap with a much higher sugar concentration than others. After you have cut out all your worst trees, you may have trouble deciding how to continue thinning your remaining trees. As sugarmakers, we don't like to cut out maple trees, even though it is necessary to achieve maximum growth on our trees. One way to distinguish between which trees to cut and which ones to leave is by testing the sap sugar concentration with a refractometer. Just a couple drips of sap will allow you to select the sweetest trees in your woods and cut out the genetically inferior trees with lower sap sugar concentration.

FIGURE 13.13. A young stand of maples that would greatly benefit from thinning. Cutting around the major crop trees at a young age will allow them to develop fuller crowns and reach tapping size sooner.

FIGURE 13.14. Analog refractometers are a good way to test individual trees for sap sugar concentration. PHOTO BY NANCIE BATTAGLIA

Game of Logging

Since cutting trees is one of the most dangerous things you'll ever do as a sugarmaker, it is well worth your time to get as much training as you can. Even if you have been cutting trees all your life, you will greatly benefit from taking a chain saw safety course such as Game of Logging. Not only will your efficiency and productivity improve, you will also be much safer in the woods. Logging is certainly not a "game," and people die or become seriously injured every year from timber-harvesting-related accidents. The Game of Logging got its name since it includes competitions among the participants on who can fell trees in the safest and most efficient manner. There are four levels in the Game of Logging: each includes a full day of training with a certified instructor. Level 1 is the introductory class and focuses on basic chain saw safety and felling techniques. With Level 2 you learn the details of maintaining your chain saw at maximum capacity, safe and efficient limbing and bucking, and how to handle springpoles and other difficult scenarios. After you have taken Level 3, you should be able to put just about any tree wherever you want to, no matter what size it is or how it is leaning (within the limits of the laws of physics). The final Level 4 is meant for professionals who want to learn how to maximize their logging activities for safety and productivity. I have taken several Game of Logging safety courses over the years and couldn't imagine cutting trees without first going through this training. Their website, www.gameoflogging.com, includes a listing of upcoming events; if you see one in your region, I highly recommend signing up for it.

FIGURE 13.15. Instructor Bill Lindloff oversees a Game of Logging participant as he cuts a notch to prepare the tree for felling. PHOTO COURTESY OF PETER SMALLIDGE

In talking about thinning, it is important to distinguish between the type of cutting I am recommending versus the cutting that normally takes place in eastern forests. The high-grading and diameter limit cuts that are often implemented under the name of forestry are more similar to mining of a non-renewable resource than sustainable management of our forest resources. Some people will encourage you to "take the best and leave the rest," since it is only profitable to cut the large, valuable trees. This is an unfortunate situation that results in many degraded stands with a preponderance of stunted, suppressed, and genetically inferior trees. Unfortunately we have all seen beautiful sugarbushes that took a century to develop devastated in a matter of days or weeks. This can happen due to greed, ignorance, or extenuating circumstances that force people to liquidate their woodland assets. Sugaring can help prevent or limit these situations by providing landowners with a steady, annual income without having to remove the largest, most valuable trees. Although there are some situations where timber harvesting will provide landowners with better financial returns, sugaring is more profitable in the vast majority of situations. See chapter 15 for a detailed analysis of the economic trade-offs between syrup and timber production.

Even though I'm not a fan of timber harvests that focus exclusively on cutting the largest, most valuable trees, there are many instances when it is appropriate to harvest mature trees from a sugarbush or other forest. Eventually all trees go into decline and die; it makes sense to harvest them before they lose all their value. Furthermore, in the later stages of development in a maturing forest, a portion of the canopy must be removed to create the appropriate light conditions to spur the development of the next cohort of trees. Regeneration cuts such as shelterwood, seed-tree harvests, and clear-cuts can all be valid silvicultural prescriptions depending on the situation at hand. Although you will probably never clear-cut or implement a seed-tree harvest in your sugarbush, at some point you may decide to do a shelterwood harvest if you want to regenerate a new stand of maples and other high-value hardwoods. Selection harvesting of individual and/or groups of trees can also be a valid management tool for uneven-aged stands, especially in

FIGURE 13.16. Group selections, in which all the trees in a small patch are harvested, are useful for establishing small patches of seedlings in an otherwise mature forest. PHOTO COURTESY OF PETER SMALLIDGE

FIGURE 13.17. In a shelterwood harvest, only the best trees are retained at a fairly wide spacing to allow enough light to reach the forest floor to stimulate regeneration. By leaving the best trees as seed sources, you are able to pass along the best genetics in the forest to the next stand of trees. PHOTO COURTESY OF PETER SMALLIDGE

FIGURE 13.18. Selection harvesting is appropriate when you want to remove just a small portion of the trees to provide enough sunlight to encourage regeneration of shade-tolerant species such as sugar maples. PHOTO COURTESY OF ROB ROUTLEDGE, SAULT COLLEGE, BUGWOOD.ORG

FIGURE 13.19. Before embarking on a regeneration harvest, it is important to have advance regeneration of sugar maples or other desirable seedlings to ensure rapid development of a new stand.

sugarbushes where you want to retain most of the large, sap-producing trees. If and when you think your woods is ready for a regeneration harvest, it is best to solicit the help of a professional forester to ensure that your goals are met to the greatest extent possible.

Invasive Species in the Sugarbush

There are no shortages of invasive species to contend with in today's world. With so many plant materials, containers, and pallets of goods being shipped around the world, a wide variety of exotic plants and insects are spreading in to sugarbushes and other woodlands. Furthermore, many of the invasive plants were deliberately introduced for wildlife or aesthetic reasons. It is very important to stay informed on current trends and issues with invasive species. After all, you can't find something if you don't know what you are looking for. Outbreaks of invasive species usually start small before they grow into uncontrollable nightmares. If you know what to look for and you spend enough time in your woods (especially during the growing season), chances are you'll be able to rectify any problems while it is still easy to do so. For instance, ripping out a few honeysuckle seedlings is no big deal, whereas trying to contend with an understory thicket of honeysuckle is time consuming, expensive, and a lot of hard work!

Fertilization and Liming Options for Your Sugarbush

As with all agricultural crops, the production potential of the land is directly tied to the long-term health of the soil. Farmers and gardeners are used to amending soil to meet the particular needs of the crops they are growing, yet it is much more difficult for sugarmakers to change the composition of forest soils. First of all, applying any type of soil amendments in a woodland setting is challenging; bringing in heavy machinery to spread fertilizers can also do serious damage to the root systems of the trees you are trying to help. Furthermore, since the area of a sugarbush tends to be quite large, it takes a substantial amount of fertilizer or lime to change soil pH and nutrient availability. Thus, even though fertilizing and liming can be of great benefit when done appropriately, few sugarmakers make this type of investment in their woodland due to the practicality and logistical challenges of doing so.

Before undertaking any type of fertilization or liming activities, it is essential to adequately test your soil and assess other characteristics of your sugarbush to determine if fertilization/liming could help. If it appears that your tree growth is stunted, it may not be the result of nutrient deficiency in the soils. It's possible that your soils are too wet or dry, or that the trees in your sugarbush are too crowded. Although there is little that can be done to change soil moisture, thinning can concentrate resources on your best trees and is often much more beneficial than fertilization. Soil testing is fairly easy to do and inexpensive, so it's worthwhile to get many samples from a wide variety of locations in your sugarbush. Limestone-based soils tend to be much more fertile and have a higher pH than granite-based soils, thus having a firm understanding of the geological formations will help you decide if fertilization and liming are right for you. Calcium is often a major limiting nutrient for sugar maples; many studies have linked higher calcium levels with increased growth and regeneration. If your soil pH tests very low and you are in an area that has been negatively affected by acid rain, liming could be one of the best investments you can make. For more information on fertilization and liming, go to www.uvm.edu/~pmrc/fertilization_brochure.pdf for an excellent article on this topic written by Tim Wilmot and Tim Perkins at UVM.

FIGURE 13.20. Closely examining the understory vegetation can be an excellent means of testing the nutrient levels in your sugarbush. If you find maidenhair fern, blue cohosh (pictured here), jack-in-the-pulpit, ginseng, foamflower, and other perennial herbs, chances are fertilizing or liming wouldn't have much effect.

Invasive plants may impair our ability to regenerate the forest and can be a huge nuisance to work around. It's not a lot of fun trying to install or maintain tubing in a woodlot infested with Japanese barberry or multiflora rose. If invasive species have gotten established in your woods, there are various types of mechanical, chemical, and cultural controls to remove them. Extensive research has been done on the most noxious species to determine the most cost-effective method of removal. To learn more about the invasive species you may find in your sugarbush and various control mechanisms, check out http://na.fs.fed.us/fhp/invasive_plants/weeds.

Sometimes the species that cause the greatest problems are actually native plants that have gotten out of control due to external factors. A perfect example of this is American beech (*Fagus grandifolia*), a common associate of sugar and red maples in northern hardwood forests. Beech can be a magnificent tree that yields

FIGURE 13.21. An entire understory being overtaken by multiflora rose. PHOTO COURTESY OF PETER SMALLIDGE

FIGURE 13.22. Once garlic mustard gets established, it spreads quickly and can quickly take over an entire understory. You may consider pulling up and eating the garlic mustard in the early spring as one method of controlling its spread, though you'll need a large appetite or good markets to make a significant dent in the population. PHOTO COURTESY OF PETER SMALLIDGE

FIGURE 13.23. This stem has been severely affected by beech bark disease and is sending up dozens of root sprouts as the main trunk dies. PHOTO COURTESY OF PETER SMALLIDGE

valuable lumber, firewood, and delicious nuts prized by both people and wildlife. However, over the past century beech trees throughout eastern North America have been decimated by an insect/disease complex called beech bark disease (BBD). It started with the introduction of an exotic scale insect to Nova Scotia around 1920 and has had a dramatic effect on northeastern hardwood forests wherever it has progressed. As the older trees die, they can send up thousands of root suckers per acre that are extremely shade-tolerant. Whereas deer love to eat maples, they rarely browse beech. Without active and deliberate management to remove the beech, sugar maples will have trouble regenerating and many sugarbushes could eventually turn into beech thickets. Thus, controlling the spread of beech is extremely important for the long-term sustainability of sugaring in the Northeast.

There are remedies that you can take to halt the spread of beech in your sugarbush.[2] If you simply cut some beech trees (especially during the dormant season), that will just exacerbate the problem and result in many more stems than were present before cutting.[3] Even though I am reluctant to use chemicals in most situations, herbicides are the most cost-effective and efficient way to control beech sprouting. Researchers with the USDA Forest Service have been conducting extensive trials of cut-stump treatment over many years, and it definitely works.[4] They recently came out with an excellent overview of four methods for controlling interfering vegetation with herbicides—it is a must read for anyone considering using herbicides to control beech and other undesirable plants. Peter Smallidge, my colleague at Cornell, has recently devoted a lot of his research and extension efforts toward chemical control of beech. Working with Peter, I've seen firsthand how spraying beech stumps with glyphosate directly after cutting will prevent them from sprouting while also killing other beech saplings that are connected to the same root system. Glyphosate is one of the safest chemicals available today, and homeowners are legally allowed to purchase a high-concentration solution and apply it on their own lands. However, you should always read the labels and be extremely cautious whenever using any chemicals!

FIGURE 13.24. Beech thickets can be a huge nuisance to deal with in sugarbushes. Not only do they inhibit maple regeneration, they make running tubing difficult and the brown leaves that rattle in the wind sound exactly like a vacuum leak. PHOTO COURTESY OF PETER SMALLIDGE

If your operation is certified organic or you don't want to use chemicals on your land (which is completely understandable), then you will have to take other measures to control beech sprouting. One cost-effective solution could be controlled grazing of goats in your sugarbush. At Cornell's Arnot Forest, they conducted an extensive "goats-in-the-woods" project for several years and determined that goats can effectively wipe out a forest understory while leaving the mature trees untouched.[5] If you are experienced with handling animals or know someone who is looking for areas to graze goats, this may be an option for you.[6] Note that you will lose everything in the understory, so before you implement this, do a careful survey of the property to make sure there aren't any plants that you wouldn't want to lose. Another possibility is flame weeding in which you apply intense heat to the base of small trees and shrubs to kill the cambium layer and prevent them from sprouting. Finally, you could just continue to mechanically cut the trees and deal with additional sprouts at each successive cutting. If you do this, I recommend starting with a high stump and cutting it a bit lower each time. It is best to do the cutting in the summer to remove as much of the biomass and starch reserves as possible.

FIGURE 13.25. When applying glyphosate to beech stumps, it is very important to wear the appropriate PPE and follow the manufacturer's guidelines on the label. PHOTO COURTESY OF PETER SMALLIDGE

Insects and diseases can kill large numbers of trees and cause major forest health issues in a short time period. There are a variety of native pests and diseases that

FIGURE 13.26. Goat damage to beech saplings in an active sugarbush. The large maple trees were untouched whereas the goats stripped away all the thin bark on the beech saplings, performing a complete kill while providing some nutrition to the animals. PHOTO COURTESY OF PETER SMALLIDGE

FIGURE 13.27. A brush saw equipped with a sharp cutting blade is an excellent tool for removing small beech whips in the understory of your sugarbush.

FIGURE 13.28. Flame weeding is a good alternative for people who don't want to cut trees or use herbicides. The hot flame kills the cambium layer, thereby preventing the trees or shrubs from resprouting. PHOTO COURTESY OF PETER SMALLIDGE

FIGURE 13.29. If you do try flame weeding, it is important to have a water source available, especially when atmospheric and environmental conditions are dry. PHOTO COURTESY OF PETER SMALLIDGE

FIGURE 13.30. Asian longhorn beetles could cause widespread mortality and wreak havoc on the maple industry if they escape from urban areas into rural forests. PHOTO COURTESY OF THE PENNSYLVANIA DEPARTMENT OF CONSERVATION AND NATURAL RESOURCES–FORESTRY ARCHIVE, BUGWOOD.ORG

FIGURE 13.31. A maple tree severely infested with Asian longhorn beetles. PHOTO COURTESY OF THE PENNSYLVANIA DEPARTMENT OF CONSERVATION AND NATURAL RESOURCES–FORESTRY ARCHIVE, BUGWOOD.ORG

affect maples, but these rarely cause widespread mortality. The most pressing concern we have in the maple industry is the Asian longhorn beetle (ALB), an invasive species from China that first appeared in New York City, Chicago, and Toronto in the 1990s and is now a major cause for concern in Massachusetts and Ohio. Eradication and control efforts have included the establishment of quarantine zones around known locations, removing all host trees within a given radius of known infestations, and injecting host trees with an insecticide called imidacloprid. ALB is commonly confused with the white-spotted pine sawyer, which looks eerily similar but has one distinctive white spot between the top of the wing covers, making identification easy. If you are unsure whether you have found an ALB or white-spotted pine sawyer, send a sample to your local authorities. It is better to report a false occurrence that an expert can validate than let an actual infestation go unchecked. For more information, check out www.asianlonghornbeetle.com and www.uvm.edu/albeetle.

Forest tent caterpillars (FTC) are probably the most destructive native pest in northeastern sugarbushes. The larvae can chew all the leaves off your sugar maples in a matter of days or weeks (red maples are not considered a host species for FTC). The trees use their stored starch reserves to put out a new set of leaves later in the summer, but these leaves are usually smaller and they won't have nearly as much time to photosynthesize and replenish starch reserves. The good news is that FTC outbreaks are cyclical and sugar maples have coevolved with these pests to withstand defoliating events from time to time. The bad news is that once an outbreak has occurred, it usually takes several years for natural forces (disease or predators) to build up and knock the FTC back to manageable levels. If there are several years of repeated defoliations coupled with other stressors, such as a persistent drought, this could cause long-term damage to your trees. Some sugarmakers spray their sugarbushes with a natural pesticide—*Bacillus thuringiensis* var. *kurstaki*—to kill the larvae as they chew on the leaves. Others choose not to spend the money and let nature take its course. If your sugarbush has been defoliated for one season, chances are you can tap your trees without any concern. However, if your trees have

FIGURE 13.32. Forest tent caterpillar larvae have just started chewing the maple leaves on this tree—a large outbreak could lead to the complete defoliation of your sugar maples in a matter of weeks. PHOTO COURTESY OF PETER SMALLIDGE

FIGURE 13.33. A five-foot-tall browse line is evident on the edge of this sugarbush where the deer have eaten all the forest vegetation within reach. Getting new seedlings established in these situations is nearly impossible without fencing. PHOTO COURTESY OF GARY GOFF

FIGURE 13.34. If you decide to plant sugar maples in an old field, you can make great use of the open space between the maples by planting one or two rotations of Christmas trees before the canopy begins to close on the maples. PHOTO COURTESY OF PETER SMALLIDGE

FIGURE 13.35. If you have high-value trees around your sugarhouse or residence that you want to protect during an FTC outbreak, you may want to apply a sticky substance such as Tanglefoot on the trunks to prevent caterpillars from traveling up to the canopy. PHOTO COURTESY OF PETER SMALLIDGE

FIGURE 13.36. A plantation of sugar maple sweet trees at Cornell University's Uihlein Forest in Lake Placid. The sweetest tree we ever found at this orchard measured 10 percent sap sugar content—we took several measurements at various time intervals just to make sure the refractometer wasn't lying to us.

been defoliated for multiple years, you may want to give them a break from tapping.

Finally, one of the biggest threats to the future of the maple industry is the overabundance of deer, another native species that has gotten out of control due to human alteration of the landscape. Sugar maple regeneration has been declining throughout eastern North America largely because of the preferential browsing by deer. Aggressive hunting will help to reduce your deer population while providing you with plenty of enjoyment and delicious venison. The cumulative mortality of deer in an area must exceed 40 percent of the herd each year to keep the deer population stable, with an even higher mortality focused on females to reduce the size and impact of the herd. If you aren't a hunter yourself, I strongly recommend getting other (responsible) people to hunt your land. In addition to the regular fall harvest season, most states allow supplemental harvesting of deer if they are causing significant damage to agricultural crops. As sugarmakers, maples are our crop trees, and if the deer continue to eat up all the maple seedlings, there won't be enough advance regeneration to replace our existing trees when they die. If you have so many deer that you can't get maple seedlings to survive in your understory, then I would certainly apply for additional harvesting permits. Even if you can't get those, try to fill as many tags in your sugarbush as possible. And don't be afraid to take the females when it is legal to do so—regenerating the future forest is much more important than trying to lure in a big buck. At Cornell's Arnot Forest, they implemented a system called Earn-a-Buck in which you have to shoot two females before you are allowed to take a buck. This program has drastically reduced the deer population, and they are now experiencing much better regeneration as a result.

Planting for the Future

There is nothing like the sight of large, beautiful sugar maples lining a country road in the Northeast. Our ancestors planted thousands of maples as a practical source of sugar and to beautify the landscape. Although we have benefited greatly from their efforts over the years, unfortunately many of these trees have been lost to development and/or road widening and maintenance. Salting of country roads in the winter is especially damaging to sugar maples and results in significant crown dieback. Just as we have benefited from those who planted maple trees in the past, I believe we have a responsibility to future generations to do the same. Even if you never expect to tap the trees yourself, you may find inspiration from Nelson Henderson's famous quote: "The true meaning of life is to plant trees, under whose shade you do not expect to sit." Indeed, I have found that most of the people who are planting sugar maples are in their 50s and 60s. They don't expect to tap the trees themselves, but they want to leave a positive legacy for their children and grandchildren to continue sugaring.

If you decide to plant maples for future syrup production, you may want to purchase trees that have been bred for higher sap sugar concentration. Whereas the sap sugar concentration in sugar maples usually hovers around 2 percent, there are genetically superior trees that have much higher concentrations. The "sweet trees" we have planted around our sugarhouse average about 4 to 5 percent sugar, yet I've taken readings as high as 7 and 8 percent sugar concentration. If all our trees were like this, we wouldn't need an RO! Not all of our sweet trees turned out to be sweet, and we have a few that always produce sap in the 2 to 3 percent range. However, by planting trees that had been grafted or cloned from known sweet trees, we have greatly improved the syrup production capacity of the open land surrounding our sugarhouse. To realize how this is possible, it's important to understand the history of the sweet tree breeding program.

In the 1960s the USDA Forest Service initiated and oversaw a program to develop genetically superior sugar maples that have higher sap sugar concentrations than ordinary trees. Together with research collaborators throughout the Northeast, they tested over 20,000 trees from many northeastern states and identified 53 that exhibited significantly higher sap sugar concentrations than all their neighbors. When doing the testing, they had to make sure that the high level of sap sweetness was not due to environmental factors, but rather

FIGURE 13.37. Collecting seeds at the sweet tree orchard in Grand Isle, Vermont.

FIGURE 13.38. A fast-growing butternut planted at The Uihlein Forest in Lake Placid. In the heart of the Adirondacks on relatively poor soils, this bare-root seedling has grown 15 feet in less than five years, even after one of the central leaders broke off when a dead tree fell on it. PHOTO BY NANCIE BATTAGLIA

something unique within the tree itself that caused the sugar content to be higher. The Forest Service then established several clonal orchards with genetically identical seedlings of the 53 sweet trees.

My predecessor at Cornell's Uihlein Forest, Lew Staats, became very interested in this program in the 1980s, and the Cornell Maple Program has been involved with sweet tree research and propagation ever since. For his MS research at SUNY-ESF, Lew conducted a provenance study of seedlings derived from the 53 selections by planting orchards on Cornell's property in Lake Placid and at a private sugarmaker's field in Pennsylvania. He tested the sugar content of these trees when they were only eight years old by inserting a hypodermic needle into the main trunk. The results were encouraging, as the average sugar content of the sweet tree orchards was approximately 30 percent higher than the sugar concentration from nearby wild trees. Of course there was tremendous variation—some trees exhibited significantly higher sweetness while others were actually below average. Ongoing research is aimed at figuring out which selections perform the best and culling out the less productive varieties.

For several years we collected seed from the clonal orchard in Grand Isle, Vermont, and performed rooted cuttings of the sweet trees planted at The Uihlein Forest with the goal of providing genetically improved seedlings to sugarmakers throughout the Northeast. The demand has always been greater than our supply; we had to limit orders to between 20 and 50 trees per person. To scale up production, in 2008 we contracted with a large nursery near Ithaca to produce fast-growing seedlings from the sweet tree seed. They were able to produce several thousand trees, but unfortunately the nursery experienced financial problems and went out of business in 2011. We have since developed agreements with two other nurseries—Forrest Keeling Nursery in Missouri and Crofters Garden Nursery in Maine—to continue growing the seedlings. Unfortunately the supply of seed has been limited from the clonal orchard so these nurseries have not been able to fill all their orders. However, when sugar maples do produce seed, it is usually a bumper crop, so following a good seed year we should be able to provide enough seed to fill the market demand. If you are interested in getting some of these seedlings for your own land, I suggest getting on their waiting list to be next in line when the bumper seed crop finally arrives.

Finally, if you have suitable soils and a relatively warm climate, you may also consider planting nut trees, such as black walnuts, butternuts, English walnuts, heartnuts, and buartnuts. Generally speaking, it is relatively easy to find tappable maple trees and much more difficult to find nut trees available for tapping. Nut trees also tend to grow significantly faster and attain tappable size much sooner than maples, so you will start to get a return on your investment more immediately. As an added bonus, you will be able to harvest delicious nuts every fall once they get to about 10 years old (or maybe even sooner). If you want to experiment with planting nut trees, I recommend starting small and eventually expanding your plantings, especially if you are located outside the natural range for these species. It is good to develop some limited successes and learn from any mistakes before you embark on a huge planting project. There are excellent resources out there to guide you in your efforts, so do plenty of research before putting any seeds or seedlings in the ground. If you take the time to do the plantings right and properly maintain the trees over time, you will soon be able to look back on your efforts and be glad that you did.

The planting process for all of the walnuts is essentially the same—simply plant the nuts in the ground in the fall and then protect and encourage them when they sprout in the spring. I recommend planting as late in the fall as possible, just before the ground freezes, to limit the amount of time squirrels have to dig them up. If there are walnut trees growing near your land, try collecting the nuts from these trees, assuming that they are healthy and appear to have good form and vigor. The vast majority of people who have walnut trees growing on their property would be glad if you helped them pick up the nuts that litter their lawn in the fall. This is how I get all my seed, and the homeowners have been thrilled that I offered to pick them up. Getting local seed that you know is adapted to your area will help ensure that your seedlings are cold-hardy and well suited to your climate. You could also purchase black walnut seedlings from several nurseries, but it is wise to first experiment with smaller trials before spending a lot of money on seedlings that might not survive.

The main concern about planting black walnut is the negative effect it can have on nearby plants. All species of *Juglans* contain a chemical in the leaves, roots, nut hulls, and twigs called juglone (5 hydroxy-1,4-naphthoquinone), though black walnuts have the highest concentrations. Some plants are considered tolerant of juglone whereas others are highly sensitive and may die if they grow too close to black walnuts. For instance, my favorite black raspberry patch is located directly under a large black walnut tree; the plants do just fine and seem to enjoy the dappled shade of the walnut canopy. On the other hand, I have seen many tomato plants either not produce fruit or die since they were growing in the vicinity of a butternut tree (and butternuts have much lower levels of juglone). Given the varied effects of juglone, I suggest doing some research on the sensitivity of nearby vegetation before planting any walnut trees. Since juglone is not water-soluble, chances are that there is little contained in the sap of the trees, though I have not been able to find any research on this. And whereas juglone can have a negative effect on some plants, the impact on people is unclear. Since we have been eating the nuts from black walnut for centuries without any ill effects, my best guess is that the syrup produced from black walnut sap is also a perfectly safe and healthy food.

Should you decide to plant black walnuts, I highly recommend first visiting the University of Missouri's Center for Agroforestry website. This contains a wealth of information on how to design, implement, and manage a black walnut orchard for nut and timber production. You can download an Excel file containing either the Black Walnut Decision Support Tool or the Black Walnut Financial Model to help you decide on tree spacing and whether it makes economic sense to spend extra money on improved varieties. These models will project out your costs and revenues over many years to determine if planting the trees for nuts and timber production will be profitable. At this time neither model includes possible revenues from tapping for syrup production (we will be working on

FIGURE 13.39. An orchard of walnut cultivars planted at the University of Missouri's Horticulture & Agroforestry Resource Center.

that), but you can imagine this could also greatly improve the cash-flow situation.

In deciding on which cultivars to plant, you should note that there isn't one "best" variety that has all the qualities breeders select for. The traits that the breeders select for include (but are not limited to) the date of leaf-out, nut weight and percent kernel, disease resistance, and bearing tendency. If you live in the Northeast, the leaf-out date is one of the most important characteristics, because temperatures below 26°F will kill emerging buds and the flowers that would eventually develop into nuts. Thus, any planting sites that are prone to late-spring frosts should utilize varieties with late leaf-out dates to protect against the cold. No matter what varieties or where you plant your trees, I recommend wide enough spacing that will allow the trees to fully develop their crowns and facilitate vehicular movement through the orchard.

Other Non-Timber Forest Products from Your Land

As the name implies, non-timber forest products (NTFPs) are basically any product you harvest from the woods besides the traditional timber products such as sawlogs, cordwood, chips, posts, poles, and so on. Maple, birch, and walnut syrups are all considered NTFPs, yet these are only a few of the many products that you can derive from your woodlands. This chapter provides a basic overview of the NTFPs that may be best suited for your sugarbushes and associated woodlands. You won't find everything there is to know on all these products; rather, my goal is to provide you with enough information so you will want to learn more and experiment with different NTFPs. Not all of the topics I describe will work for your land—you need to think about the ecology of your property as well as your time and resources to determine what will work best for you. With careful planning and some additional work, it is possible that you could develop other operations on your land that may be just as enjoyable, profitable, and sustainable as syrup production.

NTFPs are often divided up into four categories: edibles, medicinals, florals/botanicals, and specialty wood products. Table 14.1 shows the most common species harvested in each of these broad categories for the eastern United States. For this chapter, I have just focused on the NTFPs that would be most suited for most sugarbushes and associated woodlands. Over the past few decades increasing attention has been paid to them in the US and throughout the world, especially in developing countries. Harvesting of NTFPs can provide a sustainable use of the forest by generating income without cutting trees, yet there is concern that overharvesting of high-value products has threatened many native populations. Although many NTFPs can be harvested freely, others are becoming increasingly rare. You should thoroughly assess your resource before deciding on what and how much to harvest of a particular species.

Edibles

There are a variety of berries, mushrooms, nuts, and vegetables that you can cultivate or forage for on your land. Knowing when and where to look for these will greatly increase your enjoyment and satisfaction while spending time in your woods. There is nothing like going for a walk and coming back with a full belly or fresh ingredients for your next meal.

Raspberries/Blackberries (*Rubus* spp.)

If you experience major storm damage or are planning to cut a significant patch of trees in your woods, there is a good chance that raspberries and/or blackberries (*Rubus* spp.) will quickly grow up in these open areas. Many species of *Rubus* occur worldwide; the three most common on the East Coast are Allegheny blackberry (*R. allegheniensis*), American red raspberry (*R. strigosus*), and Black raspberry (*R. occidentalis*). Typically you will only have one (or occasionally two) of these species on your land. Rubus seeds lie dormant in the soil for decades, and any seeds found under a humus layer in a mature, dense forest will never germinate. However, once they are brought to the surface and have proper sunlight, water, and contact with mineral soil, they will do so quickly. You can help the cause by

TABLE 14.1: Common Non-Timber Forest Products Harvested in the Eastern United States

EDIBLES	Berries (serviceberry, elderberry, raspberry, blueberry, mulberry, paw paw, etc.) Nuts (butternut, black walnut, hazelnut, beech, etc.) Mushrooms (wild or cultivated—chanterelles, morels, chicken-of-the-woods, shiitake, oyster, etc.) Ramps/Leeks Fiddleheads
MEDICINALS	Ginseng Goldenseal Witch hazel Bloodroot Mayapple Blue cohosh Black cohosh Elderberry Slippery elm
FLORALS/ BOTANICALS	Balsam/Evergreen boughs Christmas trees Ferns Pine cones Grapevines Moss Birch bark
SPECIALTY WOOD PRODUCTS	Burls Twigs/Branches Willow for baskets Taphole maple lumber

FIGURE 14.1. Raspberry and blackberry patches often appear along roadsides and following timber harvesting. PHOTO COURTESY OF CHRIS EVANS, ILLINOIS WILDLIFE ACTION PLAN

scarifying the soil during timber harvesting; this easily occurs when harvesting occurs without a blanket of snow protecting the ground.

The wild raspberries and blackberries tend to be smaller and sweeter than the cultivated varieties. Unless they occur in great abundance, there is usually not enough to harvest for commercial sale. However, berry picking is always an enjoyable and "fruitful" activity, especially for kids. One of my greatest memories of childhood is when my father took us berry picking in the summertime. I remember getting excited every time we found a patch of berries and scarfing them down as fast as we could pick them. Of course if you have the patience and self-control not to eat them right away (unlike myself), they are easily preserved for year-round use by freezing or turning into jam.

Wild Leeks/Ramps (*Allium tricoccum*)

If you are fortunate enough to have wild leeks on your property, you probably have a *lot* of them. They are not found everywhere, but when they do occur, it is usually in large quantities. Also known as ramps (based on a similar English plant known as ramson), these perennial vegetables can cover a forest floor with a thick green blanket in April and May. Leeks die back in early summer soon after the foliage appears on the canopy trees. They have adapted to capture as much as light as possible in the short window between the time the ground thaws and when it is cast in deep shade.

Because leeks have been heavily harvested in Quebec, they are considered an endangered species in the province; you are not allowed to dig plants from the wild anymore. Thus, researchers are looking at growing leeks as an understory plant in established forests. Their results have been promising, but it still takes a long time to harvest a bulb large enough for eating or selling. If you are fortunate to have wild leeks growing on your property, take careful precautions to ensure that any harvesting you do is sustainable over the long term. Since leeks take so long to grow and mature, make sure that you don't harvest them faster than they can regenerate. If you don't currently have leeks in the understory, I recommend finding someone who will let you dig some up for transplanting. I have successfully transplanted hundreds of leeks to our sugarbush in Lake Placid. It's not hard to do, especially if you get to them in the early spring when temperatures are relatively cool and wet.

Ramps are particularly abundant in Appalachia, where several festivals are held each spring celebrating this special plant. Because it is one of the first greens to appear, people have traditionally eaten them as a spring tonic. Indeed, after going through a long winter without fresh produce, we can only imagine how much we would crave something green, nutritious, and tasty in the spring. Some of the Native American tribes relied heavily on this plant for spring sustenance. In fact, some people believe that the city of Chicago is named after a massive grove of wild leeks (Chicagou) growing along the edge of Lake Michigan where the huge metropolis now exists.

Jim Chamberlin works for the Forest Service out of the Blacksburg, Virginia, office and is one of the leading researchers on wild leeks. He is especially concerned about the overharvesting of this native plant, as it is unlikely that the vast quantities currently being sold in the marketplace can be sustained from wild populations. Jim recommends that one of the best ways to conserve the plant is to eat the greens instead of just the bulb. If you do harvest the bulbs, then be sure to eat both the bulb and the greens. However, an even better method is to cut off part of the foliage before it starts to senesce and then dry the leaves to use as a spice in cooking. You still get the entire wild leek flavor without having to dig the bulb out of the soil, and you can continue to come back to the same plant every year to gather healthy foliage. If you do have your mind set on

FIGURE 14.2. An understory of leeks with green leaves as far as the eye can see. PHOTO COURTESY OF JAMES CHAMBERLIN

FIGURE 14.3. Digging wild leeks, or ramps, is a common practice in Appalachia. Reused burlap sacks (as long as they are clean!) are a great way to harvest them from the woods. PHOTO COURTESY OF JAMES CHAMBERLIN

FIGURE 14.4. Wild leeks, or ramps, for sale at a farmers' market in New York City. The price of $3 a bunch equates to about $12 per pound, which is a standard price for wild leeks. PHOTO COURTESY OF JAMES CHAMBERLIN

picking the bulbs, wait until later in the season when the bulbs will have attained their maximum size. The bulbs are very small when the plants first emerge in the spring, and you can get a much better yield by waiting for them to mature. Finally, rather than digging up an entire mass of bulbs with a shovel, it is much better to just pull up some of the bulbs in a large clump. This way you don't disturb the forest floor, and most of the leeks remain in the ground with room to expand.

Wild Mushrooms

You may not have ever paid much attention to wild mushrooms, or know how identify the different species, but once you start looking for them, you'll find a plethora of edible, medicinal, and poisonous mushrooms growing wild in your woods! It takes a lot of studying and practice to identify which ones are good for you and—even more important—which ones could make you sick or even kill you. Some mushrooms are more easily identified than others, but I never recommend picking and eating one unless you are 100 percent certain of its identity *and* can also verify it with an experienced mushroom forager. Although the limited descriptions in this book are certainly not sufficient for you to identify or harvest any mushrooms yourself, there are a number of excellent field guides and books on identifying, harvesting, and cooking wild mushrooms. If you decide to start foraging for mushrooms, I strongly suggest studying as much as you can and going out with experienced foragers before picking and eating any wild mushrooms yourself. An important point to remember is that whenever you try a wild mushroom for the first time, you should always just have a small portion to see how your body reacts before eating any more. Another important adage to keep in mind: "There are old mushroom foragers and bold mushroom foragers, but there are no old and bold mushroom foragers." Finding wild mushrooms can be great fun and very rewarding, but always keep safety as your top concern.

There are two mushrooms that I am always looking for when out in the woods during the summer and fall. Chicken-of-the-woods, or sulphur shelf (*Laetiporus*

FIGURE 14.5. Chaga mushrooms appear to be black gnarly growths of burnt charcoal protruding from the stem of birch trees. PHOTO COURTESY OF JOSEPH O'BRIEN, USDA FOREST SERVICE, BUGWOOD.ORG

FIGURE 14.6. As their Latin name suggests, turkey tails (*Trametes versicolor*) exhibit a wide variety of colors. PHOTO COURTESY OF JOSEPH O'BRIEN, USDA FOREST SERVICE, BUGWOOD.ORG

spp.), is easily distinguished by its sulfur-yellow brackets and bright orange tops. Whenever you find one, it is usually quite large—I once came across 20 pounds growing on a single hemlock trunk. Oyster mushrooms (*Pleurotus* spp.) are another favorite that is found wild all over the world, coming in a wide variety of colors, shapes, and sizes. I often find them growing on sugar maple and aspen trees—sometimes on snags but usually on fallen logs. They are also easily cultivated on other carbon-based substrates. There are many other delicious edible mushrooms, including morels, chanterelles, porcinis, giant puffballs, truffles, honey mushrooms, and so on. But this is a book about sugaring, so you'll have to get a book on mushrooms to learn more about them.

Not only are many mushrooms good to eat, but some are also used for medical purposes. Exclusively found on birch trees, chaga (*Inonotus obliquus*) is mostly used as an anti-cancer and immune-boosting agent; several studies have proven its effectiveness in this regard. The reishi mushrooms, *Ganoderma lucidum* and *G. tsugae*, are extremely popular medicinals that have been used for centuries throughout the world, especially in Asia. *G. lucidum*, or linghzi, grows naturally on the base of hardwoods in warmer climates, whereas *G. tsugae* can often be found growing on the base of coniferous trees in northern forests. Turkey tails (*Trametes versicolor*) are

FIGURE 14.7. Chicken-of-the-woods are also known as sulpher shelf mushrooms.

FIGURE 14.8. A bounty of oyster mushrooms on a fallen maple log in the sugarbush.

often found on fallen logs and old stumps in temperate forests throughout the world. They are best known as a natural source for their anti-cancer polysaccharide PSK and are currently being evaluated for other benefits.

To extract some of the medicinal compounds present in some mushrooms, they must be boiled in hot water for an extended period of time. Many scientific studies testing the efficacy of medicinal mushrooms use this method (called decoction) to extract mushrooms' compounds. Wild Branch Botanicals is a company in Vermont that developed an even better way to extract the beneficial compounds—they simply put the mushrooms in their evaporator when making maple syrup. The "maple medicinals" they currently produce include chaga, reishi, turkey tails, and herbs such as echinacea and burdock. By putting the mushrooms or herbs in the evaporator and leaving them in there for two to four hours while the sap is boiling, they are able to produce medicinal maple syrup loaded with beneficial compounds. They concentrate the mushroom decoction at a rate of 960 grams (2.1 pounds) of dried mushrooms for every gallon of syrup produced (this equals 1,250mg per teaspoon). Wild Branch Botanicals does this as a business and a science, using standardized methods at every step, from growing and harvesting, to drying and brewing. The state health department has recently required that the medicinal syrup be produced in a certified commercial kitchen, so they teamed up with another company to make the syrup in an approved facility. They are currently looking for a partner that has a sugarhouse that also meets the certified kitchen requirements so that they can go back to making their maple medicinals with their original method—a conventional evaporator of boiling maple sap.

You could try this method in your own sugaring operation, but you have to know what you are doing and be very careful throughout the entire process. You must make sure that the mushrooms you are using are 100 percent safe and have been verified by an expert mycologist, that you clearly label the mushroom-infused syrup, and that you completely drain and clean the evaporator with the mushrooms before starting a new batch of syrup. If this is something that intrigues you, I'd recommend giving it a try, especially if you know how to identify mushrooms and understand their medicinal powers. However, you should note that the hot water (or sap) extractions do not release all the valuable compounds; other methods may be necessary to release the medicinal compounds in some mushrooms. Furthermore, as seen in the case of Wild Branch Botanicals, you should not attempt to sell any of the mushroom-infused syrup without proper licensing and approvals from your state or province. No approvals would be necessary for home use, but commercial production and sales require much more scrutiny. Finally, if you do try this, be sure to properly and fully clean out your evaporator before you go back to making regular maple syrup.

FIGURE 14.9. Oyster mushrooms grown with the totem method on aspen logs. Inside the black bags are three logs stacked on top of each other with sawdust spawn in-between.

Growing Mushrooms

If you have a large enough woodlot and are actively managing it, you will likely cut out much more wood than you could possibly burn in your evaporator or woodstove. You could either let the trees decompose naturally on the forest floor, sell the logs and firewood on the open market, or turn the logs into gourmet

FIGURE 14.10. Log-grown shiitake mushrooms command much greater prices due to their unique flavors and perceived health benefits. PHOTO COURTESY OF JOSEPH O'BRIEN, USDA FOREST SERVICE, BUGWOOD.ORG

and medicinal mushrooms. I'd highly recommend experimenting with mushroom production, as there are relatively easy ways to cultivate a wide variety of mushrooms in your woods. The most popular varieties include shiitakes (*Lentinula edodes*), oysters, maitake or hen-of-the-woods (*Grifola frondosa*), *Stropharia* spp., and lion's mane (*Hericium erinaceus*), though you can find spawn for many others as well.

I first learned about growing mushrooms when I was in graduate school studying forestry. During the winter break, I purchased some shiitake plug spawn and set out to inoculate my first logs. There were a couple of red maple and red oak trees shading my garden, so I cut them down, bucked the trees into manageable lengths, and proceeded to drill hundreds of holes in a systematic pattern on the logs. My father and I inserted the plugs, sealed them with wax, and then stacked the logs near the stream under a patch of hemlocks. We used fairly large logs, so it took a couple of years for the first mushrooms to appear, but when they eventually did it was very exciting! Optimum fruiting came three to seven years after inoculation, when we would gather dozens of shiitakes with a week or so of a soaking rainstorm. Now that the logs have mostly decomposed, mushrooms only appear on the spongy logs sporadically, so we have inoculated more logs to replace them. I never really cared much for mushrooms before growing them myself, but now I love shiitakes, especially pan-seared in butter and then caramelized with maple syrup!

The totem method is another relatively easy method of growing mushrooms. It basically involves stacking short logs on top of one another with spawn in between. Given the right conditions, mushrooms can start to appear within the first year. Last June we cut a couple of fairly large aspen trees and bucked the stem into 18" lengths. We stacked three of the short logs on top of one another with a couple inches of spawn sandwiched between the logs and placed on the bottom and top of the pile. We covered the freshly inoculated logs with a brown leaf bag, and the entire stack was placed within a black plastic garbage bag to retain moisture. We put together five of these totems and placed them on the north side of a small shed so that they would be shaded and receive additional runoff from the roof when it rained. By October we already had oyster mushrooms popping out of the top of the logs by the top layer of sawdust spawn. The entire process only took a couple of hours and the logs will produce oyster mushrooms for many years to come.

If you are at all interested in mushrooms, I highly recommend three of Paul Stamets's books, *The Mushroom Cultivator*, *Growing Gourmet and Medicinal Mushrooms*, and *Mycelium Running: How Mushrooms Can Help Save the World*. They are all fascinating books full of great information and written by one of the foremost experts in the field. I once saw Paul speak at an organic farming conference, and his message is quite compelling and inspiring. If you do a search for "TED talks Stamets," you'll see a short video of him explaining 'six ways that mushrooms can help save the world.' By taking the 18 minutes to watch him speak, you'll never look at mushrooms the same. The more you learn about edible and medicinal mushrooms, the more you will want to grow and forage for them.

Black Walnuts

Black walnuts are truly incredible trees. In addition to tapping your trees for sap and syrup production and eventually sawing them for valuable lumber, you can also harvest delicious and nutritious nuts each fall. Getting the nutmeat out of a black walnut isn't easy, but many people think it's worth the effort. After all, black walnuts are the proverbial "hard nut to crack"—and removing the husk is no easy feat, either. There are various nutcrackers on the market and tools/techniques for removing the hulls; some work better than others. If you plan on harvesting a significant amount of nuts over your lifetime, it certainly makes sense to invest in a high-quality nutcracker. I would also dedicate a pair of gloves and clothes to harvesting and processing the nuts. Black walnuts have been used as a traditional, natural dye, and they will undoubtedly stain your hands and anything you wind up touching.

The largest buyer and processor of black walnuts in the world is Hammons Products in Stockton, Missouri. They have invested in sophisticated machinery that can do all the hard work of hulling and cracking the nuts to separate out the valuable nutmeat. If you have

FIGURE 14.11. When they first fall off the tree, black walnuts looks just like big green tennis balls. Although most people wait until the nuts are ripe to obtain the nutmeat, there are traditional recipes that call for picking the green nuts when they are very small and still developing to pickle them for a healthy snack. PHOTO COURTESY OF PETER SMALLIDGE

FIGURE 14.12. Butternuts are oblong-shaped, yellowish green in color, and usually have a sticky husk. PHOTO COURTESY OF PAUL WRAY, IOWA STATE UNIVERSITY, BUGWOOD.ORG

a large number of trees and live close enough to one of their hulling locations, you may consider gathering and selling the nuts to them. In 2011, they had 190 locations spread out over 12 states throughout the Midwest to purchase and process nuts. In an average year, they purchase roughly 25 million pounds of nuts and pay between 10 and 13 cents a pound for the whole nuts. Hammons typically purchases about 65 percent of their nuts from their home state, as Missouri is the center of black walnut distribution in the United States and contains the most black walnut trees of any state.

Out of the 25 million pounds of nuts that Hammons purchases every year, they are only able to yield about 2 million pounds of nutmeat. This 7 percent recovery rate is typical of wild trees, yet there are many cultivars available that produce bumper crops every year with a high percentage of nutmeat in the hulls. The University of Missouri has been doing a lot of research and development to produce cultivars that yield high-quality, light-colored kernels with thin shells that are relatively easy to crack. Although the vast majority of black walnuts that Hammons purchases are from wild trees, they pay premium prices for walnuts that come from improved orchards. If you are going to plant walnut trees and want to eat or sell the nuts, I highly recommend planting one of the improved varieties.

Butternuts

With a name like *butternut*, it's hard to imagine that these wouldn't taste delicious. Although not usually as prolific as black walnuts, they do tend to put out heavy crops every few years. The main challenge is getting to the nuts before the squirrels take them all. As with black walnuts, don't wear nice clothes when processing them, as they will also turn your hands and clothes brown. As previously mentioned, these trees are developing cankers and dying from the fungus *Sirococcus clavigignenti-juglandacearum*. If you are able to gather nuts from healthy trees, I would plant those in the hope that some have natural resistance to butternut canker. On the other hand, if your trees are already suffering from the disease but still producing nuts, the chances of any offspring being resistant are *very* low, so you might as well eat all of those.

Beechnuts

Maples are often found with beech and yellow birch as the primary components of the beech-birch-maple northern hardwood forest. Although beech trees (*Fagus grandifolia*) have been decimated by beech bark disease (BBD) throughout most of North America, there are still millions of healthy nut-bearing trees out there, especially in regions that have yet to be impacted by BBD. Even diseased trees that are clearly dying can still produce decent crops of nuts on a periodic basis. The nuts only appear on large, mature trees and are usually out of the reach of most people, but occasionally you will find low-growing branches that are loaded with nuts, especially on roadsides and woodland edges. They tend to open after the first frost and if you aren't quick, the squirrels will beat you to all the nutmeat. If you are going to cut any mature nut-bearing beech trees in your woods, try to plan the cutting in September or October so that you can harvest the nuts from the branches once the tree is on the ground.

Fiddleheads

From late March into May (depending on your latitude and elevation), just as the leaves are beginning to pop out on deciduous trees, many ferns start to rise up from the ground after the long, cold winter. In that brief period

FIGURE 14.13. Harvesting beech nuts takes a lot of patience and luck. You must get the nuts after they have matured and before the shells have opened and released the individual kernels on the ground. PHOTO COURTESY OF BILL COOK, MICHIGAN STATE UNIVERSITY, BUGWOOD.ORG

when the fiddleheads have appeared and before they unfurl, some species of ferns can be harvested for a delicious springtime vegetable. Fiddleheads are very nutritious, providing a healthy source of vitamin A, vitamin C, iron, and potassium. People often describe the flavor as a cross between asparagus and broccoli, and you can use it in the same ways as these more common vegetables.

The window for harvesting is very short and you need to keep a careful eye on them to know when to start picking. The key to sustainably harvesting these wild vegetables is to only take two or three fiddleheads from each plant. You normally have six to eight to choose from on any given crown, so by only taking a fraction, you will allow the plant to develop normally and continue to produce year after year. I usually bring a knife to cut them—it makes harvesting a lot easier and provides for a nice clean break on the stem.

Although the fiddleheads of many fern species are technically edible, some are considered poisonous and should be avoided. The ostrich fern (*Matteuccia struthiopteris*) produces the most popular fiddleheads and is the species most people are referring to when they talk about fiddleheads. Ostrich ferns tend to grow along rivers and streams in semi-open woods. If you have a sugarbush or stand of trees along a riparian area, be sure to look for them. They have an impressive ability to spread laterally from stolons and quickly colonize an area. They can withstand heavy floodwaters and help to stabilize streambanks with their extensive root systems. I had been concerned that the heavy floods from Hurricane Irene may have devastated a favorite patch of fiddleheads in the Adirondacks, but when I went back to harvest the following spring they were as plentiful as ever. If you don't have any fiddleheads on your own property, find someone else who does. They transplant relatively easy (I've moved dozens of them in the fall) and will spread fairly rapidly once they become established.

FIGURE 14.14. Ostrich ferns are easily identified by the sterile fronds that persist throughout the winter. A recently emerged fiddlehead is harvested by cutting it off at the base with a knife.

You may just want to harvest fiddleheads for your own consumption, but if you have access to extensive areas of them (such as a partially wooded floodplain), there are good markets for this spring vegetable. Patrick von Aderkas published a great article discussing the history of the fiddlehead market in eastern North America and various attempts to commercialize the harvest.[1] Whereas most fiddleheads used to be canned for year-round use, with refrigerated trucks and modern freezers, you can now buy "fresh frozen" fiddleheads year-round in some grocery stores and various websites. Retail prices often range between \$10 and \$20 per pound, so it can be worth your while to harvest fiddleheads, especially if you are already selling maple products at farmers' markets and other venues in the late spring when fiddleheads are in season.

Medicinal Plants

There is a large and growing market for plant-based medicines, and many of these products are derived from plants found in our woods. Although this chapter highlights only ginseng and goldenseal, there are dozens of other plants used in natural and herbal medicines. If this piques your interest, I recommend pursuing additional resources to help you to discover the valuable medicines you may have growing wild in your woods. The two books that I always take with me on foraging expeditions are *Peterson's Field Guide to Medicinal Plants/Herbs: Eastern and Central North America* and *Identifying and Harvesting Edible and Medicinal Plants* by Steve Brill.

Ginseng (*Panax quinquefolius*)

American ginseng is a perennial understory herb that grows best under the shade of maple trees—it is probably the most valuable "legal" crop that you can grow on your land. In recent years wild or wild-simulated ginseng has been sold to licensed dealers for prices up to $800 or more per dry pound or $250 to $300 per pound fresh weight. Sugarbushes are the ideal place to plant ginseng since they are full of large maple trees that probably won't be cut in the foreseeable future. It requires between 8 and 12 years to grow a harvestable-sized ginseng root from seed, and most of us plan to have our woodlands intact for at least the next decade or more.

Much of my knowledge of and interest in ginseng came from working with Bob Beyfuss, the recently retired ginseng specialist for Cornell University Cooperative Extension. Bob set up 12 ginseng research plots in our sugarbush in Lake Placid to study the microclimates and forested environments in which it grows best. Although many of the test sites failed, the best-performing plot was located in a moist, well-drained, shaded valley with deep, fertile soils and about 80 percent canopy of large, mature sugar maples. These are the ideal conditions for ginseng, so if you have places like this on your property, it would be well worth your time to try growing this valuable herb.

Ginseng is a finicky plant that will only grow in certain forest types, so Bob developed a visual site assessment tool (VSA) that is useful in determining whether your property would be well suited for production of ginseng (Table 14.2).[2] It includes six categories where you enter in values for the dominant tree species, exposure, slope, soil and site surface characteristics, understory plants, and security. Going through this fairly simple exercise is a great way to identify if you could grow ginseng on your land in a profitable manner.

Having sugar maple as the dominant species yields the highest score in Bob's VSA tool. Not only do sugar maples prefer the same rich, deep, moist, well-drained soils that ginseng thrives in, but they also drop a tremendous amount of calcium in their leaves each fall. Ginseng utilizes the leaf litter as an excellent source of mulch and nutrients each year. The maples also provide

FIGURE 14.15. Bob Beyfuss, retired New York State ginseng specialist for Cornell, digging up a ginseng root in the Catskill Mountains of New York. Notice that the leaves are just starting to yellow and the berries either have fallen off naturally or have been planted. PHOTO COURTESY OF BOB BEYFUSS

FIGURE 14.16. By late summer the berries have fully ripened, making the otherwise difficult-to-find plant easy to spot in the woods. Whenever you dig a root from the wild, you are required to plant the berries to help perpetuate the species. PHOTO COURTESY OF BOB BEYFUSS

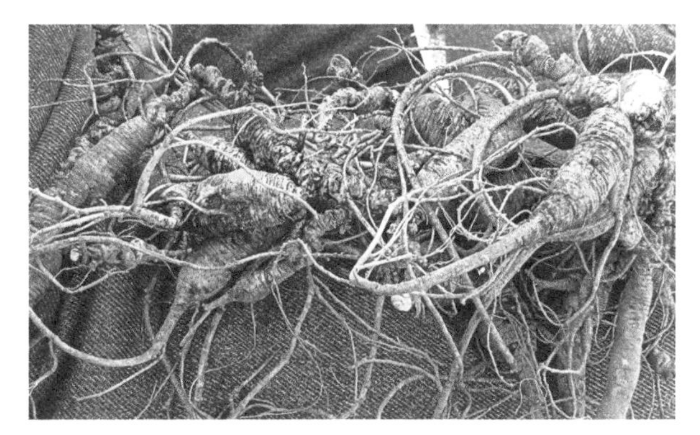

FIGURE 14.17. A collection of extremely old roots dug up in the Adirondacks. These roots are thought to contain the greatest medicinal value; some individual roots can fetch $100 or more.

TABLE 14.2: Visual Site Assessment and Grading Criteria for Potential Woodland Ginseng Growing Operation for a Northern Forest

Circle Only One Choice for Each Category	
CATEGORY A—DOMINANT TREE SPECIES (50% OR MORE OF MATURE TREES)	**POINTS**
1. Sugar maple (add additional 5 points if average circumference is greater than 60″, add 2 additional points if there is a presence of butternut); in areas south of NY, tulip poplar is equivalent in value to sugar maple as an indicator tree species	10
2. White ash or black walnut (add additional 4 points if average circumference is greater than 60″, add 2 additional points if there is a presence of butternut)	8
3. Mixed hardwoods consisting of beech, black cherry, red maple, white ash, red oak, basswood	5
4. Mixed hardwoods as above plus some hemlock and/or white pine	5
5. Red and/or white oak	3
6. Ironwood, birch, aspen	1
7. All softwoods, pine, hemlock, spruce, fir	0
Subtotal	**15**
CATEGORY B—EXPOSURE (ORIENTATION)	
1. North, east, or northeast facing	5
2. South, southeast, northwest	2
3. West, southwest	0
Subtotal	**2**
CATEGORY C—SLOPE	
1. 10–25% slope	5
2. Level	3
3. 30% or greater slope	0
Subtotal	**5**
CATEGORY D—SOIL AND SITE SURFACE CHARACTERISTICS	
1. Site dominated by mostly very large trees more than 20″ in diameter, few surface rocks, 75% of site plantable	10
2. Site dominated by medium-sized trees, 10–20″ in diameter, some surface rocks, 50% plantable	8
3. Small trees less than 10″ in diameter, very stony, 25–50% plantable	5
4. No large trees, saplings and shrubs dominate or large rock outcropping, many boulders, less than 25% tillable	3

5. Soil too rocky to plant anywhere, poorly drained, standing water present	0
Sutotal	**8**
CATEGORY E—UNDERSTORY PLANTS (SELECT HIGHEST-SCORING ONE ONLY)	
1. Reproducing population of wild ginseng	15
2. Sparse wild ginseng	10
3. Maidenhair fern, stinging nettle, rattlesnake fern, or red or white baneberry	8
4. Christmas fern, blue cohosh, red-berried elderberry, foamflower	6
5. Jack-in-the-pulpit, other ferns, trillium, bloodroot (bloodroot is a much higher-scoring indicator plant south of NY), jewelweed, mayapple, herb Robert (a type of wild geranium), true or false Solomon's seal	5
6. Wild sarsaparilla, Virginia creeper, ground nut, yellow lady's slipper, hepatica	3
7. Club moss, princess pine, bunchberry, garlic mustard, pink lady's slipper	0
8. Woody shrubs such as honeysuckle, mountain laurel, witchhazel, barberry, maple leaf viburnum, arrowwood, shrubby dogwoods, alder, lowbush or highbush blueberry, spicebush (spicebush is often found with wild ginseng in southern or midwestern sites and is considered a good indicator plant there)	0
Subtotal	**10**
CATEGORY F—SECURITY	
1. Very close to full-time residence of potential grower, with planting site within easy viewing of residence (noisy outdoor dogs housed nearby add 5 points)	10
2. Forested land less than 440 yards (0.25 mile) from grower's residence, patrolled regularly	8
3. Regularly patrolled woodlot within 1 mile of residence	3
4. Nonresident grower or remote woodlot	0
Subtotal	**10**
TOTAL SCORE (ADD POINTS FROM EACH CATEGORY'S SUBTOTAL):	50

Results:

50 points or above	Excellent site, great potential
40–50 points	Good site, do complete soil analysis
30–40 points	Fair site, test soil
Less than 30 points	Poor site, look elsewhere

Prepared by Bob Beyfuss, Cornell Cooperative Extension Agent, Greene County, NY, August 1998; revised December 2003.

the optimum shade while simultaneously bringing water to the soil surface from deep underground through a process known as hydraulic lift.

One of the best ways to determine if ginseng will grow well in your woods is to examine what is already growing on the forest floor. If you find wild ginseng growing naturally, then of course it is a great site to produce more. However, just because ginseng isn't already growing there doesn't mean it won't do well. Some of the understory plants that indicate a great site for ginseng include maidenhair fern, rattlesnake fern, Christmas fern, baneberry, blue cohosh, foamflower, and stinging nettle.

If you decide to plant ginseng, I recommend using the wild-simulated approach. The roots you eventually dig will resemble the wild ones that fetch the highest prices in the marketplace. It also requires the least amount of work and site preparation to get started. You can basically just plant seeds or rootlets in appropriate areas within your woods, monitor and tend them on a periodic basis, and eventually harvest them many years later. If you think that growing ginseng may fit in to your management plans, I recommend further reading, including Bob Beyfuss's *Practical Guide to Growing Ginseng* and *Ginseng, the Divine Root* by David Taylor.

Goldenseal (*Hydrastis canadensis*)

Chances are if you have a good site for growing ginseng, you can also try your luck with goldenseal. I had heard about goldenseal for many years without ever paying much attention to it. Then I attended the North American Agroforestry Conference in Athens, Georgia, and got to meet several people who have grown it successfully down there. Since it grows well in relatively poor soils and under long, hot summers, it will surely thrive in the more hospitable soils and climates where sugar maples are found. Once you know what the market demand and potential for goldenseal is, and how easy it is to grow, it's hard not to get excited about growing this herb.

The most cost-effective and easiest way to grow goldenseal is from seed. You can also buy cuttings from rhizomes, but those are more expensive and take more time and effort to plant. The best method I heard of was from a grower in Georgia who raises heritage turkeys. He puts the goldenseal seed in with their feed and keeps them fenced in a carefully chosen section of his woodlot. The turkeys plant the seeds in their scat, providing a simple and natural way to spread this species throughout his land.

Traditionally, much of goldenseal sold on the worldwide market has come from wild-harvested plants. Consequently, those populations have fallen drastically over time. Goldenseal is now listed in the Convention on International Trade in Endangered Species of Wild Fauna and Flora (CITES). Growing this plant will ease pressure on wild populations while producing excellent returns for your efforts. As with all new ventures, if you are interested in trying this out, first do some research to determine where you will sell the final crop and

FIGURE 14.18. Maidenhair fern is an excellent indicator of fertile, well-drained soils that may be ideal sites for planting ginseng. PHOTO COURTESY OF PETER SMALLIDGE

FIGURE 14.19. Goldenseal is a valuable perennial herb that may grow well in the understory of your sugarbush. PHOTO COURTESY OF LEW DIEHL, BUGWOOD.ORG

FIGURE 14.20. Llewellyn Butler and Grant Cavanaugh proudly standing by their birch oil still in Pennsylvania. PHOTO COURTESY OF GRANT CAVANAUGH

whether it will make economic sense for you to do so. Chances are you'll be glad you did.

Birch Oil

If you have a lot of black birch or yellow birch trees on your land, and you plan on harvesting some of them, you might consider making birch oil. The inner bark and small twigs of birch trees contain methyl salicylate, which is the active ingredient in aspirin. It used to be made in large quantities throughout Appalachia, primarily from black birch trees. Once chemists figured out how to synthesize it, the production of natural birch oil dropped off considerably. It is made by putting the bark and twigs in water and then using a still to capture the oil. Many of the stills that were used to make moonshine were also used to produce birch oil.

Decorative Items

Many plants growing in the woods can be used for decorative purposes. Some of these plants are ecologically important and should not be harvested, even though they could generate significant income. Others are invasive or harmful plants that may be competing with your more desired vegetation. Knowing what you should harvest and what is better left alone could help your bottom line while also benefiting the ecology, health, and productivity of your land.

Vines

In many forests, wild grapes, honeysuckle, and other vines can be found climbing up trees and extending their foliage well into the forest canopy. Some vines wrap themselves around smaller trees, eventually girdling them, while others produce dense foliage that cuts off much of the available sunlight to their host trees. Removing them will often improve the health and performance of the host trees, though they must be removed carefully to avoid doing more harm than good. If you have vines growing in your woodlot and wish to remove them, you may be able to use them personally or sell them to other wreath makers, especially during the holidays. Remember to keep the vines as long as you can, leaving decorative tendrils when possible, and try to weave them or sell them to another wreath maker before they dry out and become brittle. Doing this will yield a valuable product while doing a service for your sugarbush and other woodlands.

FIGURE 14.21. Grapevines can cause serious damage to your crop trees, so it is best to cut them out before they get to this stage.

Holiday Greenery

If you use tubing to collect sap, the last thing you want in your sugarbush is evergreen trees. These coniferous species provide cover and food for squirrels, so to reduce rodent damage to your tubing system, you should try to remove as many evergreen trees from your woodlot as possible. November and December are the best months to do this since you can use the high-quality boughs for holiday decorations. Balsam fir is one of the most desirable species due to its fragrance and soft needles. White pine boughs are also commonly used since they bend easily and have attractive, long needles. If these species are not available locally, people have been known to use just about any evergreen species for wreaths and other holiday decorations. If you don't have the time or desire to cut the evergreens yourself, advertise to local crafters that they can cut boughs from your woods. Even if you don't get paid anything, you will get free labor to remove the conifers that wreak havoc with tubing systems. If you do let people onto your land to cut greens, make sure they have the appropriate insurance, especially if they will be using a chain saw. You may just ask that they only cut evergreens that are small enough to use hand loppers for. It might be best to cut larger trees yourself and then let other people chop off the top limbs for greens. Note that not all evergreens will be desired for holiday decorations, so it's worth checking with the end users to see what materials they prefer before you cut too much.

Birch Bark

The bark of paper birch trees, if properly harvested and marketed, is much more valuable than the timber in a given tree. Although it is technically possible to remove some of the outer bark without killing the tree, this could lead to long-term health issues and put the tree at risk for disease and decay. If you are not careful and remove the inner bark (phloem), this could quickly kill the tree by cutting off sap flow between the branches and roots. The best time to remove the bark is just before or directly after cutting a tree. Larger trees produce the best bark, and it is worth taking the time to plan any cutting to more easily extract the bark. Late spring or early summer is generally the best time to remove bark, as it comes off relatively easily that time of year.

The price of birch bark varies greatly between $2 and $18 per square foot, depending on the quality, location, and particular vendor. Although most of the birch bark on the market is from paper birch, I have also seen the bark from young yellow birch and black birch for sale. There is usually a good market for birch bark since it is used for such a wide variety of artistic and functional products. Birch-bark canoes are one of the best-known uses for the bark, as Native Americans made excellent use of the watertight bark to line their canoes. Although very few birch-bark canoes are still produced today, those that are made command premium prices in the market. Birch bark is often used for furniture making and interior decorating of fine homes, especially in the Adirondacks. A variety of baskets and containers are made by weaving the inner bark or using the outer bark for coverings. The small branches and twigs of birch trees (especially paper birch) are also valuable for various crafts and woodworking projects.

FIGURE 14.22. In northern states and into Canada, balsam fir tends to grow like a weed in the understory of many sugarbushes. The tender branches on young trees make excellent boughs.

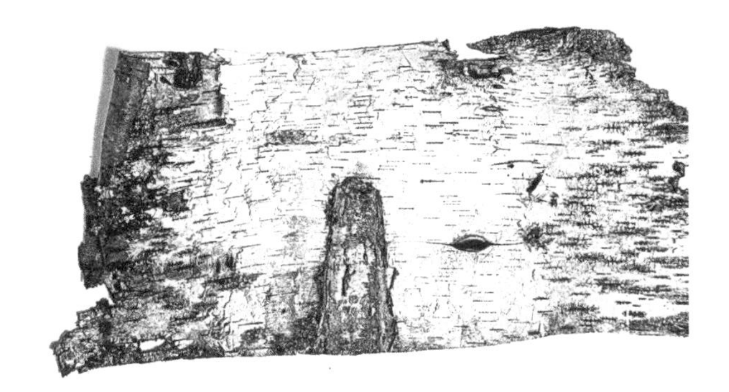

FIGURE 14.23. The bark of paper birch trees is often more valuable than the wood contained within it. If you know of trees being harvested, try to salvage the bark either directly before or after the trees are felled. PHOTO BY NANCIE BATTAGLIA

Pinecones

Although I generally advise against keeping coniferous trees in or around your sugarbush, if you do have them on your property, you may be able to collect and sell the cones to various craftspeople. Eastern white pine and Norway spruce yield some of the largest and most valuable cones in the eastern United States. Although you won't get rich selling them, I have seen them on the Internet for anywhere from 10 to 50 cents each.

FIGURE 14.24. If you have a lot of white pine or Norway spruce trees, it may be worth your time to gather the large, attractive cones. PHOTO BY NANCIE BATTAGLIA

Specialty Wood Products

The products that are derived from wood but don't end up in normal commercial distribution channels are known as specialty wood products. The two most common specialty wood products that we can derive from our sugarbushes are taphole maple lumber and burls.

Taphole Maple Lumber

For too many years we have let the timber industry dictate that the bottom 8 feet of a maple tree was basically worthless if it had been tapped for syrup production. The tapholes and associated stain columns in the wood were considered defects, so the butt logs of tapped trees were either left in the woods to rot or sold for little money in the firewood and pulpwood markets. Sugar maple is prized for its clear, white sapwood, and anything that takes away from this is considered a defect that reduces the commercial value of the logs and lumber. However, there are certain types of figured maple lumber that command premium prices, such as curly maple, birdseye maple, and tiger maple. These beautiful patterns rarely occur in maple trees and the demand is much greater than the supply, so they fetch top dollar in the marketplace. Taphole maple lumber is a beautiful and rare product that should also be highly valued—the fact that it often gets discarded for firewood represents one of the biggest wasted opportunities in the maple industry today.

While taphole maple is beautiful in itself, the fact that the holes and stain columns came from producing maple syrup is what makes this lumber extremely valuable. People love maple syrup and the story of maple sugaring. They have an image of quaint rural life, Grandpa tending the evaporator, kids gathering buckets with horse-drawn sleighs, and pure white snow everywhere. This romantic image of maple sugaring is what helps to sell maple syrup to many people and is what could also help sell taphole maple wood products to even more. Some people are drawn to it because they considered it to be "recycled"—the trees that were used for syrup are also now used for lumber. Although this isn't an accurate definition of recycling, the environmental friendly nature of taphole maple lumber and the "salvaged" image people have of it are excellent selling points.

My first venture into taphole maple lumber came in 2005 when I started working for Cornell at the maple research center in Lake Placid. I noticed that there were many maple trees in the sugarbush that were in severe decline as well as a number of spruce, fir, and pine that needed to come out. During my first summer on the job, after tearing down the old and worn-out tubing system, I contracted with a logging firm to harvest all the marked trees. I asked the loggers to sort out the first butt logs from all the maples that had been tapped. Since they would have normally sold these logs for firewood, the loggers simply reduced the stumpage payment to Cornell by the value of the logs as firewood (a few hundred dollars). They put them in a separate pile for us and we hired a local sawyer with a portable bandmill to come turn the logs into 6,000 board feet of maple lumber. At 30 cents per board foot to do the sawing, it cost about $2,300 for the entire job.

This turned out to be a very profitable venture. We used less than 1,000 board feet of the taphole lumber for wainscoting and to build the sales counter in our new education center. Without any advertising, much of the remaining lumber was sold within two years to a wide variety of interested buyers. The lumber that did not contain any tapholes went to someone who wanted a new hardwood floor made from locally harvested maples. These ordinary maple boards fetched the lowest price, $3 per board foot. The remaining boards with tapholes were sold for nearly twice that price to various people who happened to come to our sugarhouse. We told the visitors how the trees had been tapped for nearly 50 years before they eventually went into decline (from natural causes, not tapping) and no longer produced enough sap to justify keeping them in our sugarbush. People loved the story and the unique character of the wood and gladly paid premium prices for the wood. If you put the time and effort into marketing taphole maple lumber, chances are you won't be able to secure enough of it to meet the market demand.

Over the past several years I have noticed a lot more interest in and people using taphole maple lumber. An

FIGURE 14.25. When the vast majority of the outer sapwood has been impacted by tapholes and the associated stain columns, it is time to cut the tree down and turn it into valuable lumber. This tree should have been harvested 10 to 15 years earlier when it was already in decline and before the internal rot set in.

FIGURE 14.26. Portable bandmills are the best way to process tapped logs into lumber. Even if you wind up hitting an old metal spout, nail, or broken drill bit, the sawblades are fairly inexpensive and easily to replace. PHOTO COURTESY OF PETER SMALLIDGE

FIGURE 14.27. Taphole maple displayed behind the reception desk at Madava Farms in Dutchess County, New York. PHOTO COURTESY OF TYGE RUGENSTEIN

FIGURE 14.28. When the New York State Maple Producers Association recently renovated their booth at the state fairgrounds, they prominently displayed taphole maple lumber for all the paneling. PHOTO COURTESY OF THE NEW YORK STATE MAPLE PRODUCERS ASSOCIATION

Internet search for "taphole maple lumber" or "tapped maple lumber" yields a number of interesting websites with people selling taphole maple lumber and furniture. There seem to be a lot of woodworking enthusiasts who are now experimenting with this unique and storied lumber. When the New York State Maple Producers Association renovated their booth at the state fairgrounds in Syracuse, they incorporated taphole maple into all the paneling and shelving. The executive director at the time, Mary Jeanne Packer, developed an attractive poster titled A New Life for Old Maples, which is displayed at the booth and has helped to promote taphole maple lumber. It makes for a great conversation piece and adds tremendous value to your trees when they are eventually harvested. The more we can do to market taphole maple lumber to the general public, the better off our entire industry will be.

Burls

You may occasionally discover burls growing on the trees in your sugarbush and associated woodlands. Although the mechanism for how they develop is not well known, burls are caused by some type of stressor to the tree and can take on enormous size. The burls that are found on sugar maple trunks and can be turned into beautiful bowls and other specialty items. You can

FIGURE 14.29. Although burls will likely reduce the sap production potential of a tree—especially when tapping directly underneath them- they are much more valuable cut down and turned into valuable bowls, furniture, and other items.

also find burls on a number of other species—I often find them on yellow birch. If you are going to harvest them, I recommend first calling around to different woodworkers and furniture makers in the area to see who may be interested in buying them. Sometimes burls will fetch extremely high prices if you have a large, figured specimen that is in high demand. Whenever you cut a burl off a tree, always leave a couple of feet of the stem on either side to let the final user determine how to utilize it.

Striped Maple Bark

Although striped maples (*Acer pensylvanicum*) usually never grow to a tappable size, they are an excellent source of basket-making and weaving material. The attractive, green striped bark is easily removed and can be turned into beautiful baskets and other crafts. If you did ever come across striped maples that were large enough to tap and then eventually used the bark from these trees for weaving material, you might be able to develop some nice wide strips that prominently display the old tapholes. I recommend making the strips about 1.5" wide; that leaves about 0.5" of bark on either side of the taphole so it doesn't break or tear at the hole.

FIGURE 14.30. The distinctive bark of the striped maple is an excellent source of weaving material. PHOTO COURTESY OF PETER SMALLIDGE

CHAPTER 15

SYRUP OR SAWTIMBER PRODUCTION:
DETERMINING THE MOST PROFITABLE USE FOR MAPLE, BIRCH, AND WALNUT TREES[1]

Maple, birch, and walnut trees are valuable not only for the sweet sap they produce in the spring, but also for the beautiful lumber that comes from harvesting mature trees. These dual purposes can often lead to conflicts between sugarmakers who would like to tap the trees and loggers who would rather harvest them for timber. Most people consider logging and sugaring to be mutually exclusive activities since trees that are being used for syrup production cannot be harvested for lumber, and if they are eventually harvested, the previous tapping reduces the commercial value of the lumber in the first log. Considering the fact that you are reading this book, chances are you would rather utilize your trees for syrup than for sawtimber production. However, if you are also planning to tap trees on other people's land, you will likely encounter landowners who don't want to let you tap their trees due to concern over reducing sawtimber value. In fact, our survey research found that approximately half of the landowners who would otherwise want to lease their trees for sugaring have not done so because of the concern of reducing sawtimber value.

Given the prevalence of this issue, I decided to dedicate a chapter of this book toward analyzing the costs and benefits of leasing taps versus sawtimber management. It is a complex (and often times confusing) issue that requires careful consideration and thought before making any decisions. In order for you to convince a landowner that it is in his or her best interests to lease woodland for tapping, you will have to know what factors would make it more profitable for the landowner to do so. Although leasing taps for many years before eventually harvesting the tree is the most profitable management strategy for most situations, there are other situations in which immediate harvesting or leaving trees untapped will generate greater revenue for the landowner. As a sugarmaker you must know what factors influence this outcome so you can concentrate on the woodlands most suitable for sugaring.

Even when they can make more money by leasing taps than cutting their trees for timber, persuading landowners to do so is not always an easy task. Since the timber industry doesn't make money from trees that are dedicated to syrup production, naturally they would advise against tapping sawtimber-quality trees. Most foresters and loggers discourage landowners from drilling holes in their valuable maple logs, and many landowners have bought into this notion. To be fair, sometimes their advice is appropriate, as there are many circumstances when a landowner could earn greater revenues by harvesting timber than producing syrup or leasing taps. However, there are probably many more situations when the opposite is true, and determining the most profitable option must consider all of the immediate and long-term costs and revenues. This can be a difficult decision to make, and the answer will depend on many attributes of the individual trees and the stand in which they occur. By gaining a better understanding of this complex topic, you will be better able to convince landowners when leasing taps will be in their best economic interests. You will also be able to know when a landowner would be better off harvesting certain trees and therefore avoid tapping the most valuable stems. When you can offer a landowner a balanced and informed opinion, you will gain much more credibility and trust. Landowners are much more likely to want to do business with a trustworthy person who provides them a balanced and fair assessment of the options for their woodland.

NPV Calculator

The most appropriate way to determine when it makes more economic sense for a landowner to lease taps for syrup production or manage their woodland for sawtimber production is through a Net Present Value (NPV) analysis. In 2010 I developed an Excel-based NPV Calculator for leasing taps versus timber production as part of my dissertation research at Cornell. Based on the data that goes into the spreadsheet, you can compare the NPVs (from a landowner's perspective) of four distinct management options:

1. Leasing the tree for tapping and eventually harvesting it.
2. Continually leasing the tree for tapping without ever harvesting it.
3. Letting the tree grow freely (no tapping) and harvesting it in the future.
4. Harvesting the tree immediately without any tapping.

NPV analyses discount all future costs and revenues back to the present time, thereby making it possible to compare the benefits of receiving a large sum of money for cutting a tree down today with those of receiving a smaller annual payment over a period of many years. The NPV Calculator allows users to enter values for 36 variables pertaining to tree size and growth, stumpage payments, lease payments, property taxes, and financial determinants. The following pages describe each of the main variables used in this analysis, providing the possible ranges for each variable in red with the default value appearing in parentheses. After you read through these variables and understand what goes into performing this type of analysis, you should download the NPV Calculator from the Chelsea Green Publishing website, www.chelseagreen.com/NPV. I recommend plugging in the values on a variety of trees in your own sugarbush or for a tract of a land that you are hoping to lease for tapping.

Although the analysis presented here is based on individual trees, most people are concerned about management options for a stand of trees or an entire forest. The NPV Calculator was designed specifically to examine the major factors affecting the economics of leasing taps versus sawtimber management for individual trees. You must consider the specific attributes of all the trees that make up a stand to decide how to manage it. It is possible (and occasionally practiced) to not tap certain trees in a stand that is managed for syrup production. For instance, if a landowner is concerned about sawtimber value, you could flag all the trees that may produce veneer logs and forgo tapping them. Even though you would like to tap the large, healthy trees that may contain veneer logs, dedicating the trees with the highest-quality stems for timber harvesting often results in the greatest returns for the landowner. After reading this chapter, you will be better able to recognize which trees are appropriate for tapping, which should be harvested immediately, and which should be managed for long-term sawtimber production.

The main variables that you must determine values for are described below. They fall into five broad categories having to do with tree size and growth rate, stumpage rates, lease payments, property taxes, and financial determinants.

Tree Size and Growth Rate

One of the first things you must measure (or estimate) is the size of the tree and the projected growth rate over time.

Initial Tree Diameter, 6–48 (12) in (dbh)

The standard way to measure the diameter of a tree is at breast height, 4.5 feet above ground level. This analysis does not make much sense for small trees, so it is best to consider trees that are 8" dbh or more, noting that the minimum size tree anyone should tap according to conservative tapping guidelines is 10".

Diameter Growth Rate Untapped, 0–0.5 (0.15) in/yr to dbh

Individual tree growth rates will vary based on many factors, including soils, genetics, age, crown position, et cetera. Typical diameter growth rates range anywhere from 0.02 to 0.08" per year for a slow-growing tree versus 0.25 to 0.5" a year for a very fast-growing tree. You could take increment cores of healthy trees that you want to save, but it is a fairly difficult process and requires specialized tools. An easier way to estimate the growth rates of your trees is to measure the width between rings on the outer edges of trees that you cut in the sugarbush.

Initial Merchantable Height, 0–48 ft

Enter the height at which the bole of the tree reaches a point where it can no longer be used for a sawlog. This generally occurs where the diameter of the tree inside the bark equals 8". It can also be to the point where there are many large branches spreading out into the crown of the tree or where a serious defect occurs that would negate any potential sawlog value.

Initial Merchantable Height Growth Rate Untapped, 0–3 (0.5) ft/yr

Enter the rate at which the merchantable height of the tree is increasing at an annual rate (in feet). If the tree has serious defects, crotches, or large limbs on the main trunk, then it may not be capable of increasing in merchantable height. For straight, clear trees that can still add merchantable height growth, the default value has been set to 0.5 feet per year.

Reduction in Diameter Growth Due to Tapping, 0–50 (10)%

You must enter the expected reduction in diameter growth rate for tapped trees. Since trees utilize stored sugar reserves for growth and we remove a portion of these reserves when we tap trees in the spring, it makes sense that trees would grow a bit more slowly. Based on the research of Brian Chabot at Cornell, the default setting is 10 percent.

Reduction in Height Growth Due to Tapping, 0–50 (10)%

Since merchantable height is strongly correlated with diameter growth of the upper logs, you can assume the same reduction in height growth due to tapping as diameter growth.

% Decline in Annual Diameter Growth Rate, 0–3 (1)%

Unless trees are growing at an increasing rate, the amount of annual diameter growth will decline as trees get larger. This is due to the fact that if a tree puts on the same amount of wood volume, by having to distribute that same volume over a larger surface area, the annual diameter growth will be reduced by a given percentage. The default setting is 1 percent.[2]

% Decline in Annual Height Growth Rate, 0–3 (1)%

Given that merchantable height growth is strongly correlated with annual diameter growth, the default setting for this variable is also set to 1.

Stumpage Rates

The term *stumpage* refers to the money a landowner receives from a logger for the ability to cut and remove a tree (or stand of trees) from a given property. This section covers all the variables that affect how much money a tree is worth (to a landowner) when it is still standing in a woodlot.

Species: Sugar or Red Maple

Since stumpage payments vary greatly by species, if you type in "sugar" or "red" the spreadsheet will reference the appropriate log prices for sugar or red maple. Note that if you want to conduct this analysis for birch or walnut trees, you can still use this calculator—just make sure that the prices you plug in for the log prices match the species name you provide for this variable. For instance, to do the analysis for black walnut, you could enter in the appropriate log prices for black walnut in the red maple section and then type "red" for this variable to reference the black walnut prices.

Height of Tree Affected by Tapping: 3–12 (8) ft

Enter in the height at which the butt log would be affected by tapping (in feet). The default value is set to 8, as you can reasonably predict that the first 8-foot butt log will contain all the tapholes and staining. This is due to tapping higher in years of heavy snow depth with tubing systems. Also, with a typical lateral line height of 3 to 4 feet and a 30" dropline extended to maximum capacity, most of the tapholes and associated stained wood winds up being approximately 4 to 8 feet above the ground. However, if you only used buckets or bags in an area without excessive snow depths, it is possible that only the bottom 5 feet would contain tapholes and staining.

Log Scale: Doyle or Scribner or International Scale

The three commonly used log scales in the United States are Doyle, Scribner, and International. The log scale you choose will have a significant impact on the estimated board foot volume of a tree, and therefore the anticipated stumpage payments. It is important to know which log scale is used in your region and enter the appropriate one.

Sawtimber Delivered Log Prices: 0–2,000 $/MBF

You should get a price sheet from a local sawmill to input the delivered log prices for sugar and red maples of a given size (assuming no grade defects) with a diameter inside bark at the small end of 10 to 11.9", 12 to 13.9", 14 to 15.9", and 16" and greater. Note that if you wanted to run the analysis for birch or walnut trees, then you would simply input those prices in this section.

Veneer Delivered Log Prices: 1,000–12,000 $/MBF

You should also get a price sheet from a local sawmill to input the delivered log prices for veneer sugar maples with a diameter inside bark at the small end of 14", 15", 16", 18", and 20".

Number of Grades Reduced by Defects in Butt Log: 0–4 (1) Grades

Although this can be a difficult task, you will need to assess the likely condition of the butt log (on the standing tree) and grade it according to the specifications provided by your local sawmill. The data from the "Hardwood Log Grade Specifications" sidebar is taken from the A. Johnson Company in Vermont acquired in May 2013.

Note that the size and extent of the heartwood has a significant impact on grade, yet this cannot be determined until the tree has been cut down. Therefore, the amount of heartwood must be estimated by examining the outer features of the tree and making an educated guess based on previous experience.

Number of Grades Reduced by Defects in Upper Logs: 0–4 (1) Grades

The same process that is undertaken to determine the number of grade reductions for the butt log must also be done to determine the amount of grade reductions in the upper logs. Often there will be several logs in the upper portion of the stem, yet this simplified analysis assumes that all the logs have the same number of defects. Therefore, an overall average number of defects (and associated grade reductions) must be assumed for all the upper logs in a given tree.

Butt Log Market: Sawlog or Cordwood or Veneer

Sawmills have different policies and guidelines for buying and scaling tapped logs. Pricing can change rapidly based on the quality of the logs and market forces. If a tapped log will be bought as a sawlog, then you should enter "sawlog"; otherwise enter "cordwood." You can also enter "veneer" if the untapped butt log is of sufficient size and quality to be sold to a veneer mill. This drastically increases the stumpage price for the log, though you should note that only a small fraction of all logs are sold for veneer.

Value of Tapped Logs as a % of Untapped Logs: 10–100 (50)%

If a tapped log is to be sold as a sawlog, this variable allows you to specify the price that will be paid to the landowner as a fraction of the untapped price. The default value is set to 50 percent, meaning that a tapped log will sell at 50 percent of the price of an untapped log of the same size and number of defects. The figure may actually be much higher or lower, depending on the buyer, market conditions, and other factors.

Hardwood Log Grade Specifications

Prime	16″ & up, 1 defect/13 ft of length
Clear	14″ & up 1 defect/10 ft of length
Select	12″ & up, 1 defect/3 ft of length
#1	9″ butt logs, 1 defect- HM & RO
#2	10″ & up, 1 defect/2 ft length
Cordwood	Maximum 20″ dia., 20 ft length

Acceptable lengths: 8'4"–16'4"
Odd lengths are accepted in hardwood
Pine & spruce even lengths only
Please mark length on the small end of the log
Please keep 8" logs to a minimum
Maximum diameter sawlogs: 30"—either end
Diameter includes sweep, crook and large knots, and flare
Hard maple: Logs with more than ⅓ diameter in heart are reduced 1 grade
Logs with more than ⅔ diameter in heart are reduced 2 grades
Long logs pay a bonus
Hard maple, red, oak, ash, and yellow birch earn a 7% price bonus for length
Long logs are 13'4" and longer in the grades of Prime, Clear, and Select

Factors Affecting Grade and Scale

Log diameter and length, size and location of defects. Defects include:

Sweep: 1" per 4 ft of length maximum.

Rot, crook, heart size (hard maple only), black heart, mineral stain, dry logs, wormholes, ingrown bark, checking, frost cracks, shake, cat faces, knots (especially knots larger than ¼ of log diameter), seams (especially spiral seams), excess gum in cherry, hard maple, or any other species with signs of tapping may not be scaled

Stumpage Rate for Cordwood: 0–50 (20) $/MBF

Although cordwood stumpage is usually based on a per-cord basis, the nature of this spreadsheet requires that stumpages prices for cordwood be described on a board foot basis. Since there are approximately 2 cords in 1,000 board feet (MBF), and the stumpage rate for cordwood is usually $10 per cord, then you should input 2 × 10 = $20 for this variable.

Stumpage Payment Method: Scaled or Contract

Loggers will often pay landowners based on a percentage of the revenues from logs brought to a sawmill; in these cases you should type in "scaled." Other times a logger may provide landowners with the money that remains after their costs for harvesting, skidding, and hauling have been covered. For these circumstances type in "contract."

Percentage of Delivered Log Price the Landowner Receives: 10–100 (50)%

This value is used when you type in "scaled" for the stumpage payment method. It will vary based on many factors, including (but not limited to) the logger you are working with, timber quality, volume to be harvested, terrain and accessibility, market demand, distance to market, season of the year, costs of timber harvesting, landowner knowledge, and terms of the contract.

Cost of Timber Harvesting: 0–500 (150) $/MBF

If you enter "contract" in the stumpage payment method, then this variable is used to calculate stumpage payments. In this case, the cost of timber harvesting is subtracted from the delivered log price to determine the stumpage price received by the landowner.

Timber Income Tax Rate: 0–60 (15)%

The tax rate a landowner pays on timber income will vary depending on whether it is classified as ordinary income or capital gains. A landowner should input the tax rate they expect to pay on the income from selling timber.

Real Annual % Increase/Decrease in Stumpage Rate: -10 to 10 (4.5)%

Stumpage prices fluctuate over time and are largely unpredictable. However, real prices (nominal-inflation) have tended to rise at an annual rate of 4.5 percent since records started being kept in the 1960s.[3] If you expect real hardwood lumber prices to change at a different rate in the future, then you could plug in a different value.

Lease Payments

The following variables all impact how much money a landowner will receive over time for allowing someone to tap his or her trees.

Annual Lease Payment: 0–2 (0.50) $/tap/yr

Enter in the annual lease payment as a price per tap. If the compensation is in syrup rather than a direct cash payment, you can determine the cash equivalent according to the following formula:

$$\text{Lease payment} = \frac{\text{gallons received} \times \$/\text{gallon (if landowner had to purchase the syrup)}}{\#\text{ of taps leased}}$$

Real Annual % Increase/Decrease in Lease Payment: -10 to 10 (0)%

This value depends on the level of inflation and terms of the lease contract, as seen in the following formula:

$$(\%\Delta LP) = \%\text{ annual increase in lease payment (nominal value)} - \text{annual rate of inflation}$$

For example, lease contracts are often developed such that the lease payment is set at a predetermined rate such as 50 cents per year for a certain time period. Given that there is no annual increase in the lease payment, if inflation was 3 percent, then the real annual decrease in lease payment would be -3 percent. If a contract is written such that the annual lease payment increases at the rate of inflation each year, then this value would be 0.

Diameter at Which the Tree Can Be Tapped: 6–12 (10) in (dbh)

The spreadsheet is currently set at 10", but in reality, this could range anywhere from 6" to 12", depending on the mutually agreed-upon desires of the maple producer and landowner. Even though conservative tapping guidelines do not permit the tapping of a tree until it reaches 10" or 12" dbh, some maple producers will tap trees much smaller than 10" dbh, especially if the stand is crowded and thinning is not an option.

Diameter at Which to Add a Second Tap: 12–36 (18) in (dbh)

Following conservative tapping guidelines, the NPV Calculator defaults to two taps being placed in a tree once it reaches 18" dbh. However, many maple producers will put a second tap in before the tree reaches 18". The spreadsheet is not set up to include more than two taps in a tree, as this is not an acceptable practice according to conservative tapping guidelines. If you would like to explore the financial implications of adding a third tap on very large trees, then simply increase the lease payment by 50 percent.

Tapping Lease Income Tax Rate: 0–60 (15)%

Income from leasing trees should be reported as ordinary income. The tax rate a landowner pays on this revenue is based on the landowners' overall income bracket.

Property Taxes

One of the main costs of owning forestland over time is the annual property tax burden. Because states such as New York, Wisconsin, and Minnesota have preferential tax treatment for sugarbushes, it is important to consider the impact of property tax payments on the overall return from managing woodland for syrup or sawtimber production. The following variables allow landowners to determine their annual property tax payments on a per-tree basis. Note that these calculations are based on the formulas used in New York and may vary for your state.

Regular Assessment per Acre: 0–10,000 (2,000) $/acre/yr

To determine what your tax savings might be, you must first understand what they currently are. By look at the tax bill for a given parcel, you can determine the assessment of the land by utilizing the following equation:

$$\frac{\text{total assessed value} - \text{value of all structures}}{\text{total acreage}}$$

Ag Assessment per Acre: 0–1,500 (300) $/acre/yr

In some states, landowners can receive an agricultural assessment, thereby reducing their property tax payments, by producing syrup themselves or leasing taps to another producer. This value is based on the agricultural value of the land (as determined by the soil type) for New York and possibly other states.

Number of Trees per Acre: 30–500 (100) # trees/acre

This variable is used to determine the amount of taxes *per tree* that a landowner must pay each year. Although taxes are paid on a per-acre basis, to conduct the NPV analysis on an individual-tree basis, you must divide the taxes per acre by the number of trees per acre. A young, dense sugarbush could have several hundred trees per acre whereas an older, established sugarbush that has been managed may only have 50 to 80 trees per acre. This is one of the most difficult and time-consuming variables to determine the value for, especially if the data is collected scientifically and systematically (which it should be).

Uniform Percentage: 0–100 (100)%

Once an assessor estimates the value of a property, its total assessment is calculated by multiplying the market value by the uniform percentage for the municipality. The percentage used does not matter; the only requirement is that all properties in the municipality use the same value.

Property Tax Payment Rate: 0–20 (5)%

The tax rate varies widely by municipality and year—you need to determine what it is for your locale and enter the appropriate figure.

Real Annual Property Tax Payment % Increase/Decrease : -10 to 10 (2)%

Property tax rates and payments are not static yet tend to fluctuate from year to year. A user can make a prediction on how tax payments will increase or decrease over time, though it's probably safe to assume that they will go up at least 2 percent per year.

Financial Determinants

There are two financial variables that greatly influence whether it makes more sense to manage maples for syrup or sawtimber production: the discount rate and the time horizon of the investment period.

Discount Rate: 0–16 (4)%

This is one of the most important variables in any NPV analysis, especially when comparing future revenues for leasing taps and timber harvesting versus harvesting a tree immediately. The U.S. Forest Service uses a discount rate of 4 percent to evaluate long-term investments,[4] so that is the default discount rate used in the spreadsheet. However, the discount rate can vary greatly based on the desires of the landowner, so you should change it accordingly to reflect your situation. To set a discount rate, you should ask yourself the following question . . . "What is the annual rate of return (interest payment) that I could expect to get over the long term if I invested the money that I would receive by harvesting the tree immediately?"

Time Horizon of the Investment Period: 2–40 (10) years

The time horizon of the investment period is based on either (1) the number of years the landowner is planning to own a particular piece of land, or (2) the number of years until the landowner will harvest the tree in question.

Examples Using the NPV Calculator

To bring this type of analysis to real life, the following section illustrates three examples for trees you may come across in a typical woodlot. Table 15.1–15.3 contains all the variables that are held constant whereas table 15.4 contains values for the variables that may differ across the scenarios. The first example presents a "best-case" scenario for leasing taps, the second example shows a case where timber management is the obvious choice, and the third option presents a situation where the decision-making process is difficult and the outcome can easily change depending on key variables. Although this chapter only describes three examples, there are numerous scenarios that could be analyzed. You should download the NPV Calculator, enter in values appropriate to your own situations, and examine the results.

Figures 15.1 through 15.5 contain NPV data for three hypothetical scenarios of individual maple trees. Two graphs are displayed for Scenarios 2 and 3, one depicting the results for a sugar maple and the other for a red maple, all else being equal. Only one graph is displayed for Scenario 1 since the stumpage rates for sugar and red maple are the same when considering cordwood. Each graph presents NPV figures, at annual intervals depending on the time horizon of the investment period, of four different options a landowner has for utilizing an individual maple tree. This analysis is carried out over a 40-year time horizon and does not account for replacement of a tree if it is harvested.

The graphs show what the NPV for a particular management strategy would be if the time horizon for the investment period was a specific number of years. For

TABLE 15.1: Constant Variables for NPV Examples

Effect of tapping on height growth rate (ft/year)	10%
Percent reduction in annual height growth rate*	1%
Percent reduction in annual dbh growth rate*	1%
Effect of tapping on diameter growth rate	10%
Stumpage rate for cordwood ($/MBF)	20
Payment method ("scaled" or "contract")	scaled
Cost of timber harvesting ($/MBF)	150
Time horizon of investment period (years)	40
Real annual tax payment percent increase/decrease	2%
Height of tree affected by tapping (ft)	8

* Adapted from Teck and Hilt (1991)

TABLE 15.2: Sawtimber Delivered Log Prices (SDLP)

DBH Range	Untapped Sugar Maple (SDLPUSM)	Untapped Red Maple (SDLPURM)
Less than 10" dbh	SRC	SRC
Between 10 and 12" dbh	0–2,000 (325) $/MBF	0–2,000 (160) $/MBF
Between 12 and 14" dbh	0–2,000 (400) $/MBF	0–2,000 (160) $/MBF
Between 14 and 16" dbh	0–2,000 (475) $/MBF	0–2,000 (250) $/MBF
Greater than 16" dbh	0–2,000 (625) $/MBF	0–2,000 (300) $/MBF

TABLE 15.3: Veneer Prices (VP)

DBH Range	Untapped Sugar Maple Veneer (USMV)
Between 14 and 15"	1,000–12,000 (3,100) $/MBF
Between 15 and 16"	1,000–12,000 (3,600) $/MBF
Between 16 and 18"	1,000–12,000 (4,600) $/MBF
Between 18 and 20"	1,000–12,000 (5,600) $/MBF
Greater than 20"	1,000–12,000 (6,100) $/MBF

example, consider a situation where someone planned on owning a parcel of land for 10 years before selling it, or was planning on conducting a timber harvest in 10 years. You would want to know what the NPV of various management options would be, given that 10-year planning horizon. The NPVs are displayed for a 40-year time frame to show how NPVs, and thus the management decision, can change based on how long someone plans on owning a parcel of land or conducting a timber harvest. What might be the most profitable option for a 10-year planning horizon may not be for a 3-year or a 40-year planning horizon.

TABLE 15.4: Scenario Assesment Table

	Scenario 1	Scenario 2	Scenario 3
TREE SIZE AND GROWTH			
Initial tree diameter at breast height (inches)	10	18	13
dbh growth rate (inches/year)	0.1	0.15	0.12
Initial merchantable height (ft)	24	32	28
Merchantable height growth of an untapped tree (ft)	0.4	0.25	0.5
STUMPAGE PAYMENTS			
Log scale utilized (Doyle, Scribner, Int'l ¼")	Doyle	Int'l ¼"	Int'l ¼"
Number of grades reduced by defects in butt log	2	0	1
Number of grades reduced by defects in upper logs	2	0	1
Real annual % increase/decrease in stumpage rate	5%	5%	3%
% of delivered log price the landowner receives	50%	67%	50%
Value of tapped logs as a % of untapped logs	50%	10%	50%
Butt log market	sawlog	veneer	sawlog
LEASE PAYMENTS			
Initial annual lease payment ($/tap/year)	$0.50	$0.50	$0.60
Real annual lease payment % increase/decrease	0%	-2%	0%
dbh at which to add the first tap (inches)	10	12	12
dbh at which to add a second tap (inches)	16	18	18
PROPERTY TAXES			
Number of trees per acre	250	150	175
Current assessment of forestland ($/acre)	$2,000.00	$1,200.00	$1,000.00
Agricultural assessment for forestland ($/acre)	$278.00	$1,200.00	$1,200.00
Uniform percentage	62%	48%	100%
Property tax payment rate	5%	5%	3%
FINANCIAL VARIABLES			
Tapping lease income tax rate	28%	33%	28%
Timber income tax rate	28%	15%	28%
Discount rate	4%	6%	4%

Scenario 1: Small Tree with Variables Favorable to Leasing

This scenario presents a "best case" for continuous leasing of taps in comparison with timber management. As seen in figure 15.1, the NPV was only positive for the management options that included leasing. This is due in large part to the fact that (1) there was no loss in sawtimber value, since the small, slow-growing tree never developed a merchantable sawlog within the 40-year time horizon, and (2) there were large savings on property taxes by using the tree for syrup production. When a tree can only be harvested for cordwood, this revenue often falls short of the regular annual property tax payments, so managing for long-term timber production can actually yield negative returns. In fact, a recent study of Massachusetts family forest owners found that long-term timber management will produce negative returns for many landowners in that state.[5]

It is worth noting that even if the tree was growing rapidly, the lease payment was lower, or a tax break for leasing did not exist, the NPV would still be positive for leasing taps and negative for timber management. When starting with small trees that are decades away from producing a sawlog, the fact that they can generate revenue each year until they are ready to be harvested will almost always favor leasing taps over timber production. Finally, figure 15.1 displays the same information for sugar and red maples in this scenario because the lease payments and stumpage rates for cordwood are typically the same for both species.

Scenario 2: Large Veneer Tree with Variables Favorable to Harvesting

In contrast with Scenario 1, this scenario presents a clear picture of when timber harvesting is more lucrative than leasing taps. With a veneer-quality tree, the

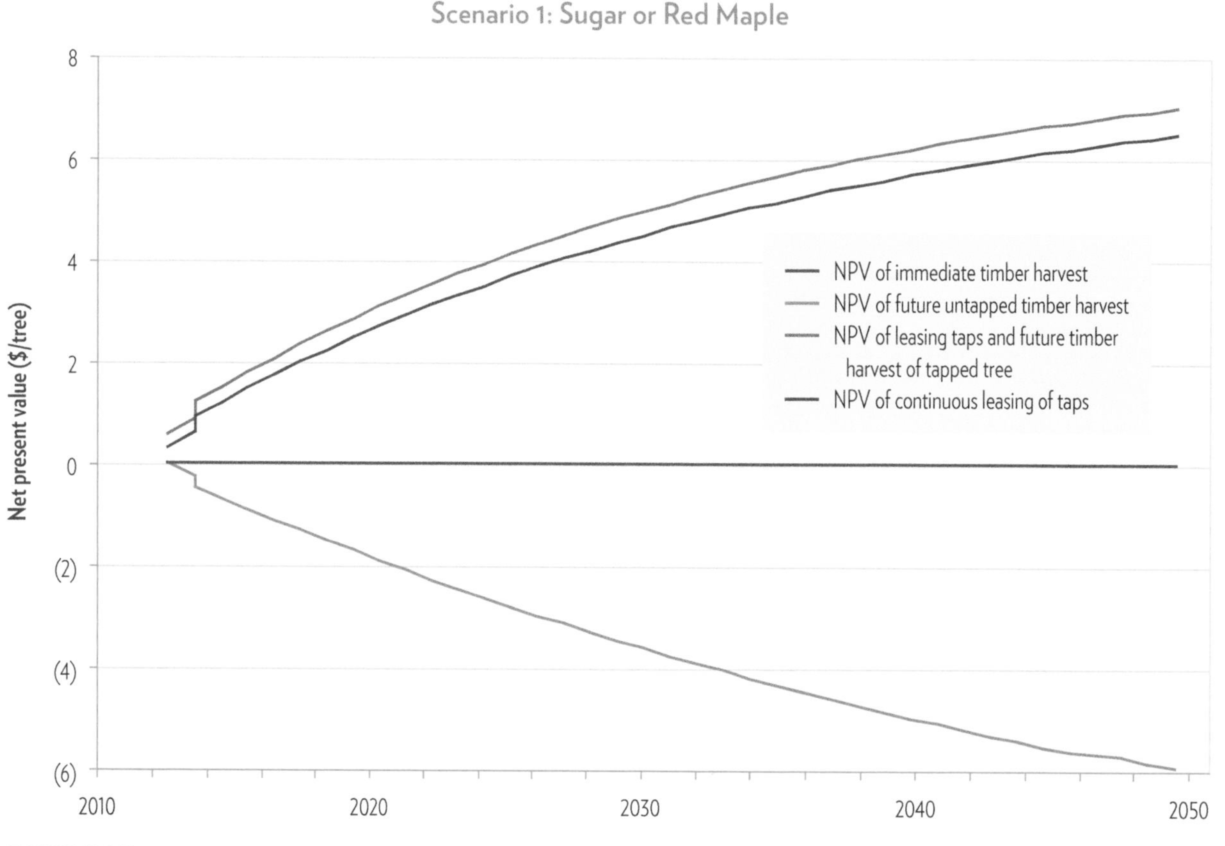

FIGURE 15.1. Net present value analysis of different management strategies for a sugar or red maple tree depicted in Scenario 1.

dilemma is not whether to lease taps or harvest timber; rather, the only question is when the tree should be cut. It becomes a matter of financial maturity depending upon the rate at which the tree is increasing in value (due to tree growth and the rise in stumpage prices) versus the alternative rate of return (the discount rate) that the landowner could earn in other investments of comparable duration, risk, liquidity, and so forth.[6] In this example, the discount rate is 6 percent whereas stumpage rates are increasing at 4.6 percent annually and the tree has a moderate growth rate of 0.12"/year. Thus, it would make sense to harvest the tree immediately or within a few years, as seen in figure 15.2. However, if the tree was growing at a much faster rate, if the discount rate was lower than 4.6 percent, or if stumpage rates were increasing at more than 6 percent annually, then it could make sense to let the tree grow for a longer time period to achieve the greatest revenues.

When considering large, merchantable trees, in order for leasing to compete with timber management, there must either be a substantial tax savings for leasing and/or a lucrative market for taphole maple lumber. Even when these conditions are met, any sugar maples that may produce veneer should be identified before a lease goes into effect and either harvested immediately or prohibited from tapping. If there is a large percentage of veneer trees in a given stand, that entire stand may be prohibited from tapping.

Since red maples are rarely used for veneer, the tree is likely to be sold for sawtimber when it is harvested, making the difference in NPVs between tapping and timber production much lower. As seen in figure 15.3, a red maple would become financially mature in approximately 20 years, at which point it would have made much less difference if the tree had been tapped (as compared with sugar maples).

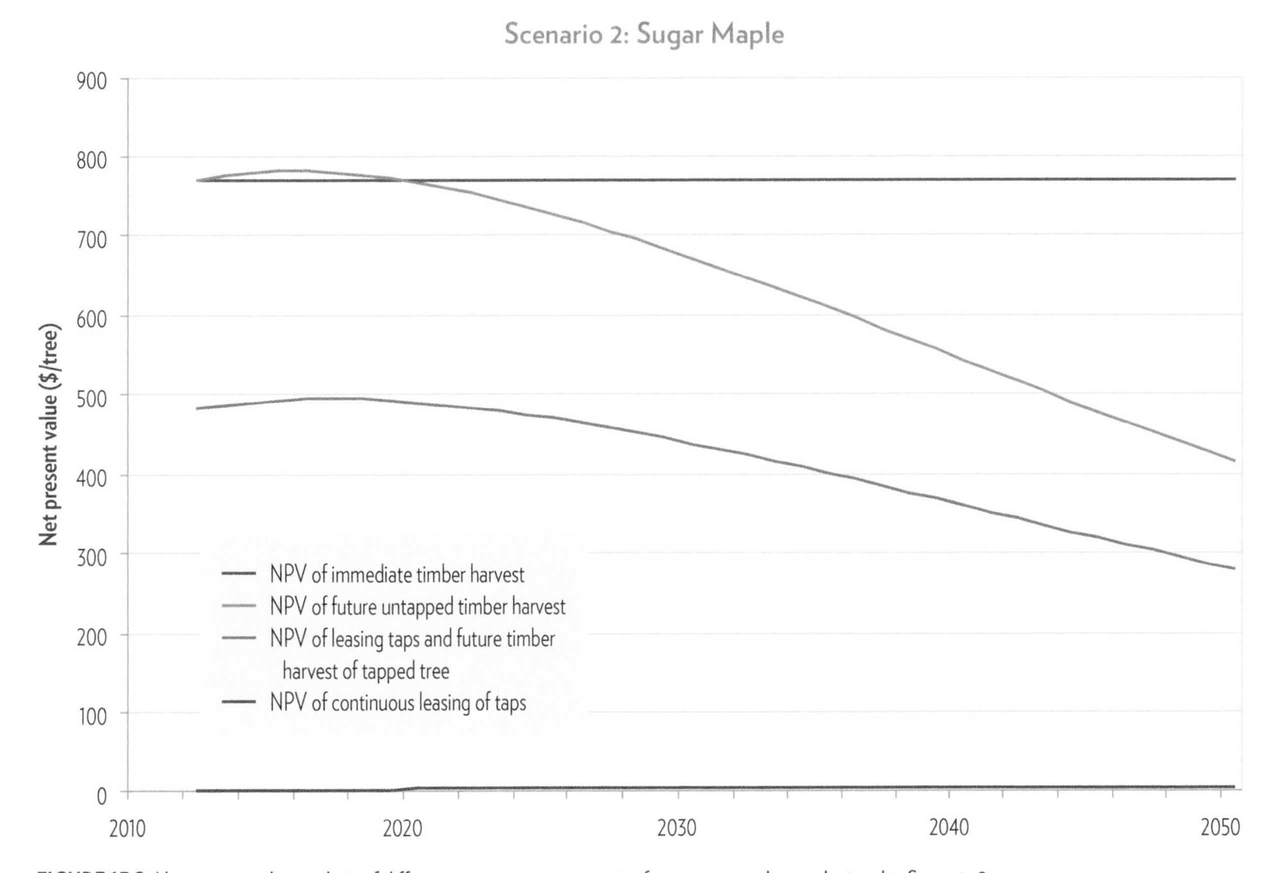

FIGURE 15.2. Net present value analysis of different management strategies for a sugar maple tree depicted in Scenario 2.

Scenario 3: Small Sawtimber Tree with Variable Outcomes

Whereas the first two scenarios have clear directives on which management strategy is more profitable, when considering a small sawtimber tree the decision is much more difficult to make, especially for sugar maples. The time horizon of the investment period should have a particularly significant impact on a landowners' decision. For sugar maples in this scenario (figure 15.4), if a landowner's planning horizon is eight years or less, the greatest returns would be from cutting immediately. However, if the landowner wants to hold on to the tree for at least 10 years, then he or she would be better off leasing taps and then harvesting a tapped tree after a minimum of 10 years. The lowest returns come from continuous leasing without ever harvesting, as the annual lease payments will never compensate for the lost stumpage value from not harvesting the mature tree.

The outcomes for this scenario are much different when considering red maples (figure 15.5). Because of the lower stumpage values, leasing and then harvesting will become the most profitable option when considering a planning horizon of at least three years. The financial returns for this management option continue to increase as the time horizon of the investment period increases, reaching a maximum after 32 years before starting to decline. Continuous leasing of red maples will actually become more profitable than long-term timber management when considering a time horizon of 23 years or greater, though it never quite reaches the value that could have been achieved with immediate timber harvesting.

For sugar and red maples, note the stark jump in NPVs once the tree reaches 14" dbh after a planning horizon of about 10 years. At this size, log prices move up a grade and the rise in NPVs reflects this jump in

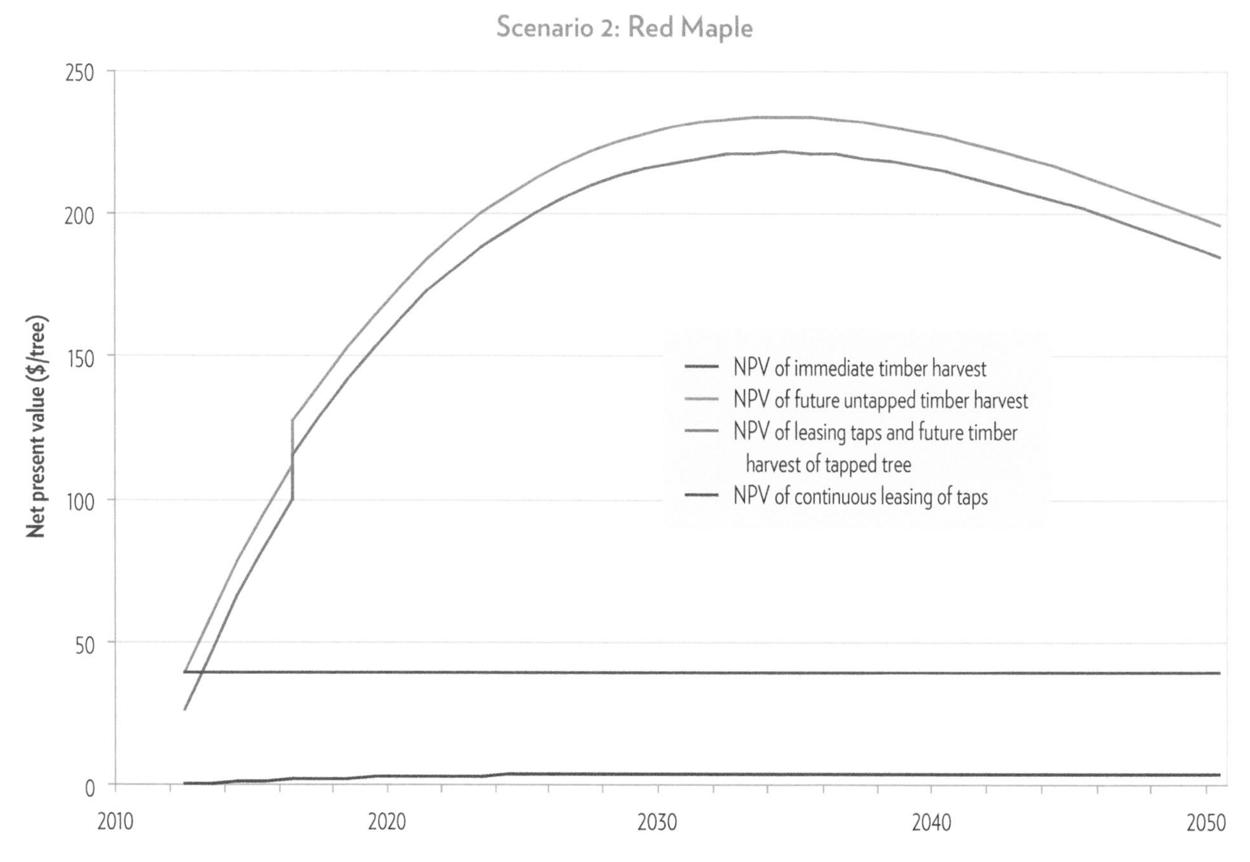

FIGURE 15.3. Net present value analysis of different management strategies for a red maple tree depicted in Scenario 2.

stumpage prices. When considering sawtimber production, the year at which this happens often coincides with maximum financial maturity (though usually at larger diameters such as 18"). After this time period, the property taxes that must be paid are greater than the accrued stumpage value from tree growth and price increases, so the NPV for timber management continually falls as the time horizon increases beyond 10 years. Finally, it is worth noting that this scenario did not include any property tax savings for leasing taps. If it did, the property tax savings would likely result in leasing taps becoming the most profitable option in all situations.

Variable Analysis

As seen from these three scenarios, the most profitable option for managing a maple tree is not always clear. Table 15.5 provides guidelines that are helpful in assessing a particular tree or stand of trees. It outlines the attributes of each variable that dictate whether an individual tree would yield greater profits through leasing taps, immediate timber production, or long-term timber management. However, just because a variable has an attribute that lends itself to a particular management strategy does not necessarily mean that using the tree for that purpose will yield the greatest profits. Rather, it is the combination of many variables—*and their weight in a particular situation*—that determines the best choice for an individual tree. The many variables involved require careful analysis each time someone decides how to manage each individual tree within a stand. The following pages contain detailed information on the most important variables, taken from a landowner's perspective.

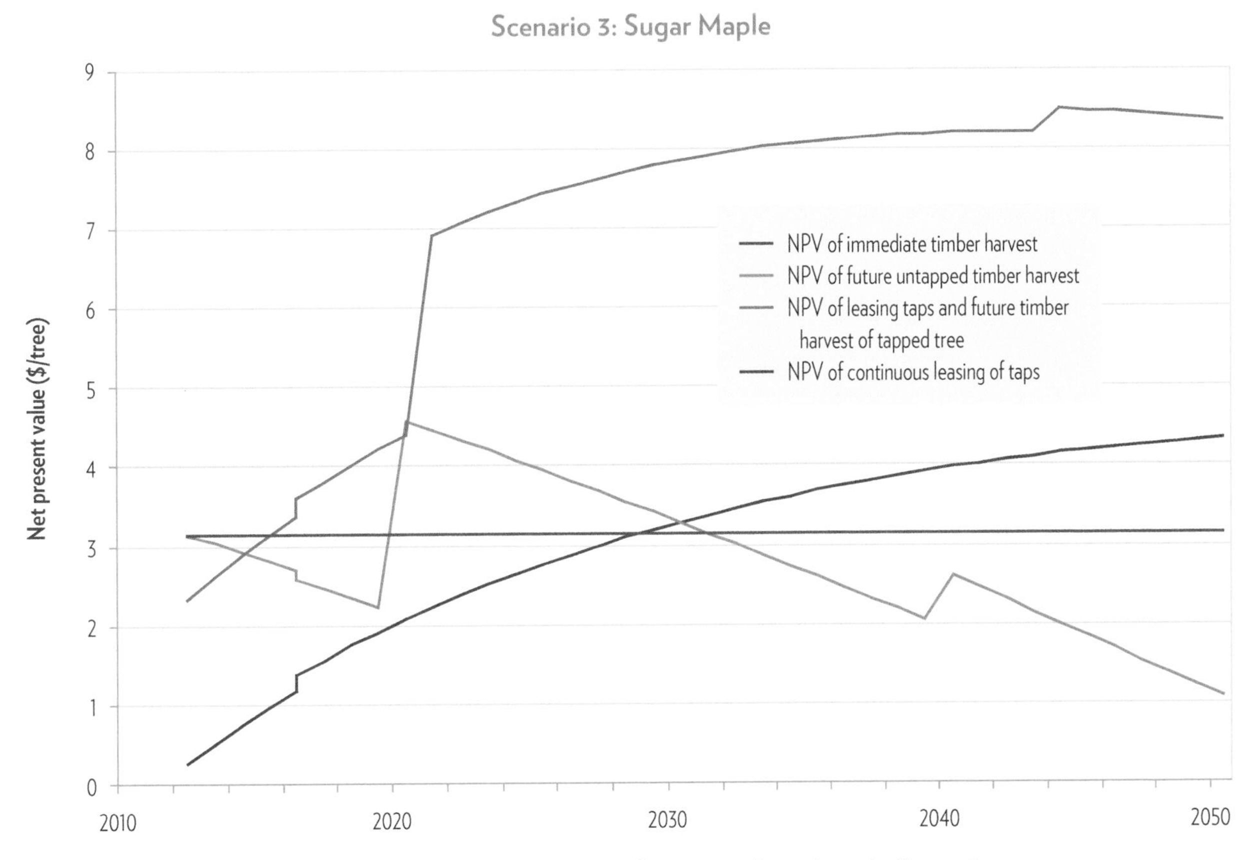

FIGURE 15.4. Net present value analysis of different management strategies for a sugar maple tree depicted in Scenario 3.

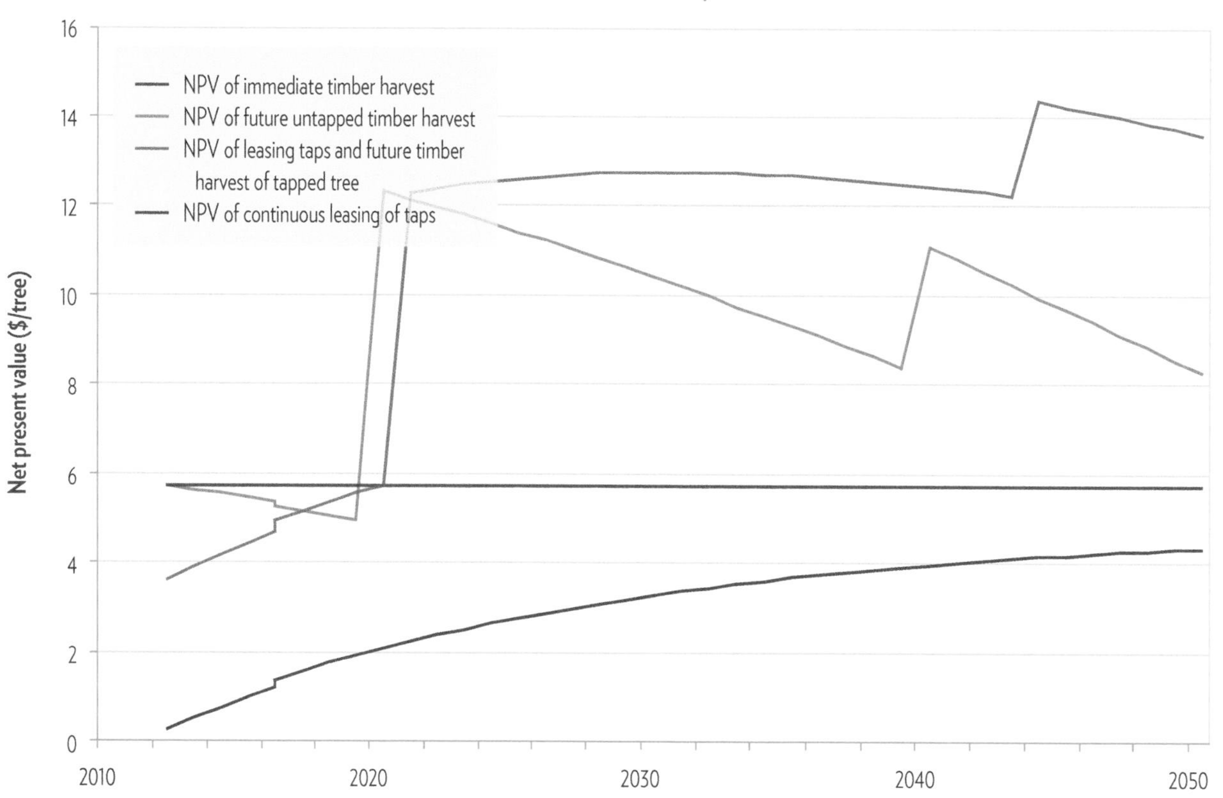

FIGURE 15.5. Net present value analysis of different management strategies for a red maple tree depicted in Scenario 3.

Species

Since the stumpage prices for red maple are much lower than those for sugar maple, landowners are better off leasing stands with a large component of red maples. The lease payments tend to be the same for sugarbushes that contain a high percentage of red maples, yet the reduction in sawtimber value is much less when tapping red maples. Although sugarmakers prefer to tap sugar maples because of the higher sugar content in the sap, many will tap red maples if they are readily available and accessible. This is especially true among large sugarmakers with reverse osmosis systems that remove about 80 percent of the water from the sap before boiling.

Tree Diameter

The smaller the initial diameter of the tree, the greater the likelihood that leasing taps will be more profitable than sawtimber management. Small-diameter trees have low stumpage values and are able to generate lease payments each year as they mature into sawtimber size. Landowners are usually paid the same lease rate whether a tree is 10", 13", or 17" dbh, so it makes more sense for landowners to lease smaller trees, capturing an annual payment without worrying as much about reducing sawtimber value.

Extent of Tapping Zone

The lower the height of the taps, the less impact on commercial lumber value loss you can expect. Landowners may request that a maple producer tap as low as possible, although in some years the snow depth may necessitate high tapping.

Growth Rates

A landowner usually receives the same lease payment no matter how fast the tree is growing, but future sawtimber

TABLE 15.5: Variables that Impact Whether Maples Should be Utilized for Syrup or Sawtimber Production

Variables	Conditions that favor leasing taps	Conditions that favor immediate cutting	Conditions that favor long term timber production
TREE SIZE AND QUALITY	Small tree (<12″ dbh) Defects in main stem Many lower branches Tree has been previously tapped Tree is a red maple	Large tree (≥18″ dbh) No defects in trunk Some dieback visible in top Tree has never been tapped Tree is a sugar maple	Medium size tree (13-16″ dbh) No defects in trunk Tall straight tree Tree has never been tapped Tree is a sugar maple
FOREST ATTRIBUTES	Dense stand of accessible maples Large percentage of red maples	Maples are widely scattered and not easily accessible	Maples are widely scattered and not easily accessible
TAPPING METHOD	Low height of tapping zone No metal objects used in collection Log buyer is familiar with tappers	Large height of tapholes/staining Metal objects used in collection No knowledge of past tapping practices	Large height of tapholes/staining Metal objects used in collection No knowledge of past tapping practices
GROWTH RATE	Slow growth rate	Slow growth rate	Fast growth rate
LOG SCALE	Doyle or Scribner	International ¼″	International ¼″
STUMPAGE RATES	Low current prices for maple Low future prices predicted Low price difference for tapped logs	High current prices for maple Uncertainty on future prices High price difference for tapped logs	Low current prices for maple High future prices predicted Low price difference for tapped logs
LEASE RATES	High lease rates above $.50/tap Includes annual fee increases	Low lease rate No annual increase in lease fees	Low lease rate No annual increase in lease fees
PROPERTY TAXES	Tax breaks available for leasing High initial assessment High tax rate	Tax breaks available for forestry Low initial assessment Low tax rate	Tax breaks available for forestry Low initial assessment Low tax rate
DISCOUNT RATE	Low discount rate Landowner wants long-term profits	High discount rate Landowner needs immediate cash	Low discount rate Landowner wants long-term profits
PLANNING	Long time horizon	Short time horizon	Medium time horizon

value is highly dependent on the size (and therefore the growth rate) of the tree. Thus, the slower a tree grows, the more likely it is that leasing taps will be more profitable than managing for sawtimber production.

Stumpage Rates

The current and future stumpage rates for maple are highly volatile and largely unpredictable. Price fluctuations have a strong impact on the timing and level of timber harvest, but it is very difficult to know what will happen to prices over time. Factors that would favor leasing include low current stumpage rates for maple logs and low expectations of future stumpage rate increases.

Log Scale

The three commonly used log scales are Doyle, Scribner, and International ¼". Since International ¼" estimates much higher volumes than Doyle or Scribner, using International ¼" favors immediate timber cutting over long-term leasing. The Doyle scale is much more likely to favor leasing taps; Scribner also has a slight advantage for leasing versus International ¼".

Log Quality

The more defects that exist in a tree, in particular the butt log, the more likely that leasing taps will be a better option. If a tree has veneer potential, the tremendous value of these logs should preclude any landowner from leasing them for tapping. However, not all maples will produce high-quality sawlogs, and poorly formed maples with many lower branches often yield the greatest amount of sap.

Lease Payments

The higher the lease payment, the greater the likelihood that maple sugaring will be more lucrative than timber management. Also, the rate at which the payment changes over time will affect profitability. The rate is often set at about 50 cents per tap per year and does not vary over time, but if there is positive inflation, then the real value of the lease payment would fall. As a sugarmaker, you want to offer a high enough payment to convince the landowner to lease taps to you while at the same time ensuring that the situation is still profitable.

Tapping Guidelines

To maximize long-term sap production, conservative tapping guidelines suggest one tap for a 10" to 17" tree and two taps for trees more than 18" dbh. However, many maple producers will start tapping trees when they are only 6" to 8", place a second tap once the trees are 12" to 14", and put in three or four taps on trees that are 18" or greater. Smaller trees will not yield as much sap as larger ones and overtapping will result in long-term declines in sap yield per taphole, but overtapping is unlikely to kill a tree. Thus, from a strictly financial perspective, a landowner should allow a maple producer to be fairly aggressive when tapping. Moreover, since a landowner is usually paid according to the number of taps put in rather than the amount of sap collected, it does not matter as much (to the landowner) how much sap the tree actually produces. However, as a sugarmaker, you should still treat the leased woods as your own and follow the same conservative tapping guidelines that you would use for your own trees. This will result in the greatest yields over time, and the landowners will likely be much happier knowing their trees are being well taken care of.

Time Horizon of the Investment Period

The length of time before a landowner plans on harvesting trees and/or selling a piece of property has a profound impact on whether or not it is more profitable to manage for syrup or timber production. If you are dealing with a landowner who desires immediate revenues, it is likely that he or she will choose to have the trees harvested, especially when considering trees that could produce high-quality sawlogs or veneer. However, if the landowner is planning on owning a property for many years, it is much more likely that leasing taps will make economic sense.

Discount Rate

The rate used to discount future cash flows to their present values is a key variable of any NPV analysis. Discounting future values is necessary to equalize the revenue streams of leasing taps (a smaller annual payment) versus managing for sawtimber production (a larger onetime payment) over time. Generally speaking, the lower the discount rate, the more profitable leasing will be. On the other hand, high discount rates often result in immediate timber harvesting generating the highest returns.

Property Taxes

In some states, such as New York and Wisconsin, landowners can qualify for reduced property tax payments by producing syrup themselves or leasing taps to another producer. The amount of tax savings will vary greatly depending on the assessed value of the land without agricultural assessment and the tax rate for that municipality. The greater the difference between the regular and reduced tax payments, the more profitable leasing will be. In deciding between leasing taps and timber management, landowners should first check to see if there is a tax advantage in their state for utilizing woodland for syrup production. I know many sugarmakers in New York who have been able to secure long-term leases at very low rates because the savings from property taxes ($3 per tap per year or more) were much higher than any possible lease payment.

Number of Trees per Acre

Conducting the NPV analysis on an individual-tree basis complicates the effect of property taxes, since those are usually calculated on a per-acre or total property basis. Although the number of trees growing on an acre has little to no effect on the overall taxes that are paid on that acre, the density of trees is directly correlated to the taxes paid *on an individual tree.* This has the potential to skew the results when there is a substantial tax savings from utilizing maple trees for syrup production. In these situations, fewer trees per acre result in more drastic differences in property taxes paid on a per-tree basis. However, a landowner will usually pay the same in property taxes for an acre of land whether 100 or 500 trees are growing on that acre. You must realize this limitation, be particularly careful when estimating the number of trees per acre, and adjust the interpretation of results accordingly.

Value of Tapped Logs as a Percentage of Untapped Logs

The value of tapped logs varies considerably in geography and time based on the current markets for standard maple lumber and the niche markets for taphole maple lumber that a particular landowner or sawmill may have. The value could also be influenced by the history of tapping and the knowledge of tapping history that the log buyer possesses. Trees that have only been tapped recently with all-plastic spouts are much more desirable than older trees that were tapped with metal spouts and/or contain nails used to support tubing. It is dangerous and costly for commercial sawmills to process logs that may contain hardware, although most sawmills are now equipped with state of the art metal detectors that will find any metal before the logs hit the sawmill. Portable bandmills can usually afford to hit metal components on an occasional basis, so if you have a bandmill yourself or could hire someone with a portable mill to saw your logs, that may be your best option. Due to the large variation in prices that is possible, it is well worth the time to explore all market options for tapped logs before they are sold.

Whether or not the timber-versus-tapping debate even makes sense is contingent on the assumption that a landowner is giving up significant sawtimber value in the first butt log by tapping. However, the market for taphole maple lumber is growing due to the interesting character and story behind the wood, so occasionally market conditions dictate that the price for tapped logs is the same as untapped logs. In these instances "you can have your cake and eat it too," generating annual income from leasing taps while eventually earning the same (or similar) stumpage prices when the trees are harvested. If this were always the case, it would not make sense to compare the two options—leasing would always generate additional revenue for the landowner without risking a significant loss in stumpage prices. In reality, I have found that many sawmills will pay lower prices for tapped logs, yet they will still saw these logs and sell the tapped lumber for a premium price. The higher prices are justified by higher handling costs when dealing with lower volumes of specialty wood and the fact that customers are willing to pay the extra price for this unique lumber. Sawmills have even higher profit margins since they are paying less for the tapped logs, so it is a win–win on both sides of the equation for them. The only one who misses out is the sugarmaker who receives a lower price for his or her logs. As sugarmakers, we need to do a better job building the market for taphole maple products and demand higher prices for our previously tapped logs.

Final Thoughts

As you can see, the decision about how to utilize maple trees is a complicated one based on a wide array of variables. When you're determining the most profitable management strategy, it takes a lot of time and knowledge to figure what course of action will yield the greatest results for the landowner. Although financial returns are important, there are many other factors that may influence a landowner's decision on what to do with his or her property.[7] Even though timber management is not the main ownership objective for the majority of landowners, most parcels eventually get cut—with the largest, most valuable trees taken in the process. As maple producers, our objective is to convince landowners to let us tap their trees until the time when they eventually conduct a timber harvest. Ideally, by providing annual income without having to harvest trees, we can stave off a potential timber harvest and help maintain the mature sugarbushes for the long term.

You may come across a sugarbush that you would really like to tap—but the landowner would rather have it cut since he needs some money right away. In these situations, you still may be able to secure access to the trees by paying the landowner a larger onetime lump payment for the stumpage/tapping rights. Consider the following situation: John owns 2,000 potential taps of mature sugar maples on a 20-acre parcel. A logger has offered him $15,000 as a stumpage payment to be able to harvest all the large, valuable trees. You may only be able to offer 50 cents per tap as an annual lease payment, so if John was looking for immediate revenue, naturally he would choose the $15,000 onetime stumpage payment over the $1,000 annual lease payment. It would take 15 years of the $1,000 lease payment just to equal the $15,000 stumpage payment, and the $1,000 he receives in year 15 really isn't worth as much as the initial $1,000 lease payment today (due to inflation and the time value of money). You may be able to strike a deal with John in which you purchase the stumpage rights for $15,000 along with the provision that you have a 15-year time frame to harvest the trees and are allowed to tap them in the meantime. By accepting this offer, John would receive the same amount of money as if he had logged it, yet he will be able to retain this mature forest for at least 15 years. It's a win–win for you and John; the only person who loses out on this deal is the logger.

The only problem with this situation is that most of us don't have $15,000 readily available to lease 2,000 taps. However, you may be able to get a bank loan to support this venture and could potentially use the timber-cutting rights as collateral for the loan. If you have an established and successful maple business, the additional 2,000 taps would make your enterprise even more profitable. You should easily be able to make your monthly loan payment with the additional profits from sugaring, and if something happened to your business and you weren't able to continue tapping the trees for syrup production, they could always be harvested for the timber value. At the end of the 15-year contract, it would probably be easy to renegotiate a new, lower lease payment to continue tapping the trees. If John didn't want to develop a new lease contract, you would still be entitled to harvest the trees, so that is a great bargaining chip to continue the tapping arrangement. Assuming you have developed a good working relationship over the 15-year time frame, the transition to a new lease agreement should go smoothly. By renewing the lease John will be able to receive an annual lease payment, and you will have secure access to the 2,000 taps for the long term. Another possibility for this situation is simply buying the 20-acre parcel from John. It's possible that John really only wants the land for hunting and would be willing to sell it if he could retain lifetime hunting rights. There are creative ways around the tapping-versus-timber conundrum—it's just a matter of figuring out what will work for you and the landowner in a given situation.

CHAPTER 16

Sugaring in a Changing Climate

It's hard to have a conversation about the future of maple syrup production without the subject of climate change coming up. Whereas there are some people who still deny that global warming is even happening, many people who accept climate change as a reality believe that it will have devastating impacts on the maple industry. Several reports have made alarming predictions about the sustainability of the maple industry as a result of global warming, and the mass media helps to perpetuate this storyline. Most reporters look for controversial stories and bad news; not being able to have Grandpa's maple syrup on your pancakes anymore certainly attracts a lot of media attention. But despite all the alarmist stories about how global warming is going to cause sugar maples to die off or render them useless for syrup production, I am much more optimistic about the future of our industry.

One way that climate change *could* affect the maple industry is by impacting the distribution and health of maple trees. Some people have predicted that maples will migrate northward and be replaced by other species that will do better in a warming climate. The other possible impact is altering the timing of sap flow and overall yields. There has been some concern that even if maples remain as one of the dominant species in the Northeast, the weather will not include enough freeze–thaw cycles during the late winter and early spring to produce marketable yields of sap. This chapter explores both of these potential impacts in significant detail and offers suggestions for how sugarmakers can adapt to a changing climate. While there is no doubt that our climate is changing, I am equally as certain that we can adapt our sugarbush management and sap collection practices to ensure a sustainable industry for the foreseeable future.

Timing of Sap Flow and Yields

According to the legends and lore of the maple industry, the sap starts flowing in maple trees sometime in late February or early March after the long, dark winter nights fade away into the warm and sunny days of early spring. While this sounds great in theory, it rarely happens this way anymore. Winter temperatures have increased even more than summer temperatures over the past century; it's not uncommon to now have perfect sugaring weather throughout January and February. Rather than a distinct winter and seasonal spring temperatures, we now have wild fluctuations in temperature from January all the way through April. It's now anyone's best guess as to when the sap will be flowing in a given year. In 2012 the only people who had good yields were those who tapped in January or early February to catch the midwinter sap flows. The season ended by mid-March for most producers in 2012 since very few operations could withstand a week of summer-like weather during the height of the traditional sugaring period. Those whose tapholes did continue to produce sap when normal sugaring weather returned in late March and early April wound up making off-flavored, "buddy" syrup due to the advancement of the buds brought upon by the extremely warm temperatures.

The early sap flows of 2012 coupled with the season-ending weather in mid-March convinced even some of the most die-hard traditionalists that they need to start tapping earlier when the weather is appropriate. In 2013, for example, there was some syrup made during

Tapping on Town Meeting Day

Maple sugaring is a lot of work—and it used to be much harder than it is today. Whereas most of the people living in the rural towns and villages of the Northeast used to be farmers, today very few people actually earn their living from the land. The tradition of waiting until after Town Meeting Day to tap in Vermont arose from the fact that farmers wanted to know how big their tax burden would be in a given year before they started sugaring. If their property tax burden was going to be high, they would set out as many taps as possible, whereas if the tax bill that year was not as cumbersome, they may only gather sap from the easiest trees to get to or forgo sugaring altogether that year. Although there is a long tradition of waiting until Town Meeting Day to tap in Vermont, rarely do the appropriate weather conditions coincide with a political artifact. By tapping your trees when the weather is suitable, no matter what the calendar says, you will be able to take advantage of early sap flows and boost your yields in most years.

the thaws in January and February, but winter held on through most of March and it wound up being more of a traditional season with heavy sap flows in April, resulting in possibly the largest syrup crop ever on record. The 2013 experience may slow the trend for earlier tapping, but it probably won't stop it altogether. The thing about climate change is that it happens whether you believe in it or not—some years the weather will be perfect and other years it will be miserable. We can't do anything about the weather, just deal with what we are given and complain about it from time to time. Sugarmakers who accept the fact that the Earth is warming and are willing to start tapping earlier will usually experience better yields than those who follow the traditional tapping dates.

There have been several research projects looking at the effect of climate on maple sap flow. Dr. Barrett Rock's lab[1] first started raising alarms about the impact of climate change when researchers discovered a correlation between increased winter temperatures and reduced yields from four northeastern states. Dr. Tim Perkins, Director of the Proctor Maple Research Center for UVM, examined long-term records from several locations throughout the Northeast and found that the sugaring season has started about a week earlier and ended 10 days ahead of schedule over the past 40 to 50 years. Although these reports may sound alarming, the data may have been influenced by the fact that most sugarmakers used to tap according to the calendar rather than the weather. Yes, the season is moving earlier, and yes, much warmer temperatures in March could reduce yields. However, simply tapping earlier when the weather is appropriate can help to boost yields in a warming world. If you wait until the "traditional" tapping time, there is a good chance your season will be cut short.

Based on evidence showing that the tapping season is moving earlier, a 2009 study[2] at Cornell set out to determine what might happen to sap flow in the northeastern United States in a warming world. Researchers modeled sap flow conditions in 2100 for the Northeast based on a wide variety of climate models. Generally speaking, they found that the season will shift earlier and that the tapping period should move into January by the end of the century. Whereas some of the southerly areas will wind up having fewer sap flow days altogether, more northerly regions (where the majority of syrup is produced) will actually have more sap flow days. Many northern locations are currently limited by the fact that it is too cold, so warmer temperatures could actually help maple producers in many regions. In other words, if I were a sugarmaker in Kentucky, I would be worried about my kids' or grandkids' ability to still tap my trees by the end of the century, but in the Adirondacks and other northerly locations, a warming world will likely benefit the overall sap flow situation. Northern producers are currently hindered by the fact that it gets so cold in the winter, so with more mild temperatures, it's likely that the late winter and early spring sap flows will be enhanced. If you want modern-day evidence of this, just consider the yields per tap in Vermont versus those in much of Quebec. Vermonters are able to produce nearly twice as much sap due in large part to more favorable weather and healthier trees south of the border.

The Importance of Vacuum Tubing

One of the main ways that sugarmakers can adapt to a changing climate is by taking advantage of the latest developments in vacuum tubing technology. Putting in new spouts every year, regularly replacing droplines, and keeping your vacuum pump on whenever the temperatures are above freezing can help compensate for unpredictable and unfavorable weather patterns. In addition, we now have a firm understanding of the role that taphole sanitation plays in determining sap yields. New, sterile spouts perform much better than previously used spouts that have been contaminated with bacteria, yeast, and mold spores. Even if your old spouts look fairly clean, chances are you would get much better yields by putting in a new spout every year. Furthermore, by utilizing new spouts and relatively clean tubing under high vacuum, you can tap early to take advantage of winter sap flows without worrying too much about tapholes drying up during the course of the traditional sugaring season.[3]

Keeping the vacuum pump running whenever the temperatures are above freezing can substantially mitigate the effects of climate change. One of the main issues we now deal with is an increase in the number of days during sugaring season when it gets too warm for sap to continuously flow out of the taphole under gravity-based systems. High vacuum allows sap to keep flowing, even without any positive pressure created by freeze–thaw events. The vacuum pump creates the pressure differential, thereby drawing sap to the taphole as long as the temperatures are above freezing. This will not kill or harm the trees, as we are still only taking a small percentage of the sap contained with a tree; as described earlier, the damage done to a tree is predominantly the result of taphole size (which can be mediated by using the 5⁄16" health spouts). The vacuum pump has no negative effects on tree health and simply makes a sugaring operation more resistant to the negative effects of climate change and unpredictable weather.

Unfortunately small-scale sugarmakers without vacuum tubing are at the mercy of the weather each year. If you rely on the freeze–thaw cycle for all of your sap flow, then there isn't anything you can do to coax the sap out of the trees. However, if you are using buckets or bags, you could at least purchase new spouts (or stainless-steel spouts that can be sterilized) before each tapping season. This will help prevent the tapholes from drying up prematurely, even though the freshest taphole with a sterile spout won't produce any sap without the right weather conditions. For anyone gathering sap without vacuum, my advice would be to tap early with new, clean spouts to catch as much of the sap flow as possible. You are much more likely to get the freezing nights that allow for pressure buildup earlier in the season, so you shouldn't necessarily wait for the perfect sugaring weather to arrive in mid-March before tapping.

Health and Distribution of Maple Trees

In 1999 Anantha Prasad and Louis Iverson developed a Climate Change Atlas[4] that predicted where the suitable habitat for many tree species would be by 2100 given varying climate scenarios. These maps painted a grave picture of the future of sugar maple in the United States. Most of the models showed the suitable habitat for sugar maple eliminated from the southern portions and greatly reduced in northern reaches of the US. These maps were intended to show where the climate would be considered most suitable for a given species, *not* where a species would be found (or not found) in the future based on a given climate forecast. However, many people did not read the "Word of Caution" statements that went along with these maps and misinterpreted the results. There was no shortage of people predicting that sugar maples would become nearly extinct from the US by the end of the 21st century.

The New England Regional Assessment Group[5] and the Union of Concerned Scientists[6] also published reports conveying serious concern about the future of maples in the Northeast. They elaborated on the negative effect climate change would have on industries that rely on sugar maples, including maple syrup, fall foliage tourism, and timber products. These reports predicted that the northern hardwood forests of New England and New York, dominated by sugar maples, red maples, birches, and beech, would be replaced by oaks (*Quercus* spp.) and hickories (*Carya* spp.) by the end of the century. The popular media ran with this story line, and its premise has become widely accepted by the general public and many scientists. However, oaks and hickories have had trouble regenerating in existing forests for several decades, raising concerns about the sustainability of these species in future forests.[7] Furthermore, there is already evidence that many oak-hickory stands are shifting to maple-dominated stands in the central hardwood region.[8]

Although climatic variables certainly influence the vegetation of a region, there are many other factors besides climate that dictate species abundance and distribution. Two recent studies[9] factor in other variables besides climate that can influence species abundance and distribution. New models include nine biological characteristics and 12 traits relating to disturbances that are expected to increase in a changing climate. With the new models, it appears less likely that maple-dominated forests will be replaced by oaks and hickories at any point in the foreseeable future. Rather than losing suitable habitat, red maples are expected to flourish and expand throughout the Northeast. The new and improved model has allowed some researchers to state that it now "seems plausible that the maple-beech-birch type will persist." Furthermore, by merely adjusting their models to focus on abundance versus presence/absence data, they reduced the predicted habitat loss of sugar maples from 90 percent to 36 percent.

In reality, many variables can affect the health of sugar maples, and it is extremely difficult to know what will happen to our forests in 50 or 100 years. One way to try to predict the future is to look at recent trends in the abundance and distribution of major tree species. By examining what has been occurring over recent decades, we can make an educated guess as to what may continue to happen over time. As part of my dissertation research I analyzed USDA Forest Service FIA (Forest Inventory & Analysis) data over the past 20 to 30 years, looking at the number of trees by diameter class for sugar maple, red maple, oaks, hickories, and beech for 25 states. Although there are many tree species growing in eastern forests besides maples, beech, oaks, and hickories, I focused solely on these dominant species since my main objective was to examine the hypothesis that maples will be replaced by oaks and hickories in the northeastern United States. Sugar and red maples were examined separately since they fill different ecological niches and have different economic values for sugaring and timber production. I lumped all oak and hickory species together to get a general sense of how the main species within the oak/hickory forest were performing. I also included beech since it is an important species found in both oak/hickory and maple/beech/birch forest types. Although it would have been useful to include other species found within these forest types (including *Betula*, *Prunus*, *Fraxinus*, *Populus*, and more), this was ancillary to my main research objective and not included due to space restraints.

If oaks and hickories will replace maples in northeastern forests, then there should be some evidence that this is already starting to take place. Similarly, if beech trees are forming dense thickets in the understory of maple-dominated stands, then the number of beech saplings should also be on the rise. Studying the abundance and distribution for these major species, stratified according to diameter class and forest type, is useful for identifying trends in forest composition over time. In particular, given that the small-diameter stems are the ones that will dominate the overstory by the end of the century, closely examining trends in saplings is one of the best ways to predict future forest composition. To do this, I compared the most recent forest inventory data with the oldest data for which electronic information was available for all states (an average of 22 years' difference). Figures 16.1–16.4

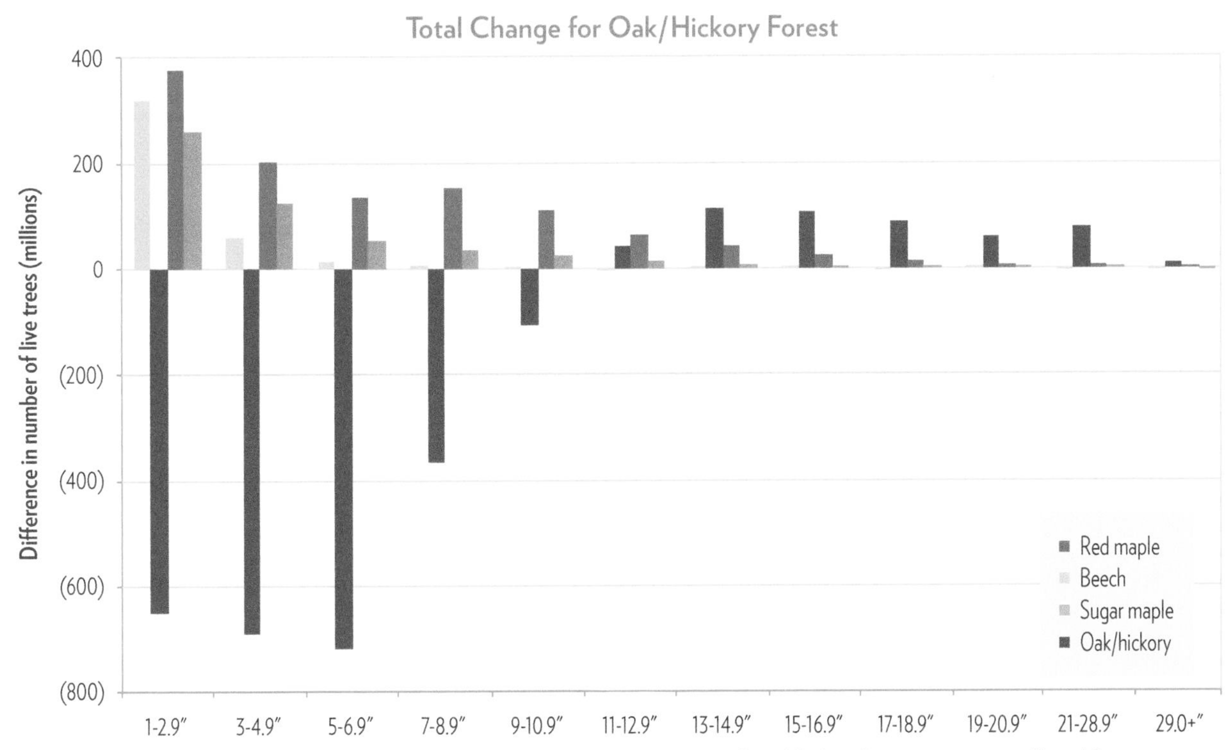

FIGURE 16.1. Difference in total number of live trees from oldest to most recent FIA survey for oak/hickory forest types in eastern United States.

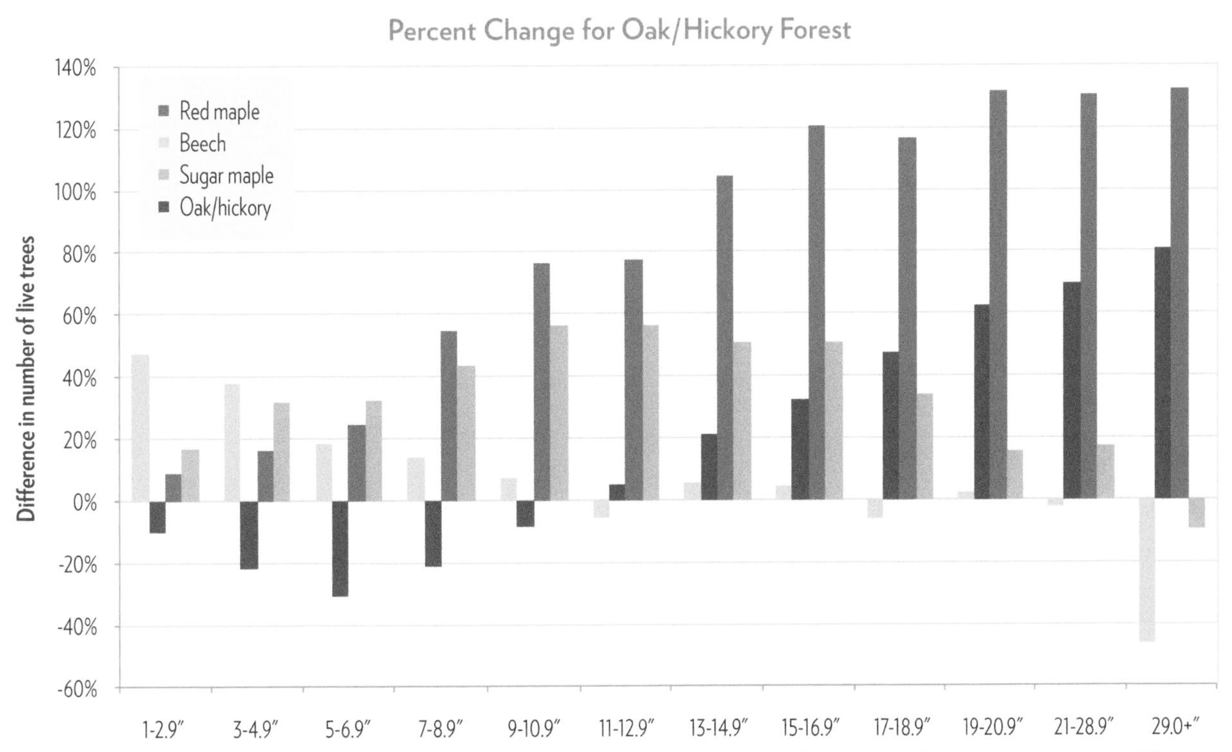

FIGURE 16.2. Percent difference in total number of live trees from oldest to most recent FIA survey for oak/hickory forest types in eastern United States.

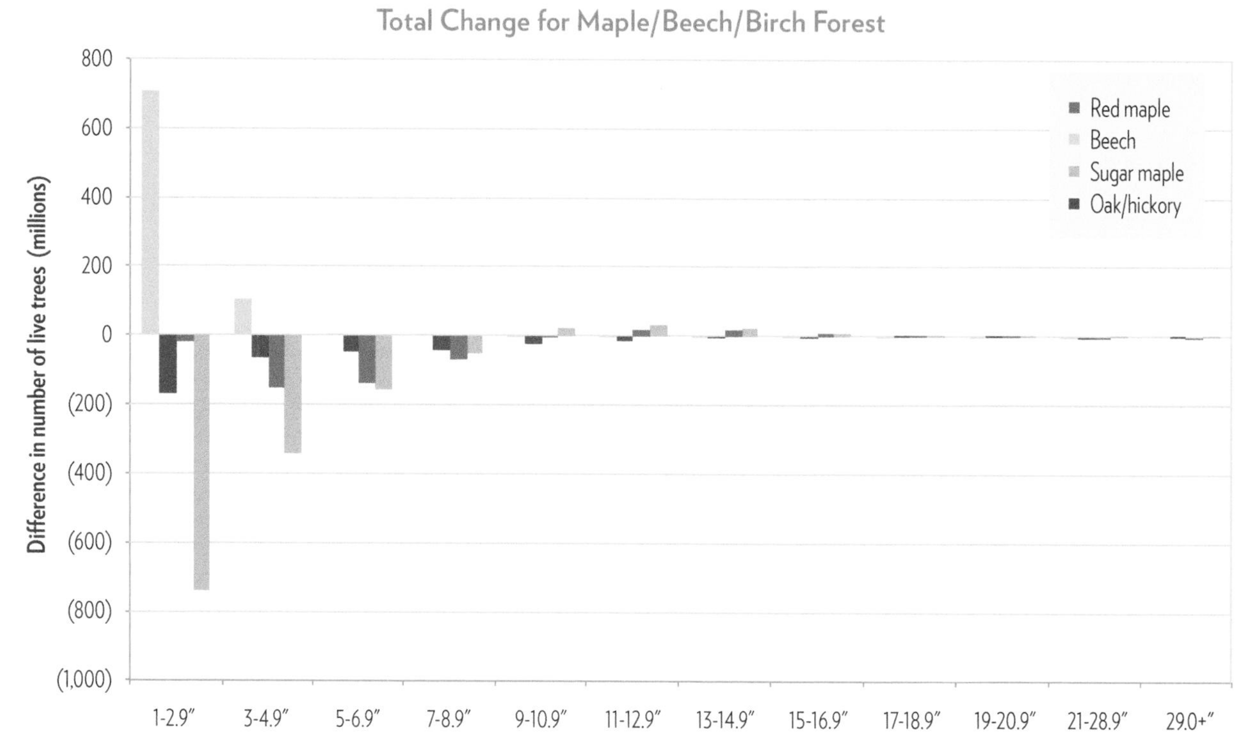

FIGURE 16.3. Difference in total number of live trees from oldest to most recent FIA survey for maple/beech/birch forest types in eastern United States.

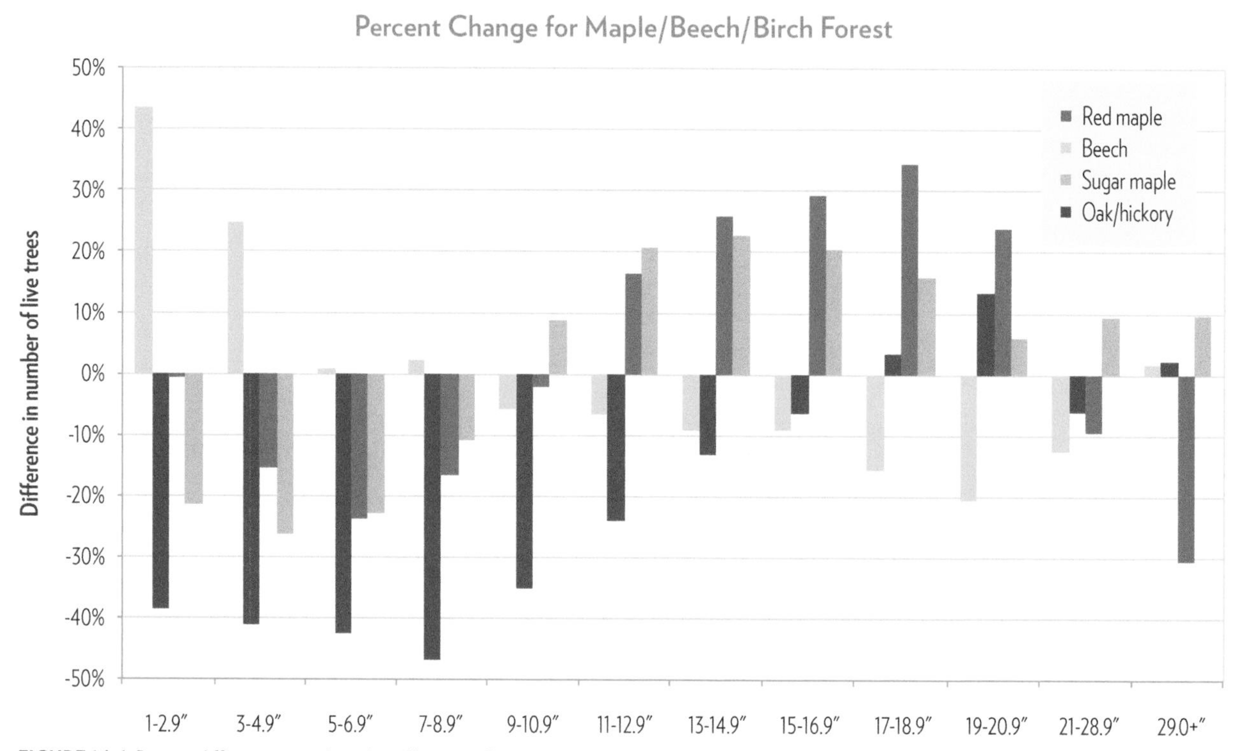

FIGURE 16.4. Percent difference in total number of live trees from oldest to most recent FIA survey for maple/beech/birch forest types in eastern United States.

present the absolute and percentage changes over that time period in the number of live trees for each of the species groups described above, stratified according to 2" diameter classes. Figures 16.1 and 16.2 display the data for oak/hickory forest types whereas figures 16.3 and 16.4 refer to maple/beech/birch forest types.

The large, established oaks and hickories are doing very well, as there are many more sawtimber-sized oaks and hickories today than there were a few decades ago. However, their presence in the understory has been falling over time, raising concerns about the sustainability of these keystone species in future forests. On the other hand, whereas shade-tolerant sugar and red maples are a small part of the overstory in oak/hickory forest types, they make up an increasingly larger portion of the understory. In fact, sugar and red maples have been rapidly increasing in abundance in oak/hickory forest types across nearly all size classes, though the changes are particularly evident in the sapling size classes. Of the species we explored, red maples have experienced the greatest overall gains in the sapling size classes and the largest percentage increases in the poletimber and small-sawtimber size classes across all forest types in the region. These trends point to increased red maple abundance in coming years, especially within oak/hickory-dominated forests. Although beech is rapidly proliferating in the understory, it has also suffered great losses in the largest size classes.

FIGURE 16.5. It is not uncommon to find a young stand of sugar and red maples beneath the overstory of oaks and hickories.

The situation is quite different in the maple/beech/birch forest types, where most of the maple syrup is produced. These forest types have experienced significant declines in sugar maple, red maple, and oak/hickory regeneration. Beech trees, not oaks or hickories, are replacing sugar maples in the understory, as there are now over 700 million additional beech trees and nearly 800 million fewer sugar maples in the 1" to 2.9" diameter class for these forest types today than there were 20 to 30 years ago. In fact, beech is the only species that has actually increased in abundance within the first four size classes. The opposite trend is seen in the larger size classes, as sugar and red maples are expanding whereas the larger beech are dying off due to beech bark disease. Sugar maple is the only species that has been increasing in every size class above 9" dbh; there are now many more potentially tappable trees on the landscape than there were just a few short decades ago. Red maples have undergone the greatest percentage increases in the small- and medium-sawtimber size classes, but have fallen in the largest size classes above 21" dbh. Oaks and hickories have experienced the greatest percentage declines in nearly all size classes, with the exception of increases in some of the larger sawtimber size classes.

These data contradict the predictions that oaks and hickories will replace maples in northeastern forests within the next century. Although there are legitimate concerns regarding sugar maple health and regeneration in certain regions, the larger size classes of sugar

FIGURE 16.6. In much of the northeastern United States, the understories of established sugarbushes are far more likely to be overtaken by beech thickets than by oaks and hickories.

maples are currently increasing throughout the eastern United States. In terms of saplings, sugar maple seems to be doing comparatively better in oak/hickory-dominated stands along the southern and western edges of its natural range than it is in the maple-dominated stands of the Northeast. Generally speaking, red maple is showing the largest increases across the region and is quickly becoming one of the dominant tree species. Larger-sized oaks and hickories are increasing as a result of growth of established trees, yet the regeneration of these species has been falling for several decades. Thus the likelihood of maple-dominated forests being replaced by oaks and hickories seems very low.

The fact that sugar maple regeneration has been increasing in the more southerly states currently dominated by oaks and hickories indicates that the problems associated with maple regeneration in the northern states may have little to do with climate change. Since there has been increased sugar maple regeneration in the southern climates and forest types that have been predicted to shift into the northeastern states, climate change in itself should not be the primary concern for the possible loss of sugar maple in the Northeast. Rather, there must be other, more important factors that have hindered sugar maple regeneration in the Northeast and assisted sugar maple advancement in the oak/

FIGURE 16.7. Red maple is becoming one of the dominant species throughout the eastern United States partly due in part to the fact that it sprouts vigorously after being cut. PHOTO COURTESY OF PETER SMALLIDGE

hickory forests of mid-Atlantic and central states. Deer browsing is a known problem that may be adversely impacting oaks and hickories more so than maples.[10] Acid precipitation has negatively affected sugar maples in many northeastern forests with limited buffering capacity. Better seed dispersal, persistent seedling banks, and rapid growth rates when light is available give maples inherent advantages over oaks and hickories in established forests. Furthermore, at the interface with spruce-fir forests at the northern edge of their range, sugar maples have been expanding their range as climate warming predicts.[11] Thus we appear to have a landscape where maples are becoming more abundant, not less.

Most forests in the eastern United States originated after clear-cutting, fire, and agricultural abandonment in the 1800s and early 1900s. Since these practices are uncommon today, regenerating oaks and hickories has become increasingly difficult. In comparison with maples, oaks and hickories have experienced more dramatic declines in the sapling size classes and greater increases in the sawtimber size classes. Thus, the immediate outlook for oaks and hickories is positive whereas the long-term trends point to lower populations of these species on the landscape. These circumstances conflict with predictions that maple/beech/birch forests will be overtaken by oaks and hickories in future

Tapping Other Species in a Changing Climate

FIGURE 16.8. Planting black walnuts provides a hedge against a changing climate and can add tremendous diversity and value to your property. Although they are concentrated in the midwestern United States, black walnuts can be planted in a wide variety of locations and climates, as seen with this plantation in Hungary. PHOTO COURTESY OF ROBERT VIDÉKI, DORONICUM KFT., BUGWOOD.ORG

Another possible adaptation to climate change could be planting and tapping other tree species, in particular black walnuts. This species can already survive in most northeastern climates, yet black walnuts currently thrive in the midwestern and Appalachian states where the summers are much warmer and drier. If you have any open fields with well-drained, fertile soils, you may consider planting them to black walnut as a hedge against a changing climate. If something were to eventually happen that wiped out all the maples, you could start producing walnut syrup as a replacement product. Nothing could ever fully take the place of maple syrup, but tapping walnut trees would certainly be better than not producing anything.

Tapping birch trees is another way of utilizing your sugaring equipment and land to produce another valuable crop in a given season. The warm weather that causes the maple season to end early is the same weather than warms the soil up and gets the birch sap flowing. By adding birch syrup to your operation, you can extend your season and make the most of the weather and resources you have available. Again, birch syrup is certainly not a substitute for maple syrup, but making it could allow you to keep sugaring and make better use of your extensive infrastructure and the expensive machinery you have already invested in.

forests. Although the climate may continue to become more favorable to the growth of established oak and hickory, the ecology of these species, in particular their limited seed dispersal mechanisms and requirement for high-light environments for establishment and growth, may prevent their natural spread across a landscape where fire, agricultural abandonment, and clear-cutting are much less prevalent.

Although red maple regeneration is down in maple/beech/birch stands, it has been expanding in the larger size classes and is also quickly becoming one of the dominant species in oak/hickory forest types. Red maple is an opportunistic, generalist species that is a heavy seeder and a prolific stump sprouter; it can also perform reasonably well in early- and late-successional stages across a wide variety of environments.[12] Fire suppression in eastern forests starting in the 20th century has hastened the presence of red maples in oak/hickory forests. Without any changes in land use, timber harvesting, and fire management practices, it is highly likely that red maple will assume an even more dominant role in future forests. Even if sugar maples start to decline in parts of the Northeast, it is highly likely that red maples will help fill the voids left in the future forest. With the use of reverse osmosis to remove most of the water before boiling, it is certainly feasible for large producers to tap red maples as a means of expanding their operations. Rather than discarding red maples as an inferior species, we should embrace them as a quality producer of sweet sap that can more easily withstand the harsh environmental conditions that we may encounter in the future.

Although persistent droughts and extreme heat events could hamper tree growth on local and/or regional scales, the long-term prospects for sugar and red maples appear favorable throughout the eastern United States. Climate change will certainly exert some influence on species abundance and distribution, but our management of forests is likely to have a much greater impact. It's important to keep in mind that drought is a much more serious problem for sugar maples than heat. We can't do anything about the weather, but we can manage our forests to make them more resilient to potential droughts. By conducting regular cutting in which less than 20 percent of the canopy is harvested, we can maintain vigorous growth in our trees without opening up the canopy too much all at once. Fully stocked forests of healthy trees are more resilient to droughts since the canopy prevents sunlight from reaching the forest floor and drying out the soil. With careful management, we can ensure that our maples stay as healthy as possible, providing an abundant supply of maples to tap for the foreseeable future, even with a changing climate.

Final Thoughts

Sugaring is highly weather-dependent; it always has been and always will be. No matter what happens to our climate over time, keep in mind that there is a big difference between weather and climate. In a warming world we will still have plenty of cold temperatures and unseasonable weather—just because we may get a long cold spell in March doesn't mean the climate isn't changing. And even if climate change were a hoax, we would still get heat waves in March some years that cause the season to end prematurely. The saying goes that the only thing predictable about sugaring is that it's unpredictable, which has always been the case. We need to take the good years with the bad ones and just do the best we can with the weather we are given.

Even though I am optimistic that we can adapt our sugaring operations to a changing climate, we should not be complacent about climate change. We all have a moral obligation to do everything we can to reduce our carbon emissions and slow the impact of global warming. Climate change is already having devastating effects throughout the world that are projected to get even worse. Even though we can adapt our sugaring practices to a warming world, there are many other people, species, and industries in which adaptation strategies are severely limited or nonexistent. By taking measures to reduce our carbon footprint, we can help ensure a more sustainable future not just for sugarmakers, but for all people and species in the world.

Finally, as much as we love maple sugaring and consider it of great importance, we must realize that if the climate ever changes to the point that we can't produce maple syrup in the Northeast, we will have much bigger problems to contend with than a lack of pure maple syrup. Maple sugaring is currently a viable industry in the mid-Atlantic states, albeit their season occurs during January and February rather than March and April. Under a worst-case scenario, the climate of the mid-Atlantic states will shift into the Northeast by 2100. I would be surprised if we aren't making more syrup in the Northeast during January and February by the end of the century, but if it gets so warm that even this isn't possible, then we will have much greater environmental and social problems to contend with. If this ever happens, our descendants will look back at photos and videos of sugaring and reminisce about the great tradition that we lost. They will criticize us for being too foolish to do anything about global warming, but the other problems in the world will be so bad that few people will be concerned about whether their pancake topping is pure maple or not—they will just be glad to have something to eat.

Further Resources for Sugarmakers

I hope that you found this book enjoyable and informative. My goal was to provide as much information as I could in a standard book format, but I couldn't cover everything. There is much to learn about sugaring; if you keep your eyes and ears open, you'll discover something new every day. This last section provides an overview of other resources that could help broaden your knowledge of tapping maple, birch, and walnut trees for syrup production. There are even more resources out there than those covered here, but I have tried to focus on the books, websites, organizations, and events that I think will be most useful for you as you develop your operation.

Books

***North American Maple Syrup Producers Manual* (*2nd edition*), Edited by Randall Heiligmann, Melvin Koelling, and Timothy Perkins**

This manual is published by The Ohio State University in cooperation with the North American Maple Syrup Council. It has served as the main reference for the maple industry for decades and contains a great deal of valuable information on all aspects of sugaring. There are some things that I purposefully didn't cover in this book because those topics are thoroughly covered in the North American manual. It is made available to sugarmakers primarily through equipment dealers and Cooperative Extension offices. The manual is currently being revised to reflect research findings and advances in technology over the past 10 years, with a third edition scheduled to be published in 2015.

***The Maple Sugar Book* by Helen and Scott Nearing**

The Nearings were famous back-to-the-landers who lived in Vermont for many years producing maple syrup as a main part of their homestead. Originally published in the 1950s, this classic book is the best resource on the history of sugaring that I have ever come across. The Nearings did a tremendous amount of research and chronicled much of the fascinating history of sugaring in North America. Since this book also explains the best practices for making syrup as of 1950, it is interesting to see what has changed and what has stayed the same over many years. I always recommend this book to anyone with a genuine interest in sugaring.

***Backyard Sugarin'* by Rick Mann**

This is a great book for those who only want to make a small amount of syrup from the trees in their backyards or neighborhoods. It contains a lot of great ideas and pictures for how to collect and process sap on a hobby scale and a limited budget. With some ingenuity and creativity, the author shows many ways in which you can produce maple syrup for yourself and your family with little investment besides your own time.

Organizations

State/Provincial Maple Producers' Associations

Joining your state or provincial maple producers' organization is an excellent way to network with other sugarmakers and become more involved with the industry. Becoming a member will also make you a more informed, effective, and efficient producer, and you will likely develop long-lasting friendships with other producers. As sugarmakers, we learn most of what we know from one another. The more sugarmakers you know, the more ideas you will be exposed to on all aspects of sugaring. Your job is to then distinguish the good ideas from the bad ones and figure what will work best for your operation. In a survey of maple producers in New York carried out in 2009, we found that members of the state association had much more progressive operations than those who did not belong to the NYSMPA. In particular, members were much more likely to use modern technology and practices in sap collection and processing, thereby allowing them to develop more efficient, profitable, and sustainable operations. We were not able to determine if these differences were due to causation or correlation, but I believe there is a little of both involved. Progressive producers who are actively engaged in their businesses are more likely to join the association, and once you have joined the association, being involved with a diverse group of producers will expose you to new ideas that will help develop your operation further.

Despite all the benefits of being a member, unfortunately only a fraction of sugarmakers belong to and actively participate in their state association. Since you have taken the time to read this book, chances are you care deeply about your sugaring operation and will soon join your association (if you aren't a member already). While simply being a member is beneficial, you will get even more out of your membership if you take an active leadership role. I have never heard of a maple producers' organization that had too many volunteers for all of the activities and roles that were available. Below is a listing of the web addresses for 12 states and four provinces that have active associations. If your state does not already have a producer organization, you can either work with others to develop your own state organization or join an existing association in a nearby state.

United States

- Maple Syrup Producers Association of Connecticut
  www.ctmaple.org
- Indiana Maple Syrup Association
  www.indianamaplesyrup.org
- Maine Maple Producers Association
  www.mainemapleproducers.com
- Massachusetts Maple Producers Association
  www.massmaple.org
- Michigan Maple Syrup Association
  www.mi-maplesyrup.com
- Minnesota Maple Syrup Producers Association
  www.mnmaple.org
- New Hampshire Maple Producers Association
  www.nhmapleproducers.com
- New York State Maple Producers Association
  www.nysmaple.com
- Ohio Maple Producers Association
  www.ohiomaple.org
- Pennsylvania Maple Syrup Producers Council
  www.pamapleassociation.com
- Vermont Maple Sugar Makers Association
  www.vermontmaple.org
- Wisconsin Maple Syrup Producers Association
  www.wismaple.org

Canada

- New Brunswick Maple Syrup Association
  http://maple.infor.ca
- Maple Producers Association of Nova Scotia
  www.novascotiamaplesyrup.com
- Ontario Maple Syrup Producers Association
  www.ontariomaple.com
- Federation of Quebec Maple Syrup Producers
  www.siropderable.ca

North American Maple Syrup Council

The North American Maple Syrup Council (NAMSC) is an international network of maple syrup producers' associations representing the 16 states and provinces that provide nearly all of the world's supply of pure maple syrup. It is a long-standing organization established in 1959 that facilitates dialogue and cooperation among the industry leaders. Elected delegates from 12 state maple producers' associations and four Canadian provinces make up their governing board. The NAMSC and the International Maple Syrup Institute (IMSI) hold a combined annual meeting each year hosted by one of the 16 member states or provinces. The NAMSC manages the NAMSC Maple Research Fund, which generates seed money through its Research Alliance Partnership Program. This unique program is the only one of its kind that provides seed money for researchers to work on a variety of issues important to the maple industry. The research findings have been instrumental in the development and improvement of the maple industry. The program is funded by voluntary donations: Maple producers and packers contribute a penny for each maple syrup container they purchase from manufacturers or equipment dealers that serve as Alliance Partners. The funds generated through this program are utilized by the NAMSC Research Fund Committee, which oversees a competitive grant program open to all maple research organizations and individuals. Producers who do not purchase their syrup containers through an Alliance Partner are encouraged to contribute a penny for every container they fill directly to the fund. Your participation in this program is welcome and encouraged. A penny per container isn't even noticeable on your bill (it's far less than 1 percent of the value of the syrup), yet all these pennies add up to help support the advancement of our industry. Research projects are reported in the *Maple Syrup Digest*, a quarterly publication that contains research articles on NAMSC-funded projects and other news of interest to sugarmakers. Membership in your local state or provincial maple producers association is the first step in supporting your industry, and through your local NAMSC delegate you are being represented on the international level at the North American Maple Syrup Council.
www.northamericanmaple.org

International Maple Syrup Institute

Whereas the NAMSC is focused more on education and research for sugarmakers, the International Maple Syrup Institute (IMSI) is geared toward marketing and promotion of pure maple products. The IMSI represents maple syrup industry stakeholders throughout the United States and Canada. The membership consists of maple syrup processors and packaging companies, state and provincial maple producer organizations, maple equipment manufacturers, research centers, and others. The goals of the IMSI are to maintain the integrity of pure maple syrup and develop standards for the maple industry to ensure that societal, economic, and environmental benefits are sustained and enhanced in the short and long term. Their website has been recently renovated and now includes a great deal of valuable information on the health benefits of maple syrup, considered by many to be the main marketing strategy for pure maple. Through the ongoing efforts of the IMSI, the maple industry has been able to develop additional markets for pure maple products and protect the pure maple name. There is always more work to be done in this arena, so if you are serious about the future development of the industry, I recommend joining the IMSI and getting involved in their activities.
www.internationalmaplesyrupinstitute.com

Cornell Maple Program

Cornell University has been conducting applied research and extension for the maple industry since the early 1900s. Cornell currently operates two research and extension facilities dedicated to syrup production—the Arnot Forest just 20 minutes south of the main campus in Ithaca, and The Uihlein Forest located in the heart of the Adirondacks near Lake Placid, New York. Working in close collaboration with county extension offices, the Cornell Maple Program offers maple schools and workshops throughout the year in nearly all parts of

the state, including the signature New York State Maple Conference in cooperation with the Vernon-Verona-Sherrill FFA. We also have a monthly webinar series that can be viewed either live or through the archives stored on our website at any time. The Cornell Maple Camp is another recent addition to our educational programming—the first one took place in 2011, and it has grown every year since then. Maple Boot Camp, as it has also been described by former attendees, provides a three-day intensive training on all aspects of maple syrup production. It combines a mixture of hands-on training with formal presentations to cover many aspects of sugarbush management, sap collection, processing, and marketing. It is geared toward people who are just starting out or who want to expand their sugaring hobby into a business.
www.cornellmaple.com

Proctor Maple Research Center

The University of Vermont has operated a maple syrup research center in Underhill since the 1940s. Dr. Tim Perkins is the current director and oversees a talented team of research and extension specialists who work on all aspects of maple syrup production. They have an excellent website with many useful publications concerning their research projects and results. In recent years they have been instrumental in determining the best designs for tubing and spouts to increase sap yields. The Proctor staff are regular presenters at maple workshops and schools throughout the Northeast, so be sure to check out their presentations to learn about the latest developments.
www.uvm.edu/~pmrc

Centre ACER

With a team of nearly 20 scientists, this is by far the largest maple research organization in the world. Although their website and trade publications are predominantly in French, they are working on getting more of their information translated to English. Furthermore, all their peer-reviewed scientific publications are in English, so

FIGURE R.1. University of Vermont technician Mark Isselhardt and research assistant Josh O'Neill at the Proctor Maple Research Center in Underhill Center, Vermont. With four identical evaporators, researchers examine the effects of processing factors (RO concentration, air injection) on syrup chemistry and flavor that occur during the transformation of sap into maple syrup. PHOTO COURTESY OF TIMOTHY PERKINS

you can still easily benefit from a lot of their work. You can also contact them with questions, as the staff are fluent in English and can easily converse via email or phone. If you happen to be in Quebec, Centre ACER also manages a subsidiary of nearly 70 people that inspects and grades all bulk maple syrup sold under the guidelines and regulations of the Federation of Quebec Maple Syrup Producers.
www.centreacer.qc.ca

Websites and Trade Publications

Sugarbush.Info

This website was developed in 2010 by Bryan Exley and a group of dedicated volunteers who are passionate about the maple industry. They have amassed a great deal of information on all aspects of maple production in a user-friendly website that is easy to navigate. It is the most comprehensive collection of electronic information available for the maple industry. As of January 2013, they have links to websites for nearly 400 sugarmakers in 25 states and provinces as well as nine birch syrup producers. There are also links to approximately 150 manufacturers, distributors, and custom fabricators of maple equipment and products. There are links for regulations in different states and provinces, roughly 30 maple events and festivals, and 20-plus links for further resources. In addition, they have a photo gallery, a glossary of maple terms, and a forum where users can post questions on a variety of topics and exchange ideas. If you haven't already registered on this site, I'd highly recommend taking the few minutes to do so and checking out all the valuable and interesting information.
www.sugarbush.info

MapleTrader.com

This website was developed and is maintained by Chris Pfeil with The Maple Guys. Chris did the maple industry a great service by developing the first online forum for maple producers to communicate and exchange ideas. The website is divided into different forums for every topic imaginable. All the exchanges are archived so you can go back to read what different people have said over the years on a given topic. There are forums dedicated to state and provincial association correspondence and happenings. As the name implies, MapleTrader.com also serves as an excellent medium for buying and selling used maple equipment. Chris does not allow commercial manufacturers or dealers to advertise; rather, the site is focused on sales of used equipment among sugarmakers. If you haven't already signed up, it is well worth the time to do so.
www.MapleTrader.com

The Maple News

This is the most well-read and widely circulated publication in the maple industry today. It comes out eight times per year with timely and relevant information on current events and happenings in the industry. A team of guest columnists provide interesting articles on their areas of expertise. There are advertisements from equipment manufacturers announcing their latest products as well as a free color classified section where subscribers can post various items for sale. For the small cost of a subscription (about 2 pints of syrup a year), getting this delivered to your door will make you a much more informed and engaged producer.
www.themaplenews.com

Special Events

Fall/Winter Maple Schools

Many state associations and Cooperative Extension agencies hold annual meetings and maple schools on Saturdays during the late fall and early winter. A calendar of all these events can usually be found in trade publications, association newsletters, and the websites listed above. The organizers of these events put a lot of time and effort into developing interesting and informative programs for all levels of producers. Many of the larger events include trade shows with representatives from the different equipment manufacturers and suppliers. I always encourage all maple producers to attend these, as they are one of the best opportunities to network with other sugarmakers and learn about the latest developments in the industry.

Summer Maple Tours

Maine, New York, Pennsylvania, Vermont, and other states and provinces usually host maple tours during the summer or fall that provide an opportunity for sugarmakers to visit other producers' operations. We are all too busy during sugaring season to get out and see how other people make syrup, but the summer or fall provides an excellent opportunity to travel around and see other operations. You don't have to be a producer in a given state to attend the tour, so I'd encourage you to break out and attend the tours in other states. The two- to three-day events make for a great mini vacation, and you are bound to come away with new ideas to implement in your own operation.

FIGURE R.2. The premier maple educational event in the United States is the New York State Maple Conference, which takes place on the first or second Friday and Saturday of January each year. It has grown to include roughly 1,000 participants and features the largest trade show and educational workshops in the maple industry.

Spring Open Houses

Soon after the sugaring season has ended, all the maple equipment manufacturers host open house events during the last weekend in April or first weekend in May. Thousands of sugarmakers descend upon the major companies in Vermont and New Hampshire to celebrate the end of the season and get deals on new equipment. Most manufacturers offer 10 percent off sales, free lunch, and educational workshops to draw people in. This is a great opportunity to visit the manufacturers, connect with other sugarmakers, and save some money on equipment.

Glenn Goodrich's One-Day Seminars

Every year on the first Saturday of November, Glenn Goodrich offers a one-day seminar on tubing installation and design at his sugarhouse in Cabot, Vermont. This is an excellent opportunity to learn from a professional tubing installer how to set up your tubing right the first time or make improvements to your existing tubing system. The first Saturday in February is also reserved for teaching people how to manage a sugarhouse so as to produce maple syrup in a cost-effective, efficient, and enjoyable manner. Glenn is a highly regarded presenter, and everyone who attends these workshops finds them entertaining and informative. You can learn more at www.goodrichmaplefarm.com.

International Maple Syrup Grading School

If you want to learn how to identify and correct off-flavors in maple syrup, I strongly encourage you to attend one of the International Maple Syrup Grading Schools. These are often held in conjunction with the IMSI and NAMSC annual meetings and cover a great deal of information over a two-day period. Through a mix of hands-on activities and interesting lectures, you will become an expert on the art and science of maple grading. Learn more at http://umaine.edu/maple-grading-school.

Notes

Chapter 1

1. Robert B. Thomas, *Farmers' Almanack*. Boston: John & Thomas Fleet, 1803.
2. Robert B. Thomas, *Farmers' Almanack*. Boston: John & Thomas Fleet, 1805.
3. An excellent book on the danger of consuming too much sugar is *Sugar Blues* by William Dufty. (New York: Grand Central Life & Style, 1986).
4. For more information, I suggest reading D. C. Allen, A. W. Molloy, R. R. Cooke, and B. A. Pendrel. "A Ten-Year Regional Assessment of Sugar Maple Mortality." In Sugar Maple Ecology and Health: Proceedings of an International Symposium. USDA FS NRS Gen. Tech. Rep. NE-261, 27–45, 1999.
5. Based on landowner survey research we carried out in 2009, go to for the full report on barriers and incentives for landowners to get involved with syrup production. M. Farrell and Richard Stedman. "Survey of the NYS Landowners: Report to the Steering Committee of the Lewis County Maple Syrup Bottling Facility." September 1, 2009, http://maple.dnr.cornell.edu/pubs/LandownerSurvey9-2009.pdf.
6. L. Li and N. P. Seeram. 2011. "Quebecol, a Novel Phenolic Compound Isolated from Canadian Maple Syrup." *Journal of Functional Foods* 3, (2011): 125–28; E. Apostolidis, L. Li, C. M. Lee, and N. P. Seeram: "In Vitro Evaluation of Phenolic-Enriched Extracts of Maple Syrup for Inhibition of Carbohydrate Hydrolyzing Enzymes Relevant to Type 2 Diabetes." *Journal of Functional Foods* 3, (2010):100–06; L. Li and N. P. Seeram: "Maple Syrup Phytochemicals Include Lignans, Coumarins, a Stilbene and Other Previously Unreported Antioxidant Phenolic Compounds." *Journal of Agricultural and Food Chemistry* 58, 11673–79.
7. My dissertation research included an analysis of the potential for increasing the production and consumption of pure maple on a statewide basis. M. Farrell, and B. Chabot. "Assessing the Growth Potential and Economic Impact of the US Maple Syrup Industry." *Journal of Agriculture, Food Systems, and Community Development* 2(2), (2012): 11–27. http://dx.doi.org/10.5304/jafscd.2012.022.009.
8. "Agriculture and Agri-Food Canada." *Canadian Maple Products Situations and Trends 2005–2006,* 19, 2006.

Chapter 2

1. For a full description of the report: M. Farrell and Richard Stedman. "Survey of the NYS Landowners: Report to the Steering Committee of the Lewis County Maple Syrup Bottling Facility." September 1, 2009, http://maple.dnr.cornell.edu/pubs/LandownerSurvey9-2009.pdf.

Chapter 4

1. D. M. van Gelderen, P. C. de Jong, and H. J. Oterdoom. *Maples of the World*. Portland, OR: Timber Press, 1994.
2. Y. Vargas-Rodriguez and W. Platt. "Remnant Sugar Maple (*Acer saccharum* subsp. *skutchii*) Populations at Their Range Edge: Characteristics, Environmental Constraints and Conservation Implications in Tropical America." *Biological*

Conservation 150(1)(2012), 111logi. http://dx.doi.org/10.1016/j.biocon.2012.03.006.

3. For a comprehensive background and history of the FIA program: "Forest Inventory and Analysis National Program," last modified June, 3, 2013, www.fia.fs.fed.us.
4. M. Farrell. "Estimating the Maple Syrup Production Potential of American Forests: An Enhanced Estimate that Accounts for Density and Accessibility of Tappable Maple Trees." *Agroforestry Systems*. Published online December 19, 2012. DOI: 10.1007/s10457-012-9584-7.
5. M. L. Crum, J. J. Zaczek, A. D. Carver, K. W. J. Willard, J. K. Buchheit, J. E. Preece, and J. C. Magnun. "Riparian Silver Maple and Upland Sugar Maple Trees Sap Sugar Parameters in Southern Illinois." In Daniel A. Yaussy, David M. Hix, Robert P. Long, and P. Charles Goebel, eds. Proceedings, 14th Central Hardwood Forest Conference, Wooster, OH, March 16–19 2004.
6. J. Kort and P. Michiels. "Maple Syrup from Manitoba Maple (*Acer negundo* L.) on the Canadian Prairies." *Forestry Chronicle* 73 (1997): 327–30.
7. Kendrick, Jenny. "Tapping the Manitoba Maple—a Prairie Cottage Industry." Statistics Canada, Catalogue no. 96-325-XPB, www.statcan.gc.ca/pub/96-328-m/2004032/4194001-eng.pdf.
8. G. Backlund and K. Backlund. *Bigleaf Sugaring: Tapping the Western Maple*. Ladysmith, BC: Backwoods Forest Management, 2004.
9. Robert H. Ruth, J. Clyde Underwood, Clark E. Smith, and Hoya Y. Yang. "Maple Sirup Production from Bigleaf Maple." Portland, OR: US Department of Agriculture, Forest Service, Pacific Northwest Forest and Range Experiment Station, 1972, www.fs.fed.us/pnw/pubs/rn181.pdf.
10. D. Bruce. "Production and Quality of Sap from Bigleaf Maple (*Acer macrophyllum* Marsh) on Vancouver Island, British Columbia." Master of science thesis, University of Victoria, 2009.
11. Philip A. Barker and D. K. Salunkhe. "Maple Syrup from Bigtooth Maple." *Journal of Forestry*, (August 1974): 491–492.
12. G. G. Naughton, W. A. Geyer, and E. Chambers. "Sugarbushing the Black Walnut." Proceedings of the annual meeting of the Northern Nut Growers Association. Columbia, MO, 2004.
13. P. Herbert. *Quarterly Bulletin of the Agricultural Experiment Station,* Michigan Agricultural College, 192.4, vii, (1924): 62–63.
14. For a detailed history of conservation efforts related to butternut canker, I suggest reading: M. E. Ostry and K. Woestel, "Spread of Butternut Canker in North America, Host Range, Evidence of Resistance Within Butternut Populations and Conservation Genetics." Proceedings of the Sixth Walnut Council Research Symposium, Lafayette, IN, July 2004, www.ncrs.fs.fed.us/pubs/gtr/gtr_nc243/gtr_nc243_114.pdf.
15. Z. Matta, E. Chambers, and G. G. Naughton. "Consumer and Descriptive Sensory Analysis of Black Walnut Syrup." *Journal of Food Science* 70(9), (2005): 610–13.
16. F. Ewers, T. Ameglio, H. Cochard, F. Beaujard, M. Martignac, M. Vandame, C. Bodet, and P. Cruiziat. "Seasonal Variation in Xylem Pressure of Walnut Trees: Root and Stem Pressures. *Tree Physiology* 21 (2001): 1123–1132.

Chapter 5

1. H. Smith and R. Lamore. "Speed of Tapping Does Not Influence Maple Sap Yields." USDA Forest Service Research Note NE-132, (1971). http://nrs.fs.fed.us/pubs/rn/rn_ne132.pdf.

Chapter 6

1. A. K. van den Berg, T. D. Perkins, M. L. Isselhardt, M-A. Godshall, and S.W. Lloyd. "Effects of Air Injection during Sap Processing on Maple Syrup Color, Chemical Composition and Flavor Volatiles." *International Sugar Journal* (111) 1321(2009): 37–42.

Chapter 7

1. US Food and Drug Administration, last modified April 3, 2013, http://www.fda.gov/Food/GuidanceRegulation/default.htm.

Chapter 8

1. C. O. Willits. "Maple Sirup Producers Manual." Agricultural Handbook No 134. Agricultural Research Service, USDA. Philadelphia, PA, 1958.
2. J. Pasto and R. Taylor. *The Economics of the Central Evaporator in Maple Syrup Production*. Bulletin 697. Pennsylvania State University Agricultural Experiment Station, 1962.

Chapter 10

1. V. Theriault. "Changes in the Quebec Maple Industry and Economic Implications for Maine and the US." Master of science thesis, University of Maine, Augusta, 2007.

Chapter 11

1. You can order a copy of the manual from the Cornell Maple Program website: Cornell Sugar Maple Research and Extension Program, last modified March 29, 2013, www.cornellmaple.com.
2. There is much more information on maple soda found on pages 93–96 in the Maple Confections Notebook : "Maple Confections," last modified October 4, 2010, http://maple.dnr.cornell.edu/pubs/confections/Confection%20Notebook%20section7.pdf.
3. J. McElroy. "Maple-Syrup Gelato Wows Italians." *Vancouver (British Columbia)*. Accessed October 6, 2012. *Province*, www.lexisnexis.com/hottopics/lnacademic.
4. To hear the full story: Steve Zind. "Vermont Beer Makers Bring Back Old-Time Maple Sap Brews." *All Things Considered*, May 15, 2012, www.npr.org/blogs/thesalt/2012/05/15/152694105/vermont-beer-makers-bring-back-old-time-maple-sap-brews.

Chapter 12

1. Sang-Hun Choe. "In South Korea, Drinks Are on the Maple Tree." *New York Times*, March 6, 2009.
2. Terazawa M. Shirakamba. "Birch Splendid Forest Biomass—Potential of Living Tree Tissues." Presented at the Tree Sap Proceedings of the 1st International Symposium on Sap Utilization, Bifuka, Japan, April 1995.
3. Olga A Zyryanova, Minoru Terazawa, Takayoshi Koike, and Vyacheslav I. Zyryanov. "White Birch Trees as Resource Species of Russia: Their Distribution, Ecophysiological Features, Multiple Utilizations." Abstract. *Eurasian Journal of Forest Research* 13(1), 2010, 25-4, doi: http://eprints.lib.hokudai.ac.jp/dspace/bitstream/2115/43853/1/EJFR13-1_004.pdf.
4. M. Terazawa. "Shirakamba Birch, Splendid Forest Biomass—Potential of Living Tree Tissues." In M. Terazawa, C. A. McLeod, and Y. Tamai, eds. *Tree Sap;* O. A. Zyryanova. "Birches as Sap Producing Species of Russia: Their Distribution, Ecophysiological Features, Utilization and Sap Productivity." Proceedings of the 1st International Symposium on Sap Utilization (ISSU), April 10–12, 1995, Bifuka, Japan. Hokkaido: Hokkaido University Press, 7–12.
5. M. Kūka, I. Čakste, F. Dimiņš, and E. Geršebeka. "Determination of Phenolic Compounds in Birch and Maple Saps." International Conference on Food Innovation, October 25–29, 2010.
6. Geun-Shik Lee, et al. "The Beneficial Effect of the Sap of *Acer mono* in an Animal with Low-Calcium Diet-Induced Osteoporosis-Like Symptoms." *British Journal of Nutrition* 100, (2009): 1011–18.

7. Y. Yeong-Min, J. Eui-Man, K. Ha-Young, C. In-Gyu, C. Kyung-Chul, and E. Jeung. 2011. "The Sap of *Acer okamotoanum* Decreases Serum Alcohol Levels after Acute Ethanol Ingestion in Rats." *International Journal of Molecular Medicine* 28, 489–95.
8. H. Yang, I. Hwang, T. Koo, H. Ahn, S. Kim, M. Park, W. Choi, H. Kang, I. Choi, K. Choi, and E. Jeung. "Beneficial Effects of *Acer okamotoanum* sap on L-NAME-Induced Hypertension-Like Symptoms in a Rat Model." *Molecular Medicine Reports* 5, (2011): 427–31.
9. A. Cochu, D. Fourmier, A. Halasz, and J. Hawari. "Maple Sap as a Rich Medium to Grow Probiotic Lactobacilli and to Produce Lactic Acid." *Letters in Applied Microbiology* 47(6), (2008): 500–07.
10. L. Lagace. "Identification of the Bacterial Community of Maple Sap by Using Amplified Ribosomal DNA (rDNA) Restriction Analysis and rDNA Sequencing." *Applied and Environmental Microbiology* 70(4), (2004): 2052–60.
11. Peter Smith. "On Tap in Vermont, Maple Sap." *The Atlantic* (April 1, 2009). www.theatlantic.com/health/archive/2009/04/on-tap-in-vermont-maple-sap/7089.
12. B. Folwell. "Sapsuckers: Tapping into a Natural Elixir." *Adirondack Life*. March–April 2010.
13. Chung et al. "The Components of Fermented Soy Sauce from Gorosoe and Bamboo Sap." *Korean Journal of Food & Nutrition*. 14 (2), (2001): 167–174.

Chapter 13

1. "Measuring Your Forest Stand." National Learning Center for Private Forest and Range Landowners, last modified April 1, 2008, http://forestandrange.org/Virtual%20Cruiser%20Vest/index.html.
2. Peter J. Smallridge and Ralph D. Nyland. "Woodland Guidelines for the Control and Management of American Beech." Cornell University Cooperative Extension ForestConnect Fact Sheet, Ithaca, New York, 2009. www2.dnr.cornell.edu/ext/info/pubs/FC%20factsheets/american%20beech%20Fact%20Sheet.pdf.
3. R. D. Nyland, A. L. Bashant, K. K. Bohn, and J. M. Verostek. "Interference to Hardwood Regeneration in Northeastern North America: Ecologic Characteristics of American Beech, Striped Maple, and Hobblebush." *Northern Journal of Applied Forestry* 23(1), (2006): 53–61.
4. J. D. Kochenderfer, J. N. Kochenderfer, and G. W. Miller. "Controlling Beech Root and Stump Sprouts Using the Cut-Stump Treatment." *Northern Journal of Applied Forestry* 23(3), (2006): 155–65; Jeffrey D. Kochenderfer, James N. Kochenderfer, Gary W. Miller. "Manual Herbicide Application Methods for Managing Vegetation in Appalachian Hardwood Forests." Gen. Tech. Rep. NRS 96. Newtown Square, PA: US Department of Agriculture, Forest Service, Northern Research Station (2012): 59. http://www.nrs.fs.fed.us/pubs/gtr/gtr_nrs96.pdf.
5. P. J. Smallidge. "Woodland Goat SARE Final Technical Report." Provided for USDA Northeast Sustainable Agriculture Research & Education. Retrieved on November 15, 2012, from www2.dnr.cornell.edu/ext/info/pubs/MapleAgrofor/Goats_in_the_Woods.finalreport.2004.pdf 15p.
6. Here is a link to an excellent fact sheet on using goats as a vegetation management tool: Brett Chedzoy. "Using Goats for Vegetation Management in the Northeast." Cornell University Cooperative Extension, 1/25/2011, www2.dnr.cornell.edu/ext/info/pubs/MapleAgrofor/Vegetation%20Management%20Using%20Goats.pdf.

Chapter 14

1. P. von Aderkas. "Economic History of Ostrich Fern, *Matteuccia struthiopteris*, the Edible Fiddlehead." *Economic Botany* 38(1), (1984): 14–23.
2. This is contained in Bob Beyfuss's manual. *Practical Guide to Growing Ginseng*. Cairo, NY: Cornell Cooperative Extension of Greene County, 1994.

Chapter 15

1. This chapter based on M. Farrell. "The Economics of Managing Maple Trees for Sawtimber Production or Leasing for Syrup Production." *Northern Journal of Applied Forestry* 29(4), (2012): 165–72.
2. Richard M. Teck and Donald Hilt. "Individual-Tree Diameter Growth Model for the Northeastern United States." USFS Northeastern Forest Experiment Station Research Paper NE-649 (1991).
3. J. E. Wagner and P. E. Sendak. "The Annual Increase of Northeastern Regional Timber Stumpage Prices: 1961–2002." *Forest Products Journal* 55(2), (2005): 36–45.
4. C. Row, H. F. Kaiser, and J. Sessions. "Discount Rate for Long-Term Forest Service Investments." *Journal of Forestry* 79(6), (1981): 367–69, 376.
5. A. W. D'Amato, P. Catanzaro, D. T. Damery, D. B. Kittredge, and K. A. Ferrare. "Are Family Forest Owners Facing a Future Where Forest Management Is Not Enough?" *Journal of Forestry* 108, (2010): 32–38.
6. S. H. Bullard, D. L. Grebner, and K. L. Belli. "Financial Maturity Concepts with Application to Three Hardwood Timber Stands." Forest and Wildlife Research Center, Mississippi State University, 2002. Last accessed October 19, 2011. http://fwrc.msstate.edu/pubs/finmaturity.pdf.
7. B. Butler. *Family Forest Owners of the United States, 2006.* US For. Serv. Gen. Tech. Rep. NRS-27 (2008).

Chapter 16

1. B. Rock and S. Spencer. *Preparing for a Changing Climate: The Potential Consequences of Climate Variability and Change, New England Regional Overview.* US Global Change Research Program, University of New Hampshire, Durham, (2001): 39–42.
2. C. B. Kinner, A. T. DeGaetano, and B. F. Chabot. "Implications of 21st Century Climate Change on Northeastern United States Maple Syrup Production: Impacts and Adaptations." *Climatic Change* 100(3), (2010): 685–702.
3. T. Wilmot. "The Timing of Tapping for Maple Sap Collection." *Maple Digest*, (June 2008): 20–28.
4. A. M. Prasad and L. R. Iverson. "A Climate Change Atlas for 80 Forest Tree Species of the Eastern United States" [database]. Northeastern Research Station, USDA Forest Service, Delaware, OH, 1999, www.fs.fed.us/ne/delaware/atlas/index.html.
5. G. Lauten, B. Rock, S. Spencer, T. Perkins, and L. Irland. "Climate Impacts on Regional Forest." In *Preparing for a Changing Climate: The Potential Consequences of Climate Variability and Change, New England Regional Overview.* US Global Change Research Program, University of New Hampshire, Durham, (2001): 32–48.
6. P. C. Frumhoff, J. J. McCarthy, J. M. Melillo, S. C. Moser, and D. J. Wuebbles. *Confronting Climate Change in the US Northeast: Science, Impacts and Solutions.* Synthesis report of the Northeast Climate Impacts Assessment (NECIA). Union of Concerned Scientists (UCS), Cambridge, MA, 2007.
7. C. W. Woodall, R. S. Morin, J. R. Steinman, and C. H. Perry. "Status of Oak Seedlings and Saplings in the Northern United States: Implications for Sustainability of Oak Forests." Proceedings of the 16th Central Hardwoods Forest Conference, (2008): 535–42.
8. J. S. Fralish and T. G. McArdle. "Forest Dynamics across Three Century-Length Disturbance Regimes in Illinois Ozark Hills." *American Midland Naturalist* 162, (2009): 418–49.
9. L. Iverson, A. M. Prasad, S. Matthews, and M. Peters. "Lessons Learned while Integrating Habitat, Dispersal, Disturbance, and Life-History Traits into Species Habitat Models under Climate Change." *Ecosystems* 14, (2011): 1005–20, http://treesearch.fs.fed.us/pubs/38757; S. Matthews, L. Iverson, A. Prasad, M. Peters, and P. Rodewald. "Modifying Climate Change Habitat Models Using Tree Species-Specific Assessments of Model Uncertainty and Life History-Factors." *Forest Ecology and Management* 262, (2011): 1460–72.

10. D. B. Kittredge and P. M. Ashton. "Impact of Deer Browsing on Regeneration in Mixed Stands in Southern New England." *Northern Journal of Applied Forestry* 12(3), (1995): 115–20.
11. B. Beckage, B. Osborne, D. Gavin, C. Pucko, T. Siccama, and T. Perkins. "An Upward Shift of a Forest Ecotone during 40 Years of Warming in the Green Mountains of Vermont, USA." *Proceedings of the National Academy of Science* 105 (2008): 4197–202.
12. M. D. Abrams. "The Red Maple Paradox: What Explains the Widespread Expansion of Red Maple in Eastern Forests?" *BioScience* 48(5), (1998): 355–64.

Index

A

B

D

E

N

O

P

Q

R

S

T

U

V

W

Y

About the Author

Michael Farrell serves as the director of Cornell University's Uihlein Forest, a maple syrup research and extension field station in Lake Placid, New York. There he taps approximately 5,000 maples, 600 birch trees, and a couple dozen black walnut and butternut trees every year. He has authored dozens of articles on maple syrup production and forest management and often presents to maple producer and landowner organizations. Michael earned his bachelor's degree in economics from Hamilton College, his master's in forestry from SUNY-ESF, and his PhD in natural resources from Cornell University.

PHOTO BY NANCIE BATTAGLIA